DISTRIBUTED MODEL PREDICTIVE CONTROL FOR PLANT-WIDE SYSTEMS

DISTRIBUTED MODEL PREDICTIVE CONTROL FOR PLANT-WIDE SYSTEMS

Shaoyuan Li and Yi Zheng
Shanghai Jiao Tong University, China

WILEY

This edition first published 2015

Registered office:
John Wiley & Sons Singapore Pte. Ltd., 1 Fusionopolis Walk, #07-01 Solaris South Tower, Singapore 138628.

For details of our global editorial offices, for customer services and for information about how to apply for permission to reuse the copyright material in this book please see our website at www.wiley.com.

Library of Congress Cataloging-in-Publication Data applied for.

A catalogue record for this book is available from the British Library.

ISBN: 9781118921562

Set in 10/12pt, TimesLTStd by SPi Global, Chennai, India
Printed and bound in Singapore by Markono Print Media Pte Ltd

1 2015

Contents

Preface

There is a class of complex plant-wide systems which are composed of many physically or geographically divided subsystems. Each subsystem interacts with some so-called neighboring subsystems by their states and inputs. The technical target is to achieve a specific global performance of the entire system.

The classical centralized control solution, which could obtain a good global performance, is often impractical for application to a plant-wide system for computational reasons and lack of error tolerance. When the centralized controller fails or a control component fails, the entire system is out of control and the control integrity cannot be guaranteed.

The distributed (or decentralized) framework, where each subsystem is controlled by an independent controller, has the advantages of error-tolerance, less computational effort, and flexibility to system structure. Thus the distributed control framework is usually adopted in this class of system, in spite of the fact that the dynamic performance of centralized framework is better. Thus, how to improve the global performance under distributed control framework is a valuable problem.

Model predictive control (MPC), as a highly practical control technology with high performance, has been successfully applied to various linear and nonlinear systems in the process industries, and is becoming more widespread. The distributed framework of MPC, distributed MPC (DMPC), is also gradually developed with the development of communication network technologies in process industries that allow the control technologies and methodologies to utilize their potentials for improving control.

For the MPC algorithm applied to the plant-wide systems, the system's architectures can be divided as follows:

1. Centralized MPC, which is a MIMO system architecture;
2. Decentralized MPC, one controller-one subsystem, but no information exchange between controllers, and
3. Distributed MPC, which assumes that each subsystem can exchange information with its neighbor's subset of other subsystems.

Since the centralized MPC is forbidden for the large-scale plant-wide system with hundreds (or thousands) of inputs and outputs variables due to its lesser flexibility, weak error tolerance and the large cost of computation, the distributed framework is usually adopted despite its lower global performance. The schematic of distributed MPC is shown in Figure 1, the whole system is composed by many spatial distributed interconnected sub-systems. Each

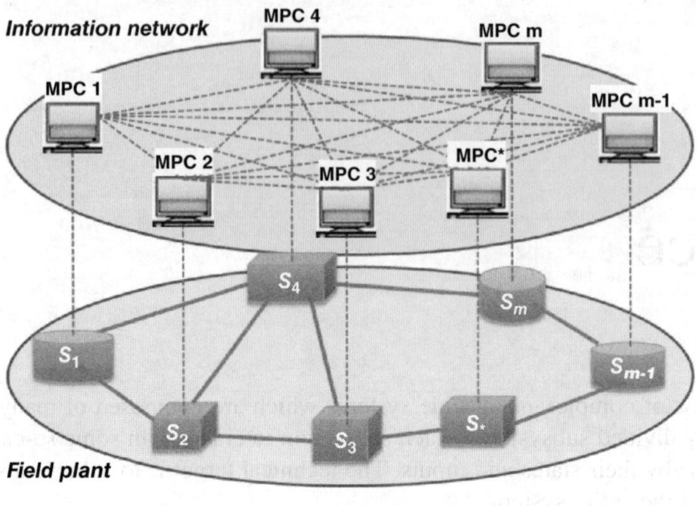

Figure 1 The schematic of distributed model predictive control

subsystem is controlled by a subsystem-based MPC and these controllers are interconnected by the network.

As mentioned before, how to improve the global performance under distributed control framework is a valuable problem. It is exactly true for the DMPC. There are many DMPC strategies and design methods in the literature, all to different ends. We have done extensive research in this topic for more than 10 years, and have proposed some strategies, e.g., the Nash optimization-based DMPC and the impacted region optimization based DMPC, etc. We found that the DMPC is definitely a useful method for large-scale plant-wide systems. Thus, we decided to write this book.

This book systematically introduces different distributed predictive control methods for plant-wide systems, including system decomposition, classification of distributed predictive control, unconstrained distributed predictive control, and the stabilized distributed predictive control with different coordinating strategies for different purposes, as well as the implementation examples of distributed predictive control. The major new contribution of this book is to show how the distributed MPCs can be coordinated efficiently for different control requirements, namely network connectivity, error tolerance, performance of entire closed-loop system, calculation speed, etc., and how to design distributed MPC. The remaining contents of this book are structured into four parts.

In the first part, we recall the main concepts and some fundamental results of the predictive control for discrete-time linear systems. The system structure model and some decomposition methods to present how to divide the entire system into interacting subsystems according to the specific control requirements is also introduced. Our intent is to provide the necessary background knowledge to understand the rest of the book.

The second part introduces the unconstrained distributed MPCs with different coordination strategies. The simplest and most practical local cost optimization based distributed MPC, Nash optimization based distributed MPC, the cooperative distributed MPC that can obtain very good performance of the entire system but each subsystem-based MPC of which requires the information of the whole system, and the networked distributed MPC with information constraints, which is a tradeoff between the two methods mentioned above. For primary readers, the major ideas and characteristics of distributed MPCs are clearly explained in a simple way without constraints.

The third part focuses on introducing the design of the stabilizing distributed MPCs with constraints for the three types of DMPCs: the local cost optimization based DMPC, the cooperative DMPC, and the networked DMPC with information constraint, respectively. The designed DMPCs can guarantee recursive feasibility and the asymptotic stability of the closed-loop system if the initial feasible solution exists.

In the last part, three practical examples are given to illustrate how to implement the introduced distributed MPC into industrial processes, they are the nonlinear networked DMPC for accelerated cooling processes in heavy plate steel mills, the speed train control with unconstrained networked DMPC, and the hierarchical DMPC for load control of a high building with multicooling resources.

In conclusion, this book tries to give a systematic overview of the latest distributed predictive control technologies to readers. We hope this book can help engineers to design control systems in their daily work or in their new projects. In addition, we believe that this book is fit for the graduate students who are pursuing their master or doctor degree in control theory and control engineering. We will be very pleased if this book is of use to you if you are interested in the control of plant-wide systems or predictive control.

Shaoyuan Li
Yi Zheng

About the Authors

Shaoyuan Li (IEEE Senior Member, 2006) is currently Professor and vice president of the School of Electronic Information and Electrical Engineering, Shanghai Jiao Tong University. He is also the discipline response person of Control Theory and Control Engineering, the vice director of Key Laboratory of Ministry of Education, the Vice President of the Chinese Association of Automation. He received his PhD in Computer and System Science from Nankai University in 1997. His research interests include predictive control, intelligent adaptive control, fuzzy intelligent control, and its applications. He has published five books and more than 200 papers in journals/conferences. Prof. Li has worked in the area of distributed model predictive control for more than 13 years.

Yi Zheng (IEEE Member, 2010) is Associate Professor at Shanghai Jiao Tong University and currently works in the School of Electronic Information and Electrical Engineering. He received his PhD in Control Theory and Engineering from Shanghai Jiao Tong University. He was with Shanghai Petrochemical Company, Ltd., Shanghai, from 2000 to 2003, GE-Global Research (Shanghai) from 2010 to 2012, and the University of Alberta from 2014 to 2015. His research interests include smart grids, model predictive control, system identification, and their applications to industrial processes. Zheng has worked in the area of distributed model predictive control for nearly 9 years.

Acknowledgement

This work was supported by the National Nature Science Foundation of China (61233004, 61221003, 61374109, 61304078), the National Basic Research Program of China (973 Program-2013CB035500), and partly sponsored by the International Cooperation Program of Shanghai Science and Technology Commission (12230709600), the Higher Education Research Fund for the Doctoral Program of China (20120073130006, 20110073110018), and the China Postdoctoral Science Foundation (2013M540364).

List of Figures

List of Tables

1

Introduction

1.1 Plant-Wide System

There is a class of systems which are composed of many interacted subsystems' industrial fields. Especially with the development of the advanced technology and the increase in the requirement of products, many new distributed processes have appeared, the processes of producing products have become more and more complex, and the scales of industrial processes have become more and more large. The automation structure for this kind of systems has changed from the traditional centralized automation system to a decentralized and centralized automation system, and then to a distributed automation system.

Correspondingly, the control algorithm and control structure for this kind of system change from centralized control and decentralized control to the distributed control system. The distributed control refers to a control system where each subsystem is controlled by an individual controller, and these controllers communicate with other subsystem-based controllers and are coordinated according to the exchanged information for obtaining good global performance or some special common goals. So far, the distributed control, especially the DMPC, has been studied and are still being studied by many scientists, and many theories and algorithms have been developed. We think it is the right time to introduce the distributed control to more students and engineers.

To make it more clear which kind of system is suitable for distributed control, we give some examples as follows.

1. **Wind power generation farm**
   In a wind turbine power generation farm, as shown in Figure 1.1, wind turbines are spatially distributed. The output wind flow rate of each wind turbine decreases with increasing generated power. It affects the input wind flow rate of the downstream wind turbines, and then their dynamics. In this way, these wind turbines interact with each other. For the automation system, each wind turbine is controlled by an individual controller. And these controllers are connected by a network (fieldbus) and are able to communicate with each other by the network.

Distributed Model Predictive Control for Plant-Wide Systems, First Edition. Shaoyuan Li and Yi Zheng.

Figure 1.1 The wind farm

2. **Multizone building temperature regulation system**
   Multizone building temperature regulation systems are a class of typical spatially distributed systems, as shown in Figure 1.2, which are composed of many physically interacted subsystems (rooms or zones) labeled as $\mathcal{S}_1$, $\mathcal{S}_2$, …, $\mathcal{S}_m$, respectively. The thermal influences between rooms of the same building occur through internal walls (the internal walls' isolation is weak) and/or door openings. A thermal meter and a heater (or air conditioner) are installed in each zone, which is used to measure and adjust the temperature of the multizone building.
3. **Distributed power network**
   Power networks are large networks consisting of a large number of components. The dynamics of the power network as a whole are the result of interactions between the individual components. The generators produce power that is injected into the network on the one side, while the loads consume power from the network on the other. If we consider each power plant, load, and station as a subsystem, it is a typical distributed system, whose subsystems interacted with each other and controlled separately.

In addition, since the number of players involved in the generation and distribution of power has increased significantly, in the near future, the number of source nodes of the power distribution network will increase even further as large-scale industrial suppliers and small-scale individual household will also start to feed electricity into the network. As a consequence, the structure of the power distribution network will change into a much more decentralized system with many generating sources and distribution agencies (Figure 1.3).

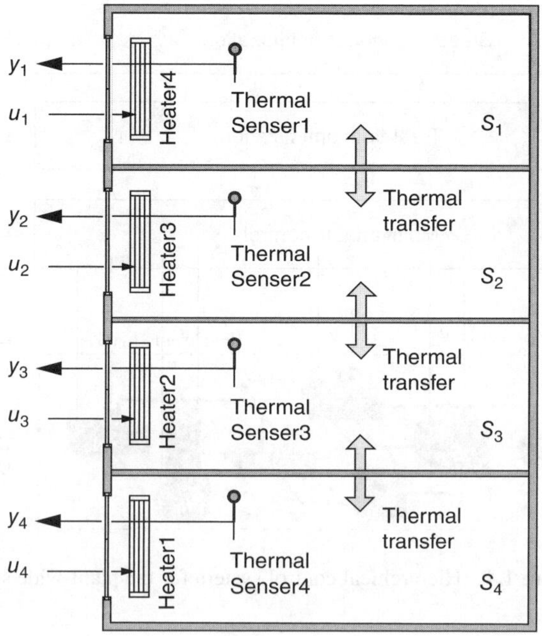

Figure 1.2 The multizone building temperature regulation system

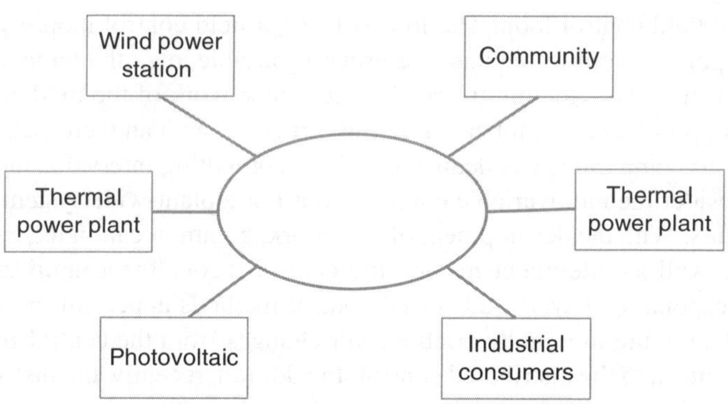

Figure 1.3 Distributed power generation power network

1.2 Control System Structure of the Plant-Wide System

The control structure is a very general concept. It includes how to schedule the controllers, and the inputs/outputs of each controller. The control system structure of the plant-wide system is shown in Figure 1.4, which is a hierarchical structure. The top layer, denoted as layer 4, is a steady economic optimization layer which is used to optimize the key process parameters, e.g., the product quantity, product quality, feeding material quality, etc. Layer 3 is a real-time optimization layer which dynamically optimizes the set-point of the multivariable layers.

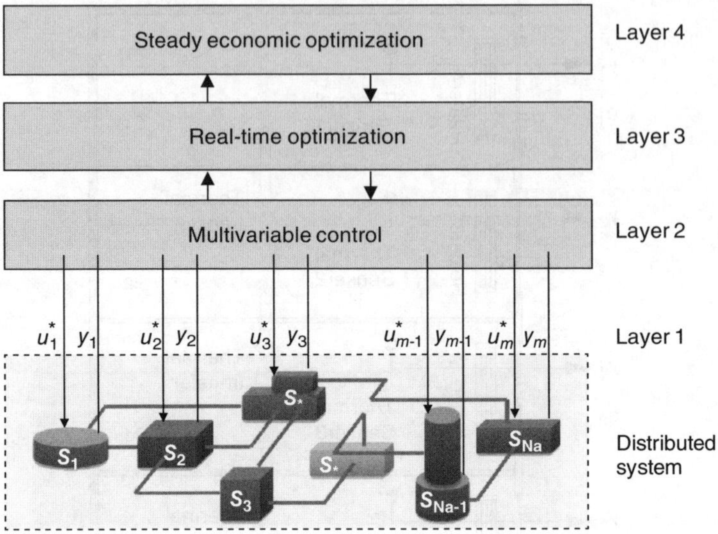

Figure 1.4 Hierarchical control system for the plant-wide system

This layer considers the dynamic economic performance and efficiency. The slow time variation of the process condition is taken into account in this layer. Below this layer is a multivariable layer which coordinates the interaction between each control loop and gives a set-point for the field control loop. The lowest layer, a field control loop layer, which is not drawn in this figure, is used to regulate the process variable, e.g., the temperature, flow rate, or pressure. In some cases, the multivariable takes some work of the field control loop layer when the control problem is complicated. In this structure, with an increase in the layer level, the information to communicate is deduced, and the computing interval is increased.

Here, we consider the multivariable control layer. For a plant-wide system, there are many inputs and outputs. With the development of a network, communication technology, and field-bus product, as well as intelligent meters, the control theory for a multivariable system is developed correspondingly. Many advanced control methods appear in the literature works, and the control structure in a multivariable layer changes from the centralized control to the decentralized control, to the distributed control. In addition, recently, the distributed structures for the real-time dynamic optimization layer and steady-state optimization layer have also appeared in the literature works. The real-time optimization layer and multivariable control loop are combined together in some cases. This is out of the scope of discussion in this book. In the following, three types of control structures, centralized control structure, decentralized control structure, and distributed control structure, in a multivariable control layer are specified to show the advantage of the distributed control framework.

1.2.1 Centralized Control

As shown in Figure 1.5, the centralized multivariable controller gets all the information of the plant-wide system, and then calculates the control law of all the inputs together, and sends the

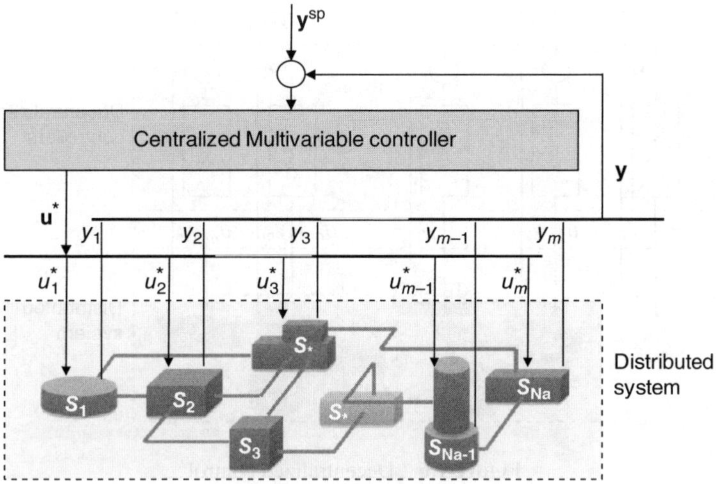

Figure 1.5 Centralized control

control signals to the actuators via the network. The control structure could achieve the best dynamic performance of the overall closed-loop system. However, since there are hundreds (or thousands) of input and output variables in a large-scale plant-wide system, the computational burden is unavoidably high if all control variables are solved together by a centralized controller during a controller period. In addition, since the information of the whole system is necessary when using centralized control, it requires that the network communication must be robust as the communication load is unavoidably high. Furthermore, under this control structure, if one of the subsystems does not work due to some fault, or we proceed with regular maintenance, the multivariable controller must be stopped, and the control of the whole system is interrupted. Thus, this control structure is not sufficiently flexible. Finally, it can be seen that if any part of the controller, actuators, sensors, networks, or control computer has a fault, the multivariable algorithm will lose its effectiveness, which means a low capability of error tolerance, which will not be expected by either the controller engineer or the owner of factories.

1.2.2 Decentralized Control and Hierarchical Coordinated Decentralized Control

Considering the less flexibility, the worse error tolerance, the large computational burden, and the heavy network communication load of centralized control, people decompose the centralized controller into many relevant small-scaled controllers, as shown in Figure 1.6. These controllers work with each other independently even when the corresponding controlled subsystems couple with each other. These classes of multivariable controllers have the advantages of simple structure, less computational burden, better error tolerance, good flexibility, and easy designing and implementation. Since the computation for obtaining the control law of the entire system is distributed to many small-scaled controllers, the computation burden of each controller is dramatically decreased. In addition, if several subsystems or controllers do not

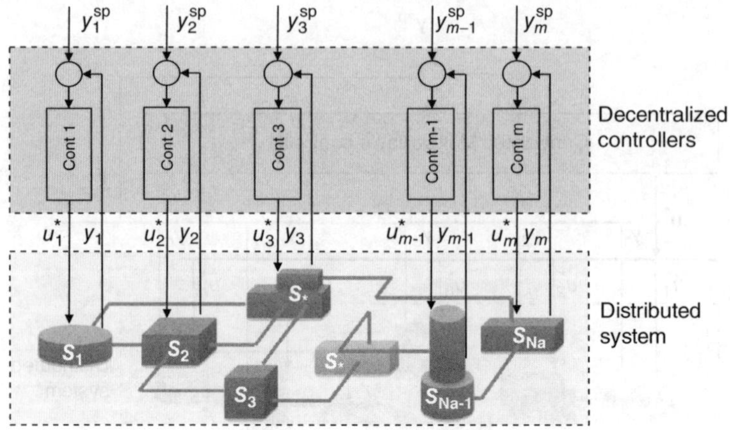

Figure 1.6 Decentralized control

work due to some fault, the other controllers are still able to work, which means good error tolerance. Furthermore, if there are some new subsystems required to be appended to or deleted from the existing plant-wide system, it needs to do nothing with the existing controllers, which means good structure flexibility.

However, since there is no communication and coordination among decentralized controllers, the controller performance is destroyed if the coupling among subsystems is strong enough. In order to avoid the degradation of the performance of the global system, one method is to enlarge the scale of each local controller, where several strong coupled subsystems are controlled by one local controller. By using this strategy, the performance of the global system could be guaranteed, but the computational burden of each local controller is increased, and the flexibility of overall system is deduced. This strategy bypasses and does not solve the problem of how to improve the global performance when strong interactions exist among the subsystems each of which is controlled by a separated controller.

Unfortunately, in most cases, strong couplings exist in the plant-wide system. Thus, people add a coordinator to coordinate each subsystem-based controller for improving the global performance of the entire plant-wide system, as shown in Figure 1.7, and is called hierarchical coordinating decentralized control. Through different coordinating algorithms, the global performance of the entire system could be significantly improved if strong interactions exist. However, all local controllers should communicate with the coordinator as the global information is necessary for the coordinator. The centralized structure appeared in the coordinator.

1.2.3 Distributed Control

Recently, with the development of computer technologies, fieldbus, network communicating technologies, and smart meter in process industries, which allows the control technologies and methodologies to utilize their potentials for improving control, the distributed control structure has appeared gradually instead of the centralized and decentralized structure for the plant-wide system. As shown in Figure 1.8, the global system is divided into many interacted subsystems, and each subsystem is controlled by a separate controller; these peer controllers communicate

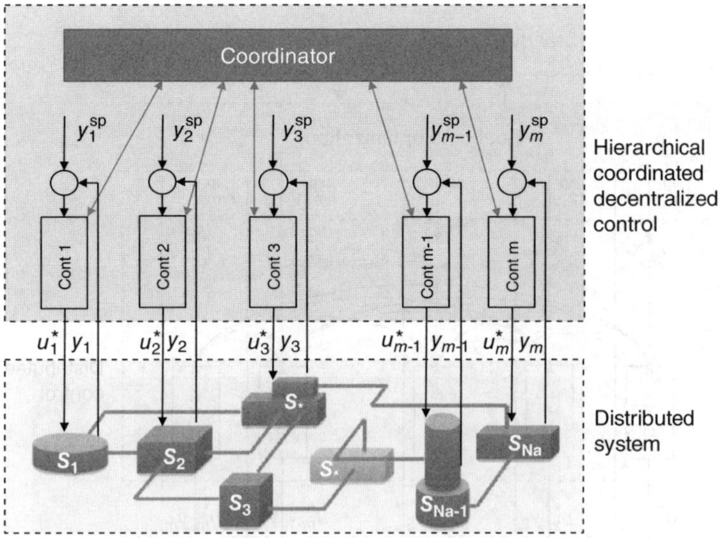

Figure 1.7 Hierarchical coordinated decentralized control

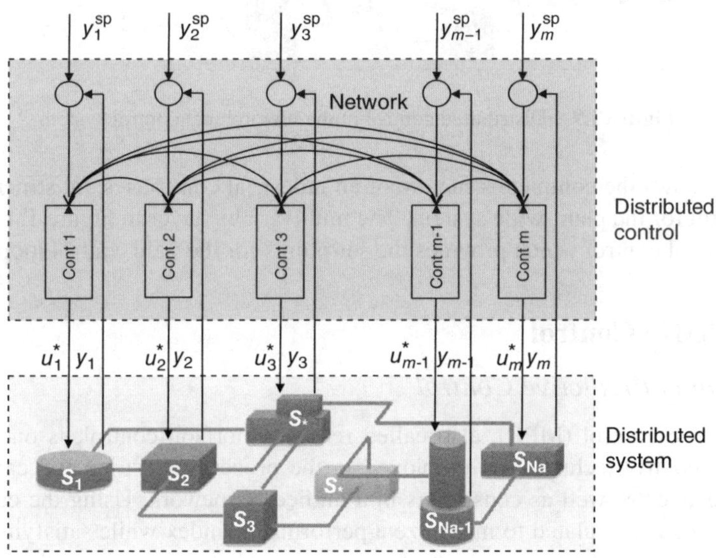

Figure 1.8 Distributed control

with each other through a network for achieving good global performance or a specifically common goal. This kind of control structure has the advantage of a decentralized control structure, e.g., high flexibility and good error tolerance, and the advantage of a centralized control structure, e.g., good global performance. In a distributed control structure, the most important problem is how to design coordinating strategies for different purposes.

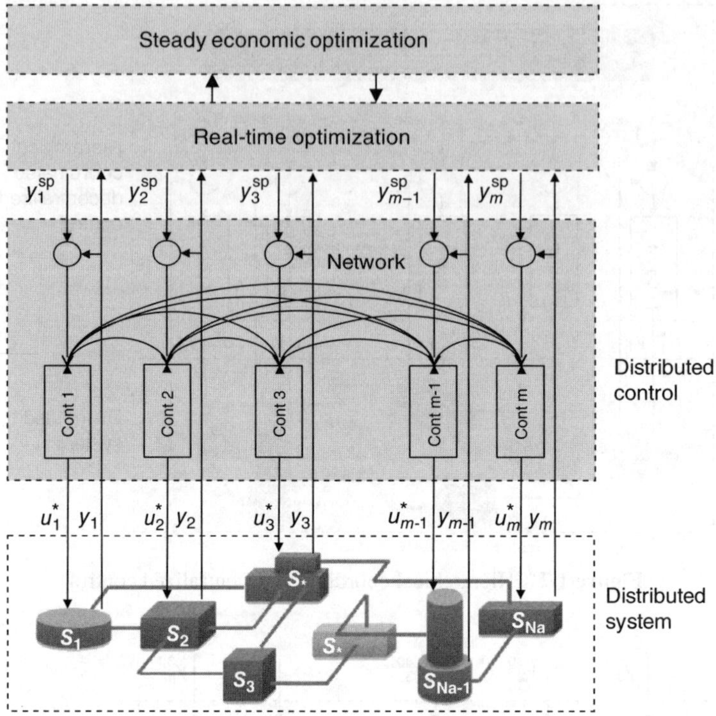

Figure 1.9 Distributed control in the hierarchical control system

Figure 1.9 shows the complete structure of an industrial control system structure with distributed control for the plant-wide system. The multivariable layer in Figure 1.4 is substituted by the distributed control which provides the set-points for the field control loops.

1.3 Predictive Control

1.3.1 What is Predictive Control

Model predictive control (MPC), also called receding horizon control, is one of the leading advanced control technologies employed in the process industries and can incorporate complex objectives as well as constraints in a unified framework. Using the current state, a control sequence is calculated to minimize a performance index while satisfying some specified constraints. Only the first element of the sequence is taken as controller output. At the next sampling time, the optimization is resolved with new measurements from the plant [1–3].

Predictive control was pioneered simultaneously by Richalet *et al.* [4] and Cutler and Ramaker [5]. The first implemented algorithms and successful applications were reported in the papers mentioned above. The use of finite impulse response models and finite step response models, which are easy to obtain for open loop stable processes, partly explains the wide acceptance especially in the hydrocarbon processing industries. Since the end of the seventies and early eighties, MPC has become the most widely applied multivariable control technique and many papers report that MPC has been applied successfully to various linear [3, 6, 7], nonlinear [8–11] systems in process industries and is becoming more widespread [3, 12, 13]. Some examples are a distillation column [6, 14], a fluidized bed catalytic cracker [15], a hydrocracker [16], a utility boiler [17], a chemical reactor [1], a transonic wind turbine [18], a pulp and paper plant [3], and a metallurgical process [12, 19–21]. Applications of MPC to faster systems were also reported, such as a mechatronic servo system [22], a power converter [23], and a robot arm [24]. This list is far from complete, but it gives an impression of the wide range of MPC applications [25].

1.3.2 Advantage of Predictive Control

Predictive control is widely recognized as a high practical control technology with high performance. It has a significant and widespread impact on industrial process control. The penetration of predictive control into industrial practice has also been helped by the following facts [2, 26]:

- Its underlying ideas are easy to understand.
- It handles multivariable control problems naturally.
- It is more powerful than proportional integral derivative (PID) control, even for single loops without constraints. It is easier to tune than PID even on "difficult" loops such as those containing long time delay.
- It is the unique control method which can deal routinely with equipment and safety constraints.
- It often obtains very small mean square error (MSE) of process variables, which allows operation closer to constraints compared with conventional control, and then frequently leads to more profitable operation.

In addition, MPC is rather a methodology than a single technique. The difference in the various methods is mainly the way the problem is translated into a mathematical formulation. However, in all methods three important items are recognizable in the design procedure: the prediction model, receding horizon optimization, and the output feedback and correction.

1.4 Distributed Predictive Control

1.4.1 Why Distributed Predictive Control

For a class of large-scale system with hundreds or thousands of inputs and outputs variables (e.g., power and energy network, large chemical processes), the classical centralized MPC, where a control agent is able to acquire the information of the global system and could obtain a good global performance, is often impractical to apply to a large-scale system for some reasons: (1) there are hundreds of inputs and outputs. It requires large computational efforts in online implementation; (2) when the centralized controller fails, the entire system is out of control and the control integrity cannot be guaranteed when a control component fails; (3) in some cases, e.g., in a multi-intelligent vehicle system, the global information is unavailable to each controller. Thus, the DMPC appears and gradually substitutes the centralized MPC.

The distributed predictive not only inherits the advantages of MPC of directly handling constraints and good optimization performance, but also has the characteristics of a distributed control framework of less computational burden, high flexibility, and good error tolerance. Using distribute predictive control, the future state information of each subsystem is able to feed into its interacted subsystem-based MPC and then satisfy the versatile control objective, e.g., large lag system, and more restrict control performance requirements.

1.4.2 What is Distributed Predictive Control

For a class of large-scale systems with hundreds or thousands of input and output variables (e.g., power and energy network, large chemical processes), as shown in Figure 1.10, the whole system is properly partitioned into several interconnected subsystems and controlled in a distributed structure. Each subsystem is controlled by a local controller, and these local controllers are interconnected by a network. If the algorithm running in each local controller is predictive control, as shown in Figure 1.10, we call the whole control the distributed predictive control. In the distributed predictive control, each local predictive control coordinates with another one by exchanging the network information. More simplified, the distributed predictive control is the distributed implementation of a set of predictive controllers, and these predictive controllers consider the feedforward information from the predictive controllers corresponding to the subsystems they interacted with.

1.4.3 Advantage of Distributed Predictive Control

The distributed predictive control not only inherits the advantages of MPC of directly handing constraints and good optimization performance, but also has the characteristics of the distributed control framework of less computational efforts, high flexibility, good error

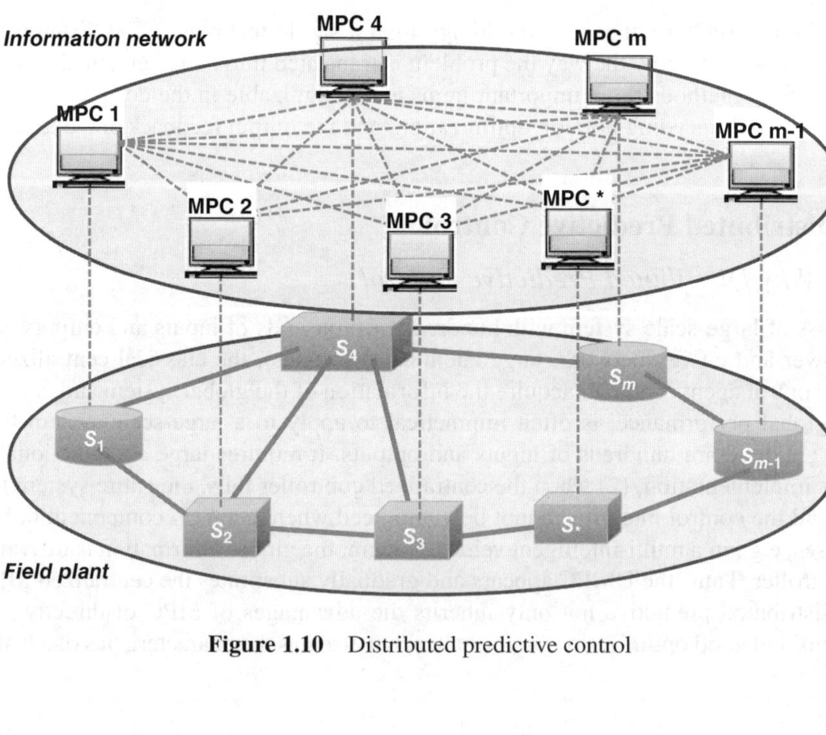

Figure 1.10 Distributed predictive control

tolerance, and no global information requirements [27, 28]. The advantages of the distributed predictive control are as follows:

- Its underlying ideas are easy to understand: the distributed predictive control is the distributed implementation of a set of predictive controllers, and these predictive controllers consider the feedforward information from the predictive controllers which corresponds to the subsystems they interacted with.
- The local predictive control can deal routinely with equipment and safety constraints.
- The local predictive control handles multivariable control problems naturally. It is more powerful than PID control, even for single loops without constraints. It is not much more difficult to tune, even on "difficult" loops such as those containing long time delay.
- It allows operation closer to constraints compared with conventional control, which frequently leads to more profitable operation.
- Since the centralized predictive control is decomposed into many small-scaled predictive controllers, the computational efforts in each small-scaled predictive control are much less than that used for solving the centralized predictive control.
- If one or several errors occur in a subsystem, the other subsystem-based predictive controllers are still able to work. There is a good error-tolerance characteristic.
- If some new subsystems are appended into the current system, it is not necessary to modify all the local predictive controls. We should only modify the predictive control whose corresponding subsystem interacts with the new added subsystems. The distributed predictive control owns high flexibility to the system structure.
- The "plug-in and plug-out" is also able to be realized if a suitable algorithm and an appropriate program are designed.

Due to these advantages, the distributed predictive control gradually takes the place of a centralized predictive control for plant-wide systems. However, as pointed out in [27–33], the optimization performance of distributed predictive control, in most cases, is not as good as that of centralized predictive control. Thus, many different coordinating strategies are proposed to solve this problem [27, 29, 31–44]. In most cases, the coordinating strategies are very important to the performance of the closed-loop systems.

1.4.4 Classification of DMPC

To improve the global performance of the DMPC, several coordination strategies have appeared in the literature, and can be classified according to the information exchange protocol needed, and to the type of cost function which is optimized [6]. There are two classes of distributed predictive control if we catalog it by the information exchange protocol.

- *Noniterative-based algorithm*: in this kind of distributed predictive control, each local predictive control communicates only once with other local predictive control within every single control period, and solves the local control law once in a control period, e.g., [34, 44–47].
- *Iterative-based algorithm*: this kind of distributed predictive control assumes that the network communication resources are abundant enough for supporting the fact that each local

predictive control communicates with other interacted local predictive control many times in a single control period. And the time cost by communicating is very little, such that it could be ignored as compared to the control period. Each local predictive control solves its optimal control law based on the presumed control sequence. Then it transforms this control law to its interacted local predictive controllers. After that, each local predictive solves the new optimal control law based on the optimal control law based on its neighbors' optimal control laws solved at the last previous iteration, and then repeats this process until the iteration broken-down conditions are satisfied, e.g., [19, 29, 37, 48, 49].

The noniterative algorithms consume less communication resources than the iterative algorithms, and have a fast computation speed in comparison to the iterative algorithms. The iterative algorithms are able to achieve a better global performance than the noniterative algorithms.

There are three kinds of DMPCs if we classify DMPCs by the cost function of each local predictive control. And the DMPCs that accommodate the same kind of cost function for each subsystem-based MPC can be solved either by the iterative algorithm or by the noniterative algorithm. We briefly review these methods as motivations for the content to be presented later in the book.

- Local cost optimization-based DMPC (LCO-DMPC): distributed algorithms where each subsystem-based controller minimizes the cost function of its own subsystem were proposed in [1–4]

$$J_i(k) = \|\mathbf{x}_i(k+N)\|^2_{\mathbf{P}_i} + \sum_{s=0}^{N-1}\left(\|\mathbf{x}_i(k+s)\|^2_{\mathbf{Q}_i} + \|\mathbf{u}_i(k+s)\|^2_{\mathbf{R}_i}\right) \tag{1.1}$$

  When computing the optimal solution, each local controller exchanges state estimation with the neighboring subsystems to improve the performance of the local subsystem. This method is simple and very convenient for implementation. An extension of this stabilizing DMPC with input constraint for nonlinear continuous systems is given in [51, 52], and a stabilizing DMPC with input and state constraints is given in [50].

  Also, an iterative algorithm for DMPC based on *Nash optimality* was developed in [1]. The whole system will arrive at Nash equilibrium if the convergence condition of the algorithm is satisfied.
- Cooperative distributed MPC (C-DMPC): to improve the global performance, distributed algorithms, where each local controller minimizes a global cost function

$$J_i(k) = \sum_{j\in \mathrm{P}} J_j(k) \tag{1.2}$$

  were proposed in [31, 37, 44, 48, 53]. In this method, each subsystem-based MPC exchanges information with all other subsystems. And some iterative stabilizing designs are proposed which take the advantages of the model of the whole system, and are used in each subsystem-based MPC. This strategy may result in a better performance but consumes much more communication resources, in comparison with the method described in (1.1).
- Networked DMPC with information constraints (N-DMPC): to balance the performance, communication cost, and the complexity of the DMPC algorithm, a novel coordination strategy was recently proposed in [19, 47, 54]. Here each subsystem-based controller minimizes

its corresponding subsystem's cost function and the cost function of the subsystems directly impacts on

$$\overline{J}_i(k) = \sum_{j \in \mathcal{P}_i} J_j(k) \tag{1.3}$$

where $\mathcal{P}_i = \{j : j \in \mathcal{P}_{-i} \text{ or } j = i\}$ is the set of subscripts of the downstream subsystems of subsystem $\mathcal{S}_i$, that is the region impacted on by subsystem $\mathcal{S}_i$. The resulting control algorithm is termed as an impacted-region cost optimization-based DMPC (ICO-DMPC) [55–57] or N-DMPC with communication constraints. It could achieve a better performance than the first method, and its communication burden is much less than the second method. Clearly, this coordination strategy as proposed in [19, 47, 54] and described in (1.3) is a preferable method to trade off the communication burden and global performance.

Some other kinds of DMPC formulations are also available in [11, 13, 29, 46, 48, 51–54, 58–64]. Among them, the methods described in [52, 62] are proposed for a set of decoupled subsystems, and the extension of [52] could handle systems with weakly interacting subsystem dynamics [51]. There is no absolute priority among these different distributed predictives. One could select different algorithms according to their purpose of employing the control system.

1.5 About this Book

This book systematically introduces the distributed predictive control with different coordination strategies for the plant-wide system, including the system decomposition, classification of distributed predictive control, unconstrained distributed predictive control, and the stabilized distributed predictive control with different coordinating strategies for different purposes, as well as the implementation examples of distributed predictive control. The major new contribution of this book is to show how the DMPCs can be coordinated efficiently for different control requirements, namely the network connectivity, error tolerance, performance of the entire closed-loop system, calculation speed, etc. This book also describes how to design DMPC. The latest theory and technologies of DMPC for coupling discrete-time linear systems are introduced in this book. The rest of this book is structured into four parts, as shown in Figure 1.11, and are organized as follows.

In the first part, Chapters 2 and 4, we recall the main concepts and some fundamental results of the predictive control for discrete-time linear systems. Some existing results about the solution and the stability of the closed-loop system under the control of MPC are provided in this part. The system model, structure model, and some decomposition methods, e.g., the relative gain array (RGA), N-step accessible matrix-based decomposition, are also introduced in this chapter to present how to divide the entire system into the interacted subsystems according to the specific control requirements. Then some coordination strategies are introduced according to the classification of coordination degree (the optimization index of each subsystem-based MPC).

Our intent is to provide only the necessary background for the understanding of the rest of the book.

In the second part, Chapters 5–7, the unconstrained DMPCs with different coordination strategies are introduced for primary readers, since the major ideas and the characteristics of

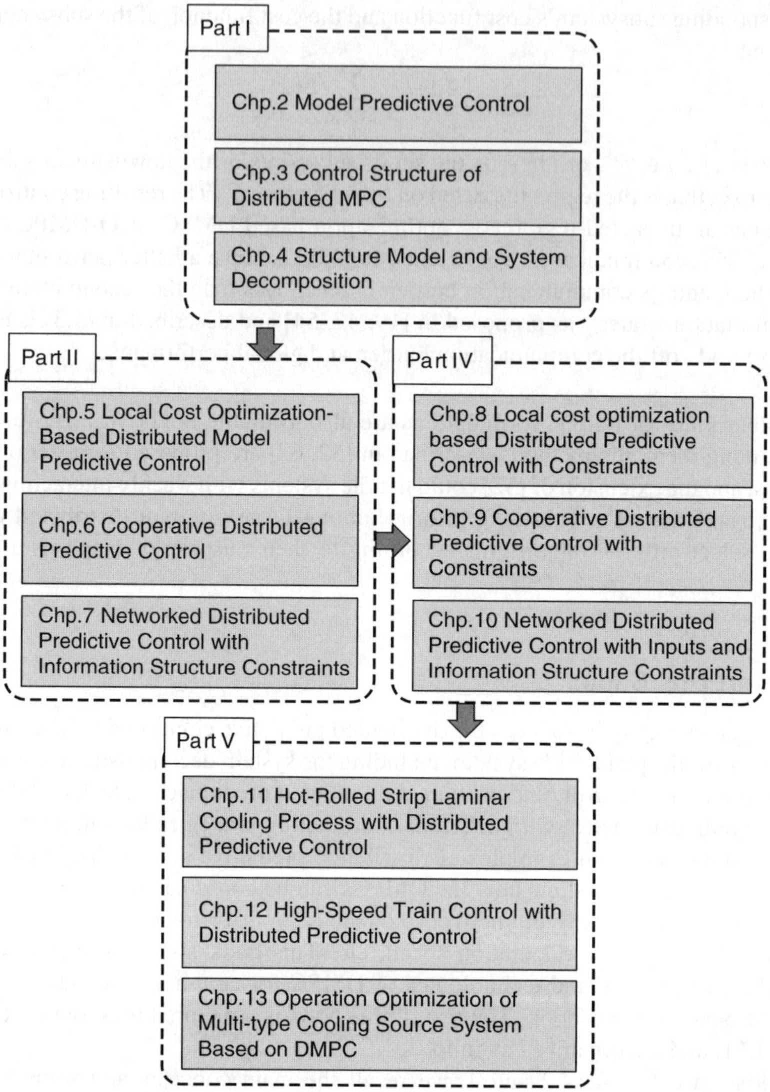

Figure 1.11 Content of this book

DMPCs can be clearly explained in a simple way without constraints. Chapter 5 presents the LCO-DMPC (the simplest and most practical one) and Nash optimization-based DMPC (the solution of which could obtain Nash optimality). Chapter 6 provides the C-DMPC which could obtain very good performance of the entire system but each subsystem-based MPC of which requires the information of the whole system. Chapter 7 introduces the N-DMPC with information constraints which is a tradeoff between the two methods mentioned above. Both iterative algorithm and noniterative algorithm for solving the optimal solution of subsystem-based MPC are given in each coordinating strategy. The predictive model, explicit solution, and stability analysis of each algorithm are also detailed in this part.

In the third part, Chapters 8–10, we focus on introducing the design methods of the stabilizing DMPCs with constraints for the advanced readers. In Chapter 8, a design method for the LCO-DMPC is developed, which is based on a dual mode scheme and is able to handle input constraints. The feasibility and stability of this method are analyzed. In addition, Chapter 9 introduces a stabilizing DMPC with constraints, in which each subsystem-based MPC optimizes the cost of whole system. The consistency constraints, which limit the error between the optimal input sequence calculated at the previous time instant, referred to as the presumed inputs, and the optimal input sequence calculated at the current time instant to within a prescribed bound, are designed and included in the optimization problem of each local predictive control. The noniterative algorithm for the related fast process is designed for solving each local predictive control. Both the feasibility and stability of this method are analyzed. Chapter 10 provides a networked distributed predictive control with inputs and information constraints, where each local predictive control optimizes not only its own performance but also that of the systems it directly impacted on. The consistency and stability constraints are designed to guarantee the recursive feasibility and the asymptotical stability of the closed-loop system if the initial feasible solution exists.

In the last part, Chapters 11–13, three practical examples are given to illustrate how to implement the introduced DMPC into the industrial process. At first, the implementation of DMPC to accelerated cooling processes in heavy plate steel mills is introduced. The control problem, the system model, the system decomposition, the control strategy, and the performance of the closed-loop system under the control of DMPC are provided. Then, different from the metallurgical process, one example of the speed train control with DMPC is presented and the technical details are also provided. Finally, a load control of a high building in Shanghai with multicooling resources system is studied, and the distributed predictive with a scheduling layer is developed and detailed in Chapter 14.

In conclusion, this book tries to give systematic and latest distributed predictive control technologies to the readers. We hope this book could help engineers to design their control systems in their daily work or in their new projects. In addition, we believe that this book is fit for the graduate students who are pursuing their master's or doctor's degree in control theory and control engineering. We will be very pleased if this book could really do something for you if you are interested in the control of a plant-wide system or predictive control.

Part One
Foundation

2

Model Predictive Control

2.1 Introduction

Predictive control appears to have been proposed independently by several people, more or less simultaneously. The pioneers were mostly industrial practitioners who implemented predictive control several years before the first publications appeared, so the publication dates do not tell the whole story.

The first description of MPC control applications was presented by Richalet *et al.* of the French company Adersa in 1976's conference [65] and later summarized in 1978's Automatica paper [4]. They described their approach as model predictive heuristic control (MPHC). The solution software was referred to as IDCOM. The emphasis of MPHC was on a control methodology which could be applied to problems too difficult to be handled by conventional PID control, but which was based on intuitive concepts and offered ease of tuning [66].

In the early 1970s, with an initial application in 1973 [66], engineers at Shell Oil developed their own independent MPC technology. Cutler and Ramaker presented details of an unconstrained multivariable control algorithm which they named dynamic matrix control (DMC) in 1979 [5]. Constraint handling, however, was still somewhat ad hoc at that time. Engineers at Shell Oil addressed this weakness by posing the DMC algorithm as a quadratic program (QP) in which input and output constraints appear explicitly, and Cutler *et al.* first described the quadratic programming solution of dynamic matrix control (QDMC) algorithm in 1983 [67]. Several years later, a more comprehensive description was published in [68]. This method emphasized optimal plant operation under constraints, and computed the control signal by repeatedly solving a linear programming (LP) problem. DMC went on to become the most well known of the commercial predictive control products [66].

All of these proposals shared the essential features of predictive control: an explicit internal model, the receding horizon idea, and computation of the control signal by optimizing predicted plant behavior. Currently predictive control has become the most popular algorithm, and the only advanced control methodology which has made a significant impact on industrial control engineering. The main reasons behind its success in these applications are as follows [26]:

- It handles multivariable control problems naturally.
- It can take account of actuator limitations.

Distributed Model Predictive Control for Plant-Wide Systems, First Edition. Shaoyuan Li and Yi Zheng.

- It allows operation closer to constraints (compared with conventional control), which frequently leads to more profitable operation.
- Control update rates are relatively low in these applications, so there is plenty of time for the necessary online computations.

As pointed out in [69], early versions of MPC and generalized predictive control did not automatically ensure stability, thus requiring tuning. This is the reason why that research in the 1990s devoted considerable attention to this topic. Research on the stability of model predictive controlled systems has now reached a relatively mature stage. The important factors for stability have been isolated and employed to develop a range of model predictive controllers that are stabilizing and differ only in their choice of the three ingredients (terminal cost, terminal constraint set, and terminal local controller) that are common to most forms of MPC [69]. These conditions are merely sufficient and several researchers [70] are seeking relaxation. Among the stabilized MPC design methods, a dual mode MPC was proposed in [71] and developed in [72, 73]. In the dual mode version, a terminal set is chosen. It solves an open loop optimal control problem solved online until all states enter the terminal set. When all the states are in their terminal set, a local feedback control law is employed to steer the states to the origin.

In this chapter, the dynamic matrix control, state space model-based predictive control, and the dual mode predictive control are introduced, since these will be used as fundamentals in the following chapters.

2.2 Dynamic Matrix Control

DMC is an algorithm based on the step response model. It applies incremental algorithms, which are very effective in removing the steady-state error. Up to now, DMC is the most widely accepted in the process industry. This section mainly refers to [2, 74].

2.2.1 Step Response Model

Suppose the system is at rest. For a linear time-invariant single-input single-output (SISO) system, let the output change for a unit input change Δu be given by

$$\{0, s_1, s_2, \dots, s_N, s_{N+1}, \dots\}$$

Here, we suppose that the system settles exactly after N steps. The step response $\{s_1, s_2, \dots, s_N\}$ constitutes a complete model of the system, which allows us to compute the system output for any input sequence,

$$y(k) = \sum_{l=1}^{N} s_l \Delta u(k-l) + s_{N+1} u(k-N-1) \tag{2.1}$$

where $\Delta u(k-l) = u(k-l) - u(k-l-1)$. Note that when $s_N = s_{N-1}$, (2.1) is equivalent to

$$y(k) = \sum_{l=1}^{N} s_l \Delta u(k-l) + s_N u(k-N) \tag{2.2}$$

Step response model (2.1) can only be used in stable processes. For a multiple-inputs and multiple-outputs (MIMO) process with m inputs and r outputs, one obtains a series of step response coefficient matrices

$$\mathbf{S}_l = \begin{bmatrix} s_{11,l} & s_{12,l} & \cdots & s_{1m,l} \\ s_{21,l} & s_{22,l} & \cdots & s_{2m,l} \\ \vdots & \vdots & \ddots & \vdots \\ s_{r1,l} & s_{r2,l} & \cdots & s_{rm,l} \end{bmatrix} \tag{2.3}$$

where $s_{ij,l}$ is the lth step response coefficient relating the jth input to the ith output.

2.2.2 *Prediction*

Suppose $y_0(k+l|k)$, $l=1,2,\ldots,P+1$, is the output prediction when the current and future control moves are kept invariant. Then, it is shown that when only the jth input is changed, other inputs remain invariant. Then considering the ith output predictions in future P sampling instants responding to the change of the jth input, yields

$$\begin{aligned} y_{ij}(k+1|k) &= y_{i,0}(k+1|k) + s_{ij,1}\Delta u(k) \\ &\vdots \\ y_{ij}(k+M|k) &= y_{i,0}(k+M|k) + s_{ij,M}\Delta u(k) + s_{ij,M-1}\Delta u(k+1|k) + \cdots \\ &\quad + s_{ij,1}\Delta u(k+M-1|k) \\ y_{ij}(k+M+1|k) &= y_{i,0}(k+M+1|k) + s_{ij,M+1}\Delta u(k) + s_{ij,M}\Delta u(k+1|k) + \cdots \\ &\quad + s_{ij,2}\Delta u(k+M-1|k) \\ &\vdots \\ y_{ij}(k+P|k) &= y_{i,0}(k+P|k) + s_{ij,P}\Delta u(k) + s_{ij,P-1}\Delta u(k+1|k) + \cdots \\ &\quad + s_{ij,P-M+1}\Delta u(k+M-1|k) \end{aligned} \tag{2.4}$$

where the notation $y_{ij}\,(k+l|k)$, $\Delta u(k+l|k)$, indicates that this estimate is based on measurements up to time k, that is, on measurements of the outputs up to $\mathbf{y}(k)$.

Writing the output predictions in a vector form, we directly obtain

$$\tilde{\mathbf{y}}_{ij}(k|k) = \tilde{\mathbf{y}}_{i,0}(k|k) + \mathbf{A}_{ij}\Delta\tilde{\mathbf{u}}_j(k|k) \tag{2.5}$$

where

$$\begin{aligned} \tilde{\mathbf{y}}_{ij}(k|k) &= [y_{ij}(k+1|k), y_{ij}(k+2|k), \ldots, y_{ij}(k+P|k)]^{\mathrm{T}} \\ \tilde{\mathbf{y}}_{i,0}(k|k) &= [y_{i,0}(k+1|k), y_{i,0}(k+2|k), \ldots, y_{i,0}(k+P|k)]^{\mathrm{T}} \\ \Delta\tilde{\mathbf{u}}_j(k|k) &= [\Delta u_j(k+1|k), \Delta u_j(k+2|k), \ldots, \Delta u_j(k+M-1|k)]^{\mathrm{T}} \end{aligned}$$

$$\mathbf{A}_{ij} = \begin{bmatrix} s_{ij,1} & 0 & \cdots & 0 \\ s_{ij,2} & s_{ij,1} & \cdots & 0 \\ \vdots & \vdots & \ddots & \vdots \\ s_{ij,M} & s_{ij,M-1} & \cdots & s_{ij,1} \\ \vdots & \vdots & \ddots & \vdots \\ s_{ij,P} & s_{ij,P-1} & \cdots & s_{ij,P-M+1} \end{bmatrix}$$

Suppose all the inputs can be changed. Then considering the ith output and applying the superposition principle yields

$$\tilde{\mathbf{y}}_i(k|k) = \tilde{\mathbf{y}}_{i,0}(k|k) + \sum_{j=1}^{r} \mathbf{A}_{ij}\Delta\tilde{\mathbf{u}}_j(k|k) \tag{2.6}$$

where

$$\tilde{\mathbf{y}}_i(k|k) = [y_i(k+1|k), y_i(k+2|k), \ldots, y_i(k+P|k)]^{\mathrm{T}}$$

Considering all the inputs and outputs yields

$$\mathbf{Y}(k|k) = \mathbf{Y}_0(k|k) + \tilde{\mathbf{A}}\Delta\mathbf{U}(k|k) \tag{2.7}$$

where

$$\mathbf{Y}(k|k) = \left[\tilde{\mathbf{y}}_1(k|k)^{\mathrm{T}} \quad \tilde{\mathbf{y}}_2(k|k)^{\mathrm{T}} \quad \cdots \quad \tilde{\mathbf{y}}_r(k|k)^{\mathrm{T}}\right]^{\mathrm{T}}$$

$$\mathbf{Y}_0(k|k) = \left[\tilde{\mathbf{y}}_{1,0}(k|k)^{\mathrm{T}} \quad \tilde{\mathbf{y}}_{2,0}(k|k)^{\mathrm{T}} \quad \cdots \quad \tilde{\mathbf{y}}_{r,0}(k|k)^{\mathrm{T}}\right]^{\mathrm{T}}$$

$$\Delta\mathbf{U}(k|k) = \left[\Delta\tilde{\mathbf{u}}_1(k|k)^{\mathrm{T}} \quad \Delta\tilde{\mathbf{u}}_2(k|k)^{\mathrm{T}} \quad \cdots \quad \Delta\tilde{\mathbf{u}}_m(k|k)^{\mathrm{T}}\right]^{\mathrm{T}}$$

$$\tilde{\mathbf{A}} = \begin{bmatrix} \mathbf{A}_{11} & \mathbf{A}_{12} & \cdots & \mathbf{A}_{1m} \\ \mathbf{A}_{21} & \mathbf{A}_{22} & \cdots & \mathbf{A}_{2m} \\ \vdots & \vdots & \ddots & \vdots \\ \mathbf{A}_{r1} & \mathbf{A}_{r2} & \cdots & \mathbf{A}_{rm} \end{bmatrix}$$

2.2.3 *Optimization*

Suppose the criterion for optimizing $\Delta\mathbf{U}(k|k)$ is to minimize the following cost function:

$$J(k) = \|\mathbf{E}(k|k)\|^2_{\tilde{\mathbf{W}}} + \|\Delta\mathbf{U}(k|k)\|^2_{\tilde{\mathbf{R}}} \tag{2.8}$$

where $\tilde{\mathbf{W}} \geq 0$ and $\tilde{\mathbf{R}} \geq 0$ are symmetric matrices

$$\mathbf{E}(k|k) = \left[\tilde{\mathbf{e}}_1(k|k)^{\mathrm{T}} \quad \tilde{\mathbf{e}}_2(k|k)^{\mathrm{T}} \quad \cdots \quad \tilde{\mathbf{e}}_n(k|k)^{\mathrm{T}}\right]^{\mathrm{T}}$$

$$\tilde{\mathbf{e}}_i(k|k) = \left[e_i\,(k+1|k) \quad e_i(k+2|k) \quad \cdots \quad e_i(k+P|k)\right]^{\mathrm{T}}$$

$$e_i(k+l|k) = y_{i,r}(k+l) - y_i(k+l|k)$$

Then, when $\tilde{\mathbf{A}}^{\mathrm{T}}\tilde{\mathbf{W}}\tilde{\mathbf{A}} + \tilde{\mathbf{R}}$ is nonsingular, minimization of (2.8) yields

$$\Delta\mathbf{U}(k) = (\tilde{\mathbf{A}}^{\mathrm{T}}\tilde{\mathbf{W}}\tilde{\mathbf{A}} + \tilde{\mathbf{R}})^{-1}\tilde{\mathbf{A}}^{\mathrm{T}}\tilde{\mathbf{W}}\mathbf{E}_0(k) \tag{2.9}$$

where

$$\mathbf{E}_0(k|k) = \left[\tilde{\mathbf{e}}_{1,0}(k|k)^{\mathrm{T}} \quad \tilde{\mathbf{e}}_{2,0}(k|k)^{\mathrm{T}} \quad \cdots \quad \tilde{\mathbf{e}}_{n,0}(k|k)^{\mathrm{T}}\right]^{\mathrm{T}}$$

$$\tilde{\mathbf{e}}_{i,0}(k|k) = \left[e_{i,0}\,(k+1|k) \quad e_{i,0}(k+2|k) \quad \cdots \quad e_{i,0}(k+P|k)\right]^{\mathrm{T}}$$

$$e_{i,0}(k+l|k) = y_{i,r}(k+l) - y_{i,0}(k+l|k)$$

$y_{i,r}\,(k+l)$ is the set-point value at the future time $k+1$ for the ith output; $y_{i,0}\,(k+lk)$ is the prediction on the ith output at future time $k+1$, when the control moves for the time k and future sampling instants are kept invariant.

At each time k, implement the following control move:

$$\Delta\mathbf{u} = \mathbf{D}\mathbf{E}_0(k) \tag{2.10}$$

where

$$\mathbf{D} = \mathbf{L}(\tilde{\mathbf{A}}^{\mathrm{T}}\tilde{\mathbf{W}}\tilde{\mathbf{A}} + \tilde{\mathbf{R}})^{-1}\tilde{\mathbf{A}}^{\mathrm{T}}\tilde{\mathbf{W}}$$

$$\mathbf{L} = \begin{bmatrix} \boldsymbol{\theta} & \mathbf{0} & \cdots & \mathbf{0} \\ \mathbf{0} & \boldsymbol{\theta} & \cdots & \mathbf{0} \\ \vdots & \vdots & \ddots & \vdots \\ \mathbf{0} & \mathbf{0} & \cdots & \boldsymbol{\theta} \end{bmatrix} \in \mathbb{R}^{m\times mM}$$

$$\boldsymbol{\theta} = \begin{bmatrix} 1 & 0 & \cdots & 0 \end{bmatrix} \in \mathbb{R}^{M}$$

A simple selection of $\tilde{\mathbf{W}}$ and $\tilde{\mathbf{R}}$ is

$$\tilde{\mathbf{W}} = \mathrm{diag}\{\mathbf{W}_1, \mathbf{W}_2, \dots, \mathbf{W}_n\}$$

$$\tilde{\mathbf{R}} = \mathrm{diag}\{\mathbf{R}_1, \mathbf{R}_2, \dots, \mathbf{R}_m\}$$

$$\mathbf{W}_i = \mathrm{diag}\{w_{i1}, w_{i2}, \dots, w_{iP}\}, \quad i \in \{1, 2, \dots, n\}$$

$$\mathbf{R}_j = \mathrm{diag}\{r_{j1}, r_{j2}, \dots, r_{jM}\}, \quad j \in \{1, 2, \dots, m\}$$

Taking $\tilde{\mathbf{R}} > 0$ guarantees the nonsingularity of $\tilde{\mathbf{A}}^{\mathrm{T}}\tilde{\mathbf{W}}\tilde{\mathbf{A}} + \tilde{\mathbf{R}}$.

2.2.4 Feedback Correction

At the initial time $k = 0$, suppose the system is in the steady state. For the start-up of DMC, we can take $y_{i,0}\ (1|0) = y_i\ (0)$, $i = 1, 2, \dots, n$. For each time $k > 0$, $y_{i,0}\ (k + l|k - 1)$ can be the basis for constructing $y_{i,0}\ (k + l|k)$ for the ith output.

Also denote

$$\varepsilon_i(k) = y_i(k) - y_i(k|k-1) \tag{2.11}$$

where

$$y_i(k|k-1) = y_{i,0}(k|k-1) + \sum_{j=1}^{m} s_{ij,1}\Delta u_j(k-1) \tag{2.12}$$

Since $\varepsilon_i\ (k)$ is the effect on the output by the unmodeled uncertainties, it can be used to predict the future prediction error, so as to compensate the predictions based on the model. In summary, we can use the following to predict $y_{i,0}\ (k + l|k)$:

$$y_{i,0}(k+1|k) = y_{i,0}(k+2|k-1) + \sum_{j=1}^{m} s_{ij,2}\Delta u_j(k-1) + f_{i,1}\varepsilon(k)$$

$$\vdots$$

$$y_{i,0}(k+M|k) = y_{i,0}(k+M+1|k-1) + \sum_{j=1}^{m} s_{ij,M+1}\Delta u_j(k-1) + f_{i,M}\varepsilon_i(k)$$

$$\vdots$$

$$y_{i,0}(k+P|k) = y_{i,0}(k+P|k-1) + \sum_{j=1}^{m} s_{ij,P+1}\Delta u_j(k-1) + f_{i,P}\varepsilon_i(k)$$

where

$$y_0(k+P|k-1) = y_{i,0}(k+P-1|k-1) \tag{2.13}$$

By summarizing the above deductions, at each time $k > 0$, $\tilde{\mathbf{y}}_0\ (k|k)$ can be calculated by

$$\tilde{\mathbf{y}}_{i,0}(k|k) = \tilde{\mathbf{y}}_{i,0}(k|k-1) + \sum_{j=1}^{m} \mathbf{A}_{ij,1}\Delta u_j(k-1) + \tilde{\mathbf{f}}_i\varepsilon_i(k) \tag{2.14}$$

where

$$\mathbf{A}_{ij,1} = \begin{bmatrix} s_{ij,2} & s_{ij,3} & \cdots & s_{ij,P+1} \end{bmatrix}^{\mathrm{T}}$$
$$\tilde{\mathbf{f}}_i = \begin{bmatrix} f_{i1} & f_{i2} & \cdots & f_{iP} \end{bmatrix}^{\mathrm{T}}$$

And $\mathbf{Y}_0\ (k|k)$ can be calculated as

$$\mathbf{Y}_0(k|k) = \mathbf{Y}_0(k|k-1) + \tilde{\mathbf{A}}_1\Delta\mathbf{U}(k|k-1) + \tilde{\mathbf{F}}\Upsilon(k) \tag{2.15}$$

where

$$\mathbf{Y}_0(k|k-1) = \begin{bmatrix} \tilde{\mathbf{y}}_{1,0}(k|k-1)^{\mathrm{T}} & \tilde{\mathbf{y}}_{2,0}(k|k-1)^{\mathrm{T}} & \cdots & \tilde{\mathbf{y}}_{n,0}(k|k-1)^{\mathrm{T}} \end{bmatrix}^{\mathrm{T}}$$
$$\tilde{\mathbf{y}}_{i,0}(k|k-1) = \begin{bmatrix} y_{i,0}\ (k+1|k-1) & y_{i,0}(k+2|k-1) & \cdots & y_{i,0}(k+P|k-1) \end{bmatrix}^{T}$$
$$\Delta\mathbf{U}(k-1) = \begin{bmatrix} \Delta u_1\ (k-1) & \Delta u_2(k-1) & \cdots & \Delta u_m(k-1) \end{bmatrix}^{\mathrm{T}}$$
$$\tilde{\mathbf{A}}_1 = \begin{bmatrix} \mathbf{A}_{11,1} & \mathbf{A}_{12,1} & \cdots & \mathbf{A}_{1m,1} \\ \mathbf{A}_{21,1} & \mathbf{A}_{22,1} & \cdots & \mathbf{A}_{2m,1} \\ \vdots & \vdots & \ddots & \vdots \\ \mathbf{A}_{r1,1} & \mathbf{A}_{r2,1} & \cdots & \mathbf{A}_{nm,1} \end{bmatrix}$$
$$\tilde{\mathbf{F}} = \begin{bmatrix} \tilde{\mathbf{f}}_1 & 0 & \cdots & 0 \\ 0 & \tilde{\mathbf{f}}_2 & \cdots & 0 \\ \vdots & \vdots & \ddots & \vdots \\ 0 & 0 & \cdots & \tilde{\mathbf{f}}_n \end{bmatrix}$$
$$\Upsilon(k) = \begin{bmatrix} \varepsilon_1\ (k) & \varepsilon_2(k) & \cdots & \varepsilon_n(k) \end{bmatrix}^{\mathrm{T}}$$

2.2.5 DMC with Constraint

In the real application of DMC, the constraints on actuator slew rates, actuator ranges, and constraints on the controlled variables usually exist. In the following, we discuss how to handle the constraint in DMC.

1. **Output constraint:** $y_{i,\min} \le y_i\ (k+l|k) \le y_{i,\max}$, $l = 1, 2, \ldots, P$.
   At each time instant k, the output prediction is (2.7). Hence, we can let the optimization problem satisfies the following constraint:

$$\mathbf{Y}_{\min} \le \mathbf{Y}_0(k|k) + \tilde{\mathbf{A}}\Delta\mathbf{U}(k|k) \le \mathbf{Y}_{\max} \tag{2.16}$$

where

$$\mathbf{Y}_{\min} = \begin{bmatrix}\tilde{\mathbf{y}}_{1,\min}^{\mathrm{T}} & \tilde{\mathbf{y}}_{2,\min}^{\mathrm{T}} & \cdots & \tilde{\mathbf{y}}_{n,\min}^{\mathrm{T}}\end{bmatrix}^{\mathrm{T}} \in \mathcal{R}^{nP}$$

$$\tilde{\mathbf{y}}_{i,\min}^{\mathrm{T}} = \begin{bmatrix}y_{i,\min} & y_{i,\min} & \cdots & y_{i,\min}\end{bmatrix}^{\mathrm{T}} \in \mathcal{R}^{P}$$

$$\mathbf{Y}_{\max} = \begin{bmatrix}\tilde{\mathbf{y}}_{1,\max}^{\mathrm{T}} & \tilde{\mathbf{y}}_{2,\max}^{\mathrm{T}} & \cdots & \tilde{\mathbf{y}}_{n,\max}^{\mathrm{T}}\end{bmatrix}^{\mathrm{T}} \in \mathcal{R}^{nP}$$

$$\tilde{\mathbf{y}}_{i,\max}^{\mathrm{T}} = \begin{bmatrix}y_{i,\max} & y_{i,\max} & \cdots & y_{i,\max}\end{bmatrix}^{\mathrm{T}} \in \mathcal{R}^{P}$$

2. **Input increment constraint:** $\Delta u_{j,\min} \le \Delta u_j\ (k+l|k) \le u_{j,\max}$. The concatenated form can be expressed as

$$\Delta\mathbf{U}_{\min} \le \Delta\mathbf{U}(k|k) \le \Delta\mathbf{U}_{\max} \tag{2.17}$$

where

$$\Delta\mathbf{U}_{\min} = \begin{bmatrix}\Delta\tilde{\mathbf{u}}_{1,\min}^{\mathrm{T}} & \Delta\tilde{\mathbf{u}}_{2,\min}^{\mathrm{T}} & \cdots & \Delta\tilde{\mathbf{u}}_{m,\min}^{\mathrm{T}}\end{bmatrix}^{\mathrm{T}} \in \mathbb{R}^{mM}$$

$$\Delta\tilde{\mathbf{u}}_{j,\min}^{\mathrm{T}} = \begin{bmatrix}\Delta u_{j,\min} & \Delta u_{j,\min} & \cdots & \Delta u_{j,\min}\end{bmatrix}^{\mathrm{T}} \in \mathbb{R}^{M}$$

$$\Delta\mathbf{U}_{\max} = \begin{bmatrix}\Delta\tilde{\mathbf{u}}_{1,\max}^{\mathrm{T}} & \Delta\tilde{\mathbf{u}}_{2,\max}^{\mathrm{T}} & \cdots & \Delta\tilde{\mathbf{u}}_{m,\max}^{\mathrm{T}}\end{bmatrix}^{\mathrm{T}} \in \mathbb{R}^{mM}$$

$$\Delta\tilde{\mathbf{u}}_{j,\max}^{\mathrm{T}} = \begin{bmatrix}\Delta u_{j,\max} & \Delta u_{j,\max} & \cdots & \Delta u_{j,\max}\end{bmatrix}^{\mathrm{T}} \in \mathbb{R}^{M}$$

3. **Input magnitude constraint** $u_{j,\min} \le u_j\ (k+l|k) \le u_{j,\max}$. The optimization problem should satisfy the following constraint:

$$\mathbf{U}_{\min} \le \mathbf{B}\Delta\mathbf{U}(k|k) + \tilde{\mathbf{u}}(k-1) \le \mathbf{U}_{\max} \tag{2.18}$$

where

$$\mathbf{U}_{\min} = \begin{bmatrix}\tilde{\mathbf{u}}_{1,\min}^{\mathrm{T}} & \tilde{\mathbf{u}}_{2,\min}^{\mathrm{T}} & \cdots & \tilde{\mathbf{u}}_{m,\min}^{\mathrm{T}}\end{bmatrix}^{\mathrm{T}} \in \mathbb{R}^{mM}$$

$$\tilde{\mathbf{u}}_{j,\min}^{\mathrm{T}} = \begin{bmatrix}u_{j,\min} & u_{j,\min} & \cdots & u_{j,\min}\end{bmatrix}^{\mathrm{T}} \in \mathbb{R}^{M}$$

$$\mathbf{U}_{\max} = \begin{bmatrix}\tilde{\mathbf{u}}_{1,\max}^{\mathrm{T}} & \tilde{\mathbf{u}}_{2,\max}^{\mathrm{T}} & \cdots & \tilde{\mathbf{u}}_{m,\max}^{\mathrm{T}}\end{bmatrix}^{\mathrm{T}} \in \mathbb{R}^{mM}$$

$$\tilde{\mathbf{u}}_{j,\max}^{\mathrm{T}} = \begin{bmatrix}u_{j,\max} & u_{j,\max} & \cdots & u_{j,\max}\end{bmatrix}^{\mathrm{T}} \in \mathbb{R}^{M}$$

$$\mathbf{B} = \text{block} - \text{diag}\{\mathbf{B}_0, \ldots, \mathbf{B}_0\}(m \text{ blocks})$$

$$\mathbf{B}_0 = \begin{bmatrix}1 & 0 & \cdots & 0\\ 1 & 1 & \ddots & \vdots\\ \vdots & \ddots & \ddots & 0\\ 1 & \cdots & 1 & 1\end{bmatrix} \in \mathbb{R}^{M\times M}$$

$$\tilde{\mathbf{u}}(k-1) = \begin{bmatrix}\tilde{\mathbf{u}}_1(k-1)^{\mathrm{T}} & \tilde{\mathbf{u}}_2(k-1)^{\mathrm{T}} & \cdots & \tilde{\mathbf{u}}_m(k-1)^{\mathrm{T}}\end{bmatrix}^{\mathrm{T}}$$

$$\tilde{\mathbf{u}}_j(k-1) = \begin{bmatrix}u_j\,(k-1) & u_j(k-1) & \cdots & u_j(k-1)\end{bmatrix}^{\mathrm{T}} \in \mathbb{R}^{M}$$

Equations (2.16)–(2.18) can be written in a uniform form as

$$\mathbf{C}\Delta\mathbf{U}(k|k) \le \mathbf{b} \tag{2.19}$$

where $\mathbf{C}$ and $\mathbf{b}$ are the matrix and vector, respectively, provided at time k. The DMC optimization problem considering these constraints can be written as follows.

Problem 2.1

$$\begin{aligned} &\min_{\Delta\mathbf{U}(k|k)} J(k) = \|\mathbf{E}(k|k)\|^2_{\tilde{\mathbf{W}}} + \|\Delta\mathbf{U}(k|k)\|^2_{\tilde{\mathbf{R}}} \\ &\text{s.t.} \quad \mathbf{C}\Delta\mathbf{U}(k|k) \le \mathbf{b} \end{aligned} \tag{2.20}$$

Problem (2.20) is a quadratic optimization problem. The feedback law solution to the constrained quadratic optimization problem is, in general, nonlinear. In MATLAB MPC Toolbox, for the DMC optimization problem for constrained systems, one can adopt "cmpc."

The above constraint DMC algorithm can be summarized as follows.

Algorithm 2.1 DMC Algorithm

Step 0. Obtain step response coefficient matrix (2.3). Choose $\tilde{\mathbf{F}}$.
Step 1. At $k = 0$,

- measure $\mathbf{y}(0)$;
- determine $\tilde{\mathbf{y}}_r(0)$;
- choose $\mathbf{y}_0\ (i|0)$, $i \in \{1, 2, \ldots, P\}$ and construct $\mathbf{Y}_0\ (0|0)$
- solve Problem 2.1 to calculate $\Delta\mathbf{u}(0)$;
- implement $\mathbf{u}(0) = \mathbf{u}(0) + \Delta\mathbf{u}(0)$.

Step 2. At time $k > 0$,

- measure $\mathbf{y}(k)$;
- determine $\tilde{\mathbf{y}}_r(k)$;
- use (2.11) to calculate $\Upsilon(k)$;
- use (2.15) to calculate $\mathbf{Y}_0(k|k)$;
- solve Problem 2.1 to calculate $\Delta\mathbf{u}(k)$;
- implement $\mathbf{u}(k) = \mathbf{u}(k-1) + \Delta\mathbf{u}(k)$.

Step 3. At time $k+1$, let $k+1 \to k$, and go to Step 2.

2.3 Predictive Control with the State Space Model

In the above section, we introduced the DMC algorithm, which is based on the step response model. In this section, we will present the MPC algorithm which is based on the state space model, since, so far, this kind of MPC is a most discussed and studied MPC method in the literature works. And this part of content mainly refers to [2, 26, 74].

For the basic formulation of predictive control and making the formulation useful in the real world, we shall assume that:

- the plant model is linear time invariant,
- there are measurement disturbances in the plant model. Handling feed forwarding measurement disturbances is very important in distributed predictive control,
- the output variables can be measured,
- the real plant is governed by the same equations as the model, although, in most cases, this is not really true,

- the cost function is quadratic,
- the cost function does not penalize particular values of the input vector $\mathbf{u}(k)$, but only the input increments vector, $\Delta\mathbf{u}(k)$, which coincides with that used in the majority of the predictive control literature,
- constraints are in the form of linear inequalities.

2.3.1 System Model

Consider a dynamic time-invariant linear system with measurement disturbances, which is express as

$$\begin{cases} \mathbf{x}(k+1) = \mathbf{A}\mathbf{x}(k) + \mathbf{B}\mathbf{u}(k) + \mathbf{E}\mathbf{d}(k) \\ \mathbf{y}(k) = \mathbf{C}\mathbf{x}(k) \end{cases} \tag{2.21}$$

where $\mathbf{x}(k) \in \mathbb{R}^{n_x}$, $\mathbf{y}(k) \in \mathbb{R}^{n_y}$, $\mathbf{u}(k) \in \mathbb{R}^{n_u}$, and $\mathbf{d}(k) \in \mathbb{R}^{n_d}$ are state, output, input, and measurable disturbances vector, respectively. **A**, **B**, **C**, **D**, and **E** are system coefficient matrices.

There are several methods to obtain the model of (2.21), and two of them are introduced as follows.

1. **Black-box model:** Most commonly, it is obtained by performing tests on the plant, which involve injecting known signals, such as steps, multi-sines, pseudorandom, or others, at the plant inputs, and recording the resulting plant outputs and the measurable disturbances. Then the linear time-invariant models can be obtained by using the system identification techniques, which range from simple curve fitting to sophisticated statistically based methods. Models obtained in this way are "black-box" models, which represent only the input–output behavior of the plant, and carry no information about its internal structure [75–77]. Although the subspace identification method could directly obtain the state space model, the state of the identified model by the subspace method usually cannot reflect a real physical state variable in an industrial plant.
2. **First-principle model:** When a first-principle nonlinear model of the plant is available, the linear time-invariant model can be obtained by the following methods:
   - One way we can do is to approximate these first-principle models by Taylor series. However, in some cases, if the first-principle model contains nonsmooth elements such as switches and look-up tables, this method is not appropriate.
   - Another very effective procedure is to use the nonlinear model to generate simulation data by injecting particular signals, such as steps, multi-sines, pseudorandom, or others, to the first-principle nonlinear model, and then to apply system identification techniques to get the black-box model using these data, as if they have been obtained from the plant itself.

The first-principle model refers to a model in which the equations are obtained from knowledge of the underlying physical, chemical, and thermodynamic processes. For motion control, in most cases, the equations describe that the nonlinear dynamics of an aircraft are accurate, and not very complicated. For complex industrial processes, such first-principle dynamic models are much more complex and expensive to develop, but this is being done increasingly commonly.

2.3.2 Performance Index

The cost function $J(k)$ penalizes deviations of the prediction outputs $\hat{\mathbf{y}}(k+i|k)$, $i = 1, 2, \ldots, P$, from a reference trajectory $\mathbf{y}_r(k+i|k)$, $i = 1, 2, \ldots, P$. We define the cost as

$$J(k) = \sum_{i=1}^{P} ||\hat{\mathbf{y}}(k+i|k) - \mathbf{y}_r(k+i|k)||^2_{\mathbf{Q}_i} + \sum_{i=1}^{M} ||\Delta\mathbf{u}(k+i-1|k)||^2_{\mathbf{R}_i} \tag{2.22}$$

where $\mathbf{Q}_i > 0$ and $\mathbf{R}_i > 0$ are weighting coefficient matrices, respectively. The prediction horizon has length P, but we do not necessarily start penalizing deviations of $\mathbf{y}$ from a set-point immediately (if $P > 1$), because there may be some delay between applying an input and seeing any effect. Thus, the reference trajectory $\mathbf{y}_r\ (k+i|k)$, $i = 1, 2, \ldots, P$, may depend on measurements made up to time k; in particular, its initial point may be the output measurement $\mathbf{y}(\mathrm{k})$. But it may also be a set-point, or some other predetermined trajectory. M, $M < P$, is the control horizon, and that $\Delta\mathbf{u}(k+i|k) = 0$ for $i > M$.

2.3.3 Prediction

To get the optimal control law, the first step is to deduce the relationship between the output and input concatenate version $\mathbf{Y}(k)$ and $\mathbf{U}(k)$. From (2.21), we can obtain that

$$\begin{aligned}
\hat{\mathbf{x}}(k+1|k) &= \mathbf{Ax}(k|k) + \mathbf{Bu}(k|k) + \mathbf{Ed}(k) \\
&= \mathbf{Ax}(k|k) + \mathbf{B}\Delta\mathbf{u}(k|k) + \mathbf{Bu}(k-1|k-1) + \mathbf{Ed}(k) \\
\hat{\mathbf{x}}(k+2|k) &= \mathbf{A}^2\mathbf{x}(k|k) + \mathbf{ABu}(k|k) + \mathbf{Bu}(k+1|k) \\
&\quad + \mathbf{AEd}(k) + \mathbf{E}\hat{\mathbf{d}}(k+1) \\
&= \mathbf{A}^2\mathbf{x}(k|k) + \mathbf{AB}(\mathbf{u}(k-1|k-1) + \Delta\mathbf{u}(k|k)) \\
&\quad + \mathbf{B}((\mathbf{u}(k-1|k-1) + \Delta\mathbf{u}(k|k) + \Delta\mathbf{u}(k+1|k)) \\
&\quad + \mathbf{AEd}(k) + \mathbf{E}\hat{\mathbf{d}}(k+1) \\
&= \mathbf{A}^2\mathbf{x}(k|k) + (\mathbf{A}+\mathbf{I})\mathbf{B}\Delta\mathbf{u}(k|k) + \mathbf{B}\Delta\mathbf{u}(k+1|k) \\
&\quad + \mathbf{AEd}(k) + \mathbf{E}\hat{\mathbf{d}}(k+1) \\
&\quad + (\mathbf{A}+\mathbf{I})\mathbf{Bu}(k-1|k-1) \\
&\;\;\vdots \\
\hat{\mathbf{x}}(k+P|k) &= \sum_{l=1}^{P} \mathbf{A}^l\mathbf{x}(k|k) + \sum_{i=1}^{P}\sum_{j=1}^{i} \mathbf{A}^{j-1}\mathbf{B}\Delta\mathbf{u}(k+i-1|k) \\
&\quad + \sum_{h=1}^{P} \mathbf{A}^{h-1}\mathbf{E}\hat{\mathbf{d}}(k+h-1|k) \\
&\quad + \sum_{i=1}^{P} \mathbf{A}^{i-1}\mathbf{Bu}(k-1|k-1)
\end{aligned} \tag{2.23}$$

where $\Delta\mathbf{u}(k+l|k)=0$, when $l \geq M$. Define that

$$\mathbf{Y}(k) = \left[\mathbf{y}(k+1|k)^{\mathrm{T}} \quad \mathbf{y}(k+2|k)^{\mathrm{T}} \quad \cdots \quad \mathbf{y}(k+P|k)^{\mathrm{T}}\right]^{\mathrm{T}}$$

$$\mathbf{X}(k) = \left[\mathbf{x}(k+1|k)^{\mathrm{T}} \quad \mathbf{x}(k+2|k)^{\mathrm{T}} \quad \cdots \quad \mathbf{x}(k+P|k)^{\mathrm{T}}\right]^{\mathrm{T}}$$

$$\mathbf{U}(k) = \left[\mathbf{u}(k|k)^{\mathrm{T}} \quad \mathbf{u}(k+1|k)^{\mathrm{T}} \quad \cdots \quad \mathbf{u}(k+M-1|k)^{\mathrm{T}}\right]^{\mathrm{T}}$$

$$\mathbf{W}(k) = \left[\mathbf{d}(k|k)^{\mathrm{T}} \quad \mathbf{d}(k+1|k)^{\mathrm{T}} \quad \cdots \quad \mathbf{d}(k+P-1|k)^{\mathrm{T}}\right]^{\mathrm{T}}$$

And

$$\mathbf{H} = \begin{bmatrix} \mathbf{A} \\ \mathbf{A}^2 \\ \vdots \\ \mathbf{A}^P \end{bmatrix}^{\mathrm{T}}$$

$$\mathbf{G} = \begin{bmatrix} \mathbf{B} & \mathbf{0} & \cdots & \mathbf{0} \\ (\mathbf{A}+\mathbf{I})\,\mathbf{B} & \mathbf{B} & \cdots & \mathbf{0} \\ \vdots & \vdots & \cdots & \vdots \\ \sum_{i=1}^{M} \mathbf{A}^{i-1}\mathbf{B} & \sum_{i=1}^{M-1} \mathbf{A}^{i-1}\mathbf{B} & \cdots & \mathbf{B} \\ \sum_{i=1}^{M+1} \mathbf{A}^{i-1}\mathbf{B} & \sum_{i=1}^{M} \mathbf{A}^{i-1}\mathbf{B} & \cdots & (\mathbf{A}+\mathbf{I})\mathbf{B} \\ \vdots & \vdots & \cdots & \vdots \\ \sum_{i=1}^{P} \mathbf{A}^{i-1}\mathbf{B} & \sum_{i=1}^{P-1} \mathbf{A}^{i-1}\mathbf{B} & \cdots & \sum_{i=1}^{P-M+1} \mathbf{A}^{i-1}\mathbf{B} \end{bmatrix}$$

$$\mathbf{F} = \begin{bmatrix} \mathbf{B} \\ (\mathbf{A}+\mathbf{I})\,\mathbf{B} \\ \vdots \\ \sum_{i=1}^{M} \mathbf{A}^{i-1}\mathbf{B} \\ \sum_{i=1}^{M+1} \mathbf{A}^{i-1}\mathbf{B} \\ \vdots \\ \sum_{i=1}^{P} \mathbf{A}^{i-1}\mathbf{B} \end{bmatrix}$$

$$\mathbf{V} = \begin{bmatrix} \mathbf{E} & \mathbf{0} & \cdots & \mathbf{0} \\ \mathbf{AE} & \mathbf{E} & \cdots & \mathbf{0} \\ \vdots & \vdots & \cdots & \vdots \\ \mathbf{A}^{P-1}\mathbf{E} & \mathbf{A}^{P-2}\mathbf{E} & \cdots & \mathbf{E} \end{bmatrix}$$

$$\mathbf{T} = \begin{bmatrix} \mathbf{C} & \mathbf{0} & \cdots & \mathbf{0} \\ \mathbf{0} & \mathbf{C} & \ddots & \vdots \\ \vdots & \ddots & \ddots & \mathbf{0} \\ \mathbf{0} & \cdots & \mathbf{0} & \mathbf{C} \end{bmatrix}$$

Then, the concatenated predictive model can be expressed as

$$\begin{cases} \mathbf{X}(k|k) = \mathbf{H}\mathbf{x}(k|k) + \mathbf{G}\Delta\mathbf{U}(k|k) + \mathbf{F}\mathbf{u}(k-1|k-1) + \mathbf{V}\mathbf{W}(k|k) \\ \mathbf{Y}(k|k) = \mathbf{T}\mathbf{X}(k|k) \end{cases} \tag{2.24}$$

It should be noted here that we assume that the state variables $\mathbf{x}(k)$ at very time instant k can be measured. However, in a real industrial process, many state variables cannot be measured directly. In these cases, an observer can be employed to estimate the current state according to the system model (2.21), the measured output in time instant k, measurable disturbances at time instant $k-1$, and input at time instant $k-1$, as shown in Figure 2.1.

Then, the predictive model with an observer becomes

$$\begin{cases} \mathbf{Y}(k|k) = \mathbf{T}\mathbf{H}\hat{\mathbf{x}}(k|k) + \mathbf{T}\mathbf{G}\Delta\mathbf{U}(k|k) + \mathbf{T}\mathbf{F}\mathbf{u}(k-1) + \mathbf{T}\mathbf{V}\mathbf{W}(k|k) \\ \hat{\mathbf{x}}(k|k) = \mathbf{A}\hat{\mathbf{x}}(k|k-1) + \mathbf{B}\mathbf{u}(k-1) + \mathbf{E}\mathbf{d}(k-1) + \mathbf{L}(\hat{\mathbf{y}}(k) - \hat{\mathbf{y}}(k|k-1)) \end{cases} \tag{2.25}$$

There are many methods provided in the literature works that readers can refer to as the starting point for the design of matrix $\mathbf{L}$, e.g., the Kalman filter, Lunberg method, etc.

2.3.4 Closed-Loop Solution

Define that

$$\mathbf{Y}_r(k) = \left[\mathbf{y}_r(k+1|k)^{\mathrm{T}} \quad \mathbf{y}_r(k+2|k)^{\mathrm{T}} \quad \cdots \quad \mathbf{y}_r(k+P|k)^{\mathrm{T}}\right]^{\mathrm{T}},$$
$$\mathbf{Q} = \text{block} - \text{diag}\left\{\mathbf{Q}_1 \ \mathbf{Q}_2 \ \cdots \ \mathbf{Q}_P\right\},$$
$$\mathbf{R} = \text{block} - \text{diag}\left\{\mathbf{R}_1 \ \mathbf{R}_2 \ \cdots \ \mathbf{R}_M\right\}.$$

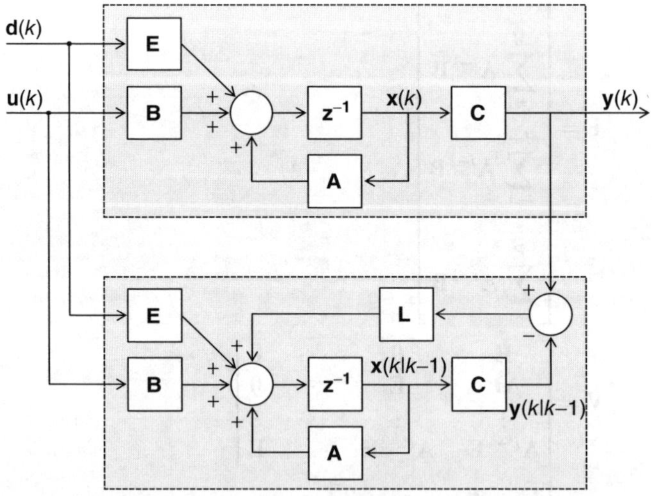

Figure 2.1 A state observer with measurable disturbances

Then, the optimization problem of MPC in each control period becomes

$$\min_{\Delta\mathbf{U}(k|k)} J(k) = ||\mathbf{THx}(k|k) + \mathbf{TFu}(k-1|k-1) + \mathbf{TVW}(k|k) + \mathbf{TG}\Delta\mathbf{U}(k|k) - \mathbf{Y}_r(k|k)||^2_{\mathbf{Q}} + ||\Delta\mathbf{U}(k|k)||^2_{\mathbf{R}} \tag{2.26}$$

which is a standard quadratic programming problem. Let

$$\mathbf{Z}(k) = \mathbf{Hx}(k|k) + \mathbf{TFu}(k-1|k-1) + \mathbf{TVW}(k|k) - \mathbf{Y}_r(k|k) \tag{2.27}$$

The explicit solution of (2.26) can be deduced as

$$\Delta\mathbf{U}(k|k) = ((\mathbf{TG})^{\mathrm{T}}\mathbf{QTG} + \mathbf{R})^{-1}(\mathbf{TG})^{\mathrm{T}}\mathbf{QZ}(k) \tag{2.28}$$

Thus, in every control period, the manipulated variable should be

$$\mathbf{u}(k) = \mathbf{D}((\mathbf{TG})^{\mathrm{T}}\mathbf{QTG} + \mathbf{R})^{-1}(\mathbf{TG})^{\mathrm{T}}\mathbf{QZ}(k) + \mathbf{u}(k-1) \tag{2.29}$$

where $\mathbf{D} = \begin{bmatrix}\mathbf{I} & 0 & \cdots & 0\end{bmatrix}$.

2.3.5 *State Space MPC with Constraint*

In the real application of MPC, the constraints on actuator slew rates, actuator ranges, and constraints on the controlled variables usually exist. We use the following equations to express those constraints.

1. **Output constraint:** $\mathbf{y}_{\min} \leq \mathbf{y}(k+l|k) \leq \mathbf{y}_{\max}$.
   At each optimization cycle, the output prediction can be calculated by (2.24). Hence, we can let the optimization problem satisfy the following constraint:

$$\mathbf{Y}_{\min} \leq \mathbf{T}(\mathbf{Hx}(k|k) + \mathbf{G}\Delta\mathbf{U}(k|k) + \mathbf{Fu}(k-1|k-1) + \mathbf{VW}(k|k)) \leq \mathbf{Y}_{\max} \tag{2.30}$$

   or

$$\mathbf{Y}_{\min} \leq \mathbf{T}(\mathbf{HA}\hat{\mathbf{x}}(k|k-1) + \mathbf{Bu}(k-1) + \mathbf{Ed}(k-1) + \mathbf{L}(\hat{\mathbf{y}}(k) - \hat{\mathbf{y}}(k|k-1)) + \mathbf{G}\Delta\mathbf{U}(k|k) + \mathbf{Fu}(k-1|k-1) + \mathbf{VW}(k|k)) \leq \mathbf{Y}_{\max} \tag{2.31}$$

   where

$$\mathbf{Y}_{\min} = \begin{bmatrix}\mathbf{y}^{\mathrm{T}}_{\min} & \mathbf{y}^{\mathrm{T}}_{\min} & \cdots & \mathbf{y}^{\mathrm{T}}_{\min}\end{bmatrix}^{\mathrm{T}}$$
$$\mathbf{Y}_{\max} = \begin{bmatrix}\mathbf{y}^{\mathrm{T}}_{\max} & \mathbf{y}^{\mathrm{T}}_{\max} & \cdots & \mathbf{y}^{\mathrm{T}}_{\max}\end{bmatrix}^{\mathrm{T}}$$

2. **Input increment constraint:** $\Delta\mathbf{u}_{\min} \leq \mathbf{u}(k+l|k) - \mathbf{u}(k+l-1|k) \leq \mathbf{u}_{\max}$.
   We can let the optimization problem satisfy the following constraint:

$$\Delta\mathbf{U}_{\min} \leq \Delta\mathbf{U}(k|k) \leq \Delta\mathbf{U}_{\max} \tag{2.32}$$

   where

$$\Delta\mathbf{U}_{\min} = \begin{bmatrix}\Delta\mathbf{u}^{\mathrm{T}}_{\min} & \Delta\mathbf{u}^{\mathrm{T}}_{\min} & \cdots & \Delta\mathbf{u}^{\mathrm{T}}_{\min}\end{bmatrix}^{\mathrm{T}}$$
$$\Delta\mathbf{U}_{\max} = \begin{bmatrix}\Delta\mathbf{u}^{\mathrm{T}}_{\max} & \Delta\mathbf{u}^{\mathrm{T}}_{\max} & \cdots & \Delta\mathbf{u}^{\mathrm{T}}_{\max}\end{bmatrix}^{\mathrm{T}}$$

3. **Input magnitude constraint:** $\mathbf{u}_{\min} \le \mathbf{u}(k+l|k) \le \mathbf{u}_{\max}$.
   We can let the optimization problem satisfy the following constraint:

$$\mathbf{U}_{\min} \le \mathbf{S}\Delta\mathbf{U}(k|k) + \mathbf{u}(k-1) \le \mathbf{U}_{\max} \tag{2.33}$$

   where

$$\mathbf{U}_{\min} = \begin{bmatrix} \mathbf{u}_{\min}^{\mathrm{T}} & \mathbf{u}_{\min}^{\mathrm{T}} & \cdots & \mathbf{u}_{\min}^{\mathrm{T}} \end{bmatrix}^{\mathrm{T}}$$
$$\mathbf{U}_{\max} = \begin{bmatrix} \mathbf{u}_{\max}^{\mathrm{T}} & \mathbf{u}_{\max}^{\mathrm{T}} & \cdots & \mathbf{u}_{\max}^{\mathrm{T}} \end{bmatrix}^{\mathrm{T}}$$
$$\mathbf{S} = \begin{bmatrix} \mathbf{I} & 0 & \cdots & 0 \\ \mathbf{I} & \mathbf{I} & \ddots & \vdots \\ \vdots & \ddots & \ddots & 0 \\ \mathbf{I} & \cdots & \mathbf{I} & \mathbf{I} \end{bmatrix} \in \mathbb{R}^{Mn_u \times Mn_u}$$

Equations (2.30)–(2.33) can be written in a uniform form as $\mathbf{C}_u \, \Delta\mathbf{U}(k|k) \le \mathbf{b}$, where $\mathbf{C}_u$ and $\mathbf{b}$ are the matrix and vector known at time k, respectively. The MPC optimization problem considering these constraints can be written as follows.

Problem 2.2

$$\min_{\Delta\mathbf{U}(k|k)} J(k) = \|\mathbf{Y}(k|k) - \mathbf{Y}_r(k|k)\|_{\mathbf{Q}}^2 + \|\Delta\mathbf{U}(k|k)\|_{\mathbf{R}}^2$$
$$\text{s.t.} \;\; \mathbf{C}_u \Delta\mathbf{U}(k|k) \le \mathbf{b} \tag{2.34}$$

Problem (2.34) is a quadratic optimization problem. The feedback law solution to the constrained quadratic optimization problem is, in general, nonlinear. In the MATLAB MPC Toolbox, for an MPC optimization problem for constrained systems, one can adopt "cmpc."

The above constraint DMC algorithm can be summarized as follows.

Algorithm 2.2 State Space MPC algorithm

Step 0. Obtain system model (2.21). Choose $\mathbf{L}$.
Step 1. At $k = 0$,

- measure $\mathbf{y}(0)$;
- determine $\mathbf{Y}_r(0)$;
- solve Problem 2.2 to calculate $\Delta\mathbf{u}(0)$;
- implement $\mathbf{u}(0) = \mathbf{u}(0) + \Delta\mathbf{u}(0)$.

Step 2. At time $k > 0$,

- measure $\mathbf{y}(k)$;
- determine $\mathbf{Y}_r(k)$;
- solve Problem 2.2 to calculate $\Delta\mathbf{u}(k)$;
- implement $\mathbf{u}(k) = \mathbf{u}(k-1) + \Delta\mathbf{u}(k)$.

Step 3. Let at time $k+1$. Let $k+1 \to k$, go to Step 2.

2.4 Dual Mode Predictive Control

From the above, it is obvious that, compared with other control technologies, MPC has powerful abilities to deal with the constraints. However, the early version of MPC for process control did not address stability theoretically so that they are not automatically stabilizing. A common method to obtain the stability of the early version of MPC is to increase the length of the control horizon, which indicates the stability property of an infinite horizon predictive control scheme. Until the 1990s, the stability theory of MPC was established [69], and the theoretical framework of that guarantees feasibility and stability, and is now well understood.

A method to obtain a stable MPC was proposed in [71]. The proposed strategy, which is denominated as dual mode control, is based on the inclusion of an inequality constraint that forces the terminal state to lie in a terminal positive-invariant region Ω at the end of the prediction horizon. When the state is outside the terminal region, the control action is provided by the conventional MPC with the above constraint. Once the state is in Ω, the controller is switched to a linear previously defined controller [71]. In this section, we will introduce the dual mode predictive control.

2.4.1 *Invariant Region*

Consider the following linear time-invariant system:

$$\begin{cases} \mathbf{x}(k+1) = \mathbf{A}\mathbf{x}(k) + \mathbf{B}\mathbf{u}(k) \\ \mathbf{y}(k) = \mathbf{C}\mathbf{x}(k) \end{cases} \tag{2.35}$$

where $\mathbf{x} \in \mathbb{R}^{n_x}$, $\mathbf{u} \in \mathcal{U} \subset \mathbb{R}^{n_u}$, and $\mathbf{y} \in \mathbb{R}^{n_y}$ are, respectively, the state, input, and output vectors, and $\mathcal{U}$ is the feasible set of the input $\mathbf{u}$, which is used to bound the input according to the physical constraints on the actuators, the control requirements, or the characteristics of the plant.

Assumption 2.1 For system (2.35), there exists a state feedback $\mathbf{u}(k) = \mathbf{K}\mathbf{x}(k)$ such that the closed-loop system $\mathbf{x}(k+1) = \mathbf{A}_c\, \mathbf{x}(k)$ is asymptotically stable, where $\mathbf{A}_c = \mathbf{A} + \mathbf{B}\mathbf{K}$.

This assumption presumes that each subsystem is able to be stabilized by a feedback control $\mathbf{K}\mathbf{x}$. The control gain $\mathbf{K}$ can be obtained by the linear matrix inequality (LMI) or linear-quadratic regulator (LQR) technology.

Lemma 2.1 *Under Assumption 2.1, for any positive scalar* c *and* $\mathbf{P} > 0$*, the set*

$$\Omega(c) = \{\mathbf{x} \in \mathbb{R}^{n_x} : \|\mathbf{x}\|_{\mathbf{P}} \leq c\}$$

is a positive invariant region of attraction for the closed-loop system $\mathbf{x}(k+1) = \mathbf{A}_c\, \mathbf{x}(k)$. *Additionally, there exists a small enough positive scalar* ε *such that* $\Omega(\varepsilon)$ *is in the feasible input set* $\mathcal{U} \in \mathbb{R}^{n_u}$ *for all* $\mathbf{x} \in \Omega(\varepsilon)$.

Proof. Under Assumption 2.1, there is a positive definite matrix $\mathbf{Q}$ such that

$$\mathbf{A}_c^{\mathrm{T}}\mathbf{P}\mathbf{A}_c + \mathbf{Q} = \mathbf{P} \tag{2.36}$$

Consider the function $V(k) = \|\mathbf{x}(k)\|_{\mathbf{P}}^2$. The time difference of $V(k)$ along the trajectories of the closed-loop system $\mathbf{x}(k+1) = \mathbf{A}_c\, \mathbf{x}(k)$ can be evaluated as

$$\begin{aligned}\Delta V(k) &= \mathbf{x}^{\mathrm{T}}(k)\mathbf{A}_c^{\mathrm{T}}\mathbf{P}\mathbf{A}_c\mathbf{x}(k) - \mathbf{x}^{\mathrm{T}}(k)\mathbf{P}\mathbf{x}(k)\\ &= \mathbf{x}^{\mathrm{T}}(k)(\mathbf{A}_c^{T}\mathbf{P}\mathbf{A}_c - \mathbf{P})\mathbf{x}(k)\\ &\le -\mathbf{x}^{\mathrm{T}}(k)\mathbf{Q}\mathbf{x}(k)\\ &\le 0\end{aligned} \tag{2.37}$$

which holds for all $\mathbf{x}(k) \in \Omega(c)\backslash\{0\}$. This implies that all trajectories of the closed-loop system that starts inside $\Omega(c)$ will remain inside and converge to the origin.

The existence of an $\varepsilon > 0$ such that $\mathbf{Kx} \in \mathcal{U}$ for all $\mathbf{x} \in \Omega(\varepsilon)$ follows from the fact that $\mathbf{P}$ is positive definite, which implies that the set $\Omega(\varepsilon)$ shrinks to the origin as ε decreases to zero. This completes the proof.

2.4.2 MPC Formulation

1. **Performance Index**
   More specifically, the performance index of dual mode MPC at each time instant k is defined as

$$J(k) = \|\mathbf{x}(k+N|k)\|_{\mathbf{P}}^2 + \sum_{l=0}^{N-1}(\|\mathbf{x}(k+l|k)\|_{\mathbf{Q}}^2 + \|\mathbf{u}(k+l|k)\|_{\mathbf{R}}^2) \tag{2.38}$$

   where $\mathbf{Q} = \mathbf{Q}^{\mathrm{T}} > 0$, $\mathbf{R} = \mathbf{R}^{\mathrm{T}} > 0$, and $\mathbf{P} = \mathbf{P}^{\mathrm{T}} > 0$. The matrix $\mathbf{P}$ is chosen to satisfy the *Lyapunov* equation

$$\mathbf{A}_c^{\mathrm{T}}\mathbf{P}\mathbf{A}_c - \mathbf{P} = -\overline{\mathbf{Q}} \tag{2.39}$$

   where $\overline{\mathbf{Q}} = \mathbf{Q} + \mathbf{K}^{\mathrm{T}}\mathbf{R}\mathbf{K} > 0$.
2. **Predictive Model**
   From (2.35), the prediction of the system l-step ahead state can be deduced easily, and is given as

$$\hat{\mathbf{x}}(k+l+1|k) = \mathbf{A}^{l}\mathbf{x}(k) + \sum_{i=1}^{l}\mathbf{A}^{i-1}\mathbf{B}\mathbf{u}(k+l-i|k) \tag{2.40}$$

3. **Optimization Problem**
   In the optimization problem of each subsystem-based MPC, the terminal state constraint set can be set to be $\mathbf{x}(k+N|k) \in \Omega(\varepsilon)$. Suppose that at some time k_0, $\mathbf{x}(k_0) \in \Omega(\varepsilon)$. Then, by Lemma 2.1, stabilization can be achieved if the controller employs its static feedback controller $\mathbf{Kx}(k)$ after the time instant k_0.Thus, the objective of the MPC law is to drive the state to the set $\Omega(\varepsilon)$. Once all states have reached these sets, they switch to the feedback

control for stabilization. Such switching from an MPC law to a terminal controller, once the state reaches a suitable neighborhood of the origin, is referred to as the dual mode MPC [71, 78].

In what follows, we formulate the optimization problem for the dual mode MPC.

Problem 2.3 Consider system (2.35). Let $\varepsilon > 0$ be as specified in Lemma 2.1. Let the update time be $k \geq 1$. Given $\mathbf{x}(k)$, find the control sequence $\mathbf{u}(k+l-1|k) : \{1, 2, \ldots, N\} \to \mathcal{U}$ that minimizes

$$J(k) = \|\mathbf{x}(k+N|k)\|_{\mathbf{P}}^2 + \sum_{l=0}^{N-1} (\|\mathbf{x}(k+l|k)\|_{\mathbf{Q}}^2 + \|\mathbf{u}(k+l|k)\|_{\mathbf{R}}^2) \tag{2.41}$$

subject to the following constraints:

$$\mathbf{u}(k+l-1|k) \in \mathcal{U},\ l = 1, \ldots, N \tag{2.42}$$

$$\mathbf{x}(k+N|k) \in \Omega(\varepsilon) \tag{2.43}$$

2.4.3 Algorithms

Before stating the dual mode MPC algorithm, we make the following assumption to facilitate the initialization phase.

Assumption 2.2 At initial time k_0, there exists a feasible control $\mathbf{u}(k_0 + l - 1|k_0) \in \mathcal{U}$, $l = 1, 2, \ldots, N$, for system (2.35), such that the solution to the system $\mathbf{x}(k_0 + l|k_0) = \mathbf{A}\mathbf{x}(k_0 + l - 1|k_0) + \mathbf{B}\mathbf{u}(k_0 + l - 1|k_0)$, $l = 1, 2, \ldots, N$, satisfies $\mathbf{x}(N + k_0) \in \Omega_i(\varepsilon)$ and results in a bounded cost $J(k_0)$.

Assumption 2.2 bypasses the difficult task of actually constructing an initially feasible solution. In fact, finding an initially feasible solution for many optimization problems is often a primary obstacle, whether or not such problems are used in a control setting. As such, many centralized implementations of MPC also assume that an initially feasible solution is available [50, 51].

Algorithm 2.3 Dual Mode MPC Algorithm
The dual mode MPC law at every time instant k is constructed as follows:

Step 1 If $\mathbf{x}(k) \in \Omega(\varepsilon)$, then apply the terminal controller $\mathbf{u}(k) = \mathbf{K}\mathbf{x}(k)$; else go to Step 2.
Step 2 Solve Problem 2.3 for $\mathbf{u}(k|k)$, and apply $\mathbf{u}(k|k)$;
Step 3 Let $k + 1 \to k$; repeat Step 1.

Algorithm 2.3 gives a method to solve the dual mode predictive control. In the next subsection, it will be shown that the dual mode predictive control policy drives the state $\mathbf{x}(k+l)$ to $\Omega(\varepsilon)$ in a finite number of updates. As a result, if $\Omega(\varepsilon)$ is chosen sufficiently small, then MPC can be employed for all time without switching to a terminal controller. Of course, in this case, instead of asymptotic stability at the origin, we can only drive the state toward the small set $\Omega(\varepsilon)$.

The analysis in the next subsection shows that the dual mode predictive control algorithm is feasible at every update and is stabilizing.

2.4.4 *Feasibility and Stability*

2.4.4.1 Feasibility

The main result of this section is that, provided that an initially feasible solution is available, for system (2.35) and at any time $k \geq 1$, $\mathbf{u}(\cdot|k) = \mathbf{u}^f(\cdot|k)$ is a feasible control solution to Problem 2.3. This feasibility result refers that, at any update $k \geq 1$, the control and state pair $(\mathbf{u}^f(\cdot|k), \mathbf{x}^f(\cdot|k))$ satisfy the control constraint (2.42) and the terminal state constraint (2.43).

Under Assumption 2.2, we can assume that there is a feasible solution at time $k-1$, which means the existence of the control law $\mathbf{u}(k-1+l-1|k-1)$ and the state $\mathbf{x}(k-1+l|k-1)$, $l = 1, 2, \ldots, N$. At time instant k, define

$$\begin{cases} \mathbf{u}^f(k+l-1|k) = \mathbf{u}(k+l-1|k-1), & l = 1, 2, \ldots, N-1 \\ \mathbf{u}^f(k+N-1|k) = \mathbf{K}\mathbf{x}(k+N-1|k-1) \end{cases} \tag{2.44}$$

According to (2.35) and (2.44), we have

$$\begin{cases} \mathbf{x}^f(k+l|k) = \mathbf{x}(k+l-1|k-1), & l = 1, 2, \ldots, N-1 \\ \mathbf{x}^f(k+N-1|k) = \mathbf{A}_c\mathbf{x}(k+l-1|k-1) = \mathbf{A}_c\mathbf{x}^f(k+l-1|k) \end{cases} \tag{2.45}$$

Since $\mathbf{x}(k+l-1|k-1)$, $l = 1, 2, \ldots, N$, is a feasible solution at time $k-1$, it means

$$\mathbf{x}(k+N-1|k-1) \in \Omega(\varepsilon).$$

Considering that $\lambda_{\max}(\mathbf{A}_c) < 1$, we have

$$\begin{aligned} &\|\mathbf{x}^f(k+N-1|k)\|_{\mathbf{P}}^2 \\ &= \|\mathbf{A}_c\mathbf{x}(k+l-1|k-1)\|_{\mathbf{P}}^2 \\ &\leq \|\mathbf{x}(k+l-1|k-1)\|_{\mathbf{P}}^2 \\ &\leq \varepsilon^2 \end{aligned}$$

Thus $\mathbf{x}^f(k+N-1|k)$ satisfied the terminal constraint (2.43).

In addition, since $\mathbf{x}(k+N-1|k-1) \in \Omega(\varepsilon)$, $\mathbf{K}\mathbf{x}(k+N-1|k-1)$ should be in the set of $\mathcal{U}$ by the definition of ε.

From above, we can conclude that if there is a feasible solution for Problem 2.3 at time $k-1$, then we can find at least one feasible solution at time k for Problem 2.3. The recursive feasibility of the dual mode predictive control is guaranteed.

2.4.4.2 Stability

By Algorithm 2.3, when $\mathbf{x}(k)$ enters $\Omega(\varepsilon)$, the terminal controllers take over to keep it in there and stabilize the system at the origin. Therefore, it remains to show that if $\mathbf{x}(0) \in \mathcal{X}\backslash\Omega(\varepsilon)$, then by the application of Algorithm 2.3, the state of system (2.35) is driven to the set $\Omega(\varepsilon)$ in finite time.

Define the nonnegative function for

$$V(k) = \|\mathbf{x}(k+N|k)\|_{\mathbf{P}}^2 + \sum_{l=0}^{N-1}(\|\mathbf{x}(k+l|k)\|_{\mathbf{Q}}^2 + \|\mathbf{u}(k+l|k)\|_{\mathbf{R}}^2) \tag{2.46}$$

In what follows, we will show that, for any $k \geq 0$, if $\mathbf{x}(k) \in \mathcal{X}\backslash\Omega(\varepsilon)$, then there exists a constant $\eta \in (0, \infty)$ such that $V(k) \leq V(k-1) - \eta$.

$$
\begin{aligned}
&V(k) - V(k-1) \\
&= \|\mathbf{x}(k+N|k)\|_{\mathbf{P}}^2 + \sum_{l=0}^{N-1}(\|\mathbf{x}(k+l|k)\|_{\mathbf{Q}}^2 + \|\mathbf{u}(k+l|k)\|_{\mathbf{R}}^2) \\
&\quad - \|\mathbf{x}(k-1+N|k-1)\|_{\mathbf{P}}^2 - \sum_{l=0}^{N-1}(\|\mathbf{x}(k-1+l|k-1)\|_{\mathbf{Q}}^2 + \|\mathbf{u}(k-1+l|k-1)\|_{\mathbf{R}}^2) \\
&\leq \|\mathbf{x}^{\mathrm{f}}(k+N|k)\|_{\mathbf{P}}^2 + \sum_{l=0}^{N-1}(\|\mathbf{x}^{\mathrm{f}}(k+l|k)\|_{\mathbf{Q}}^2 + \|\mathbf{u}^{\mathrm{f}}(k+l|k)\|_{\mathbf{R}}^2) \\
&\quad - \|\mathbf{x}(k-1+N|k-1)\|_{\mathbf{P}}^2 - \sum_{l=0}^{N-1}(\|\mathbf{x}(k-1+l|k-1)\|_{\mathbf{Q}}^2 + \|\mathbf{u}(k-1+l|k-1)\|_{\mathbf{R}}^2) \\
&= \|\mathbf{x}^{\mathrm{f}}(k+N|k)\|_{\mathbf{P}}^2 + \|\mathbf{x}^{\mathrm{f}}(k+N-1|k)\|_{\mathbf{Q}}^2 + \|\mathbf{u}^{\mathrm{f}}(k+N-1|k)\|_{\mathbf{R}}^2 \\
&\quad - \|\mathbf{x}(k-1+N|k-1)\|_{\mathbf{P}}^2 - \|\mathbf{x}(k-1|k-1)\|_{\mathbf{Q}}^2 - \|\mathbf{u}(k-1|k-1)\|_{\mathbf{R}}^2
\end{aligned}
\tag{2.47}
$$

Substituting (2.44), (2.45), and (2.39) into (2.47) yields

$$
\begin{aligned}
&V(k) - V(k-1) \\
&= \|\mathbf{A}_c\mathbf{x}(k-1+N|k-1)\|_{\mathbf{P}}^2 + \|\mathbf{x}(k-1+N|k-1)\|_{\mathbf{Q}}^2 + \|\mathbf{K}\mathbf{x}(k-1+N|k-1)\|_{\mathbf{R}}^2 \\
&\quad - \|\mathbf{x}(k-1+N|k-1)\|_{\mathbf{P}}^2 - \|\mathbf{x}(k-1|k-1)\|_{\mathbf{Q}}^2 - \|\mathbf{u}(k-1|k-1)\|_{\mathbf{R}}^2 \\
&= \|\mathbf{x}(k-1+N|k-1)\|_{\mathbf{P}}^2 \\
&\quad - \|\mathbf{x}(k-1+N|k-1)\|_{\mathbf{P}}^2 - \|\mathbf{x}(k-1|k-1)\|_{\mathbf{Q}}^2 - \|\mathbf{u}(k-1|k-1)\|_{\mathbf{R}}^2 \\
&= -\|\mathbf{x}(k-1|k-1)\|_{\mathbf{Q}}^2 - \|\mathbf{u}(k-1|k-1)\|_{\mathbf{R}}^2 \\
&< 0
\end{aligned}
\tag{2.48}
$$

Thus, for any $k \geq 0$, if $\mathbf{x}(k) \in \mathcal{X}\backslash\Omega(\varepsilon)$, there is a constant $\eta \in (0, \infty)$ such that $V(k) \leq V(k-1) - \eta$. It then follows that there exists a finite time k' such that $x(k') \in \Omega(\varepsilon)$. This concludes the proof of stability of dual mode predictive control.

We have now established the feasibility and the stability of the resulting closed-loop system. That is, if an initially feasible solution could be found, subsequent feasibility of the algorithm is guaranteed at every update, and the resulting closed-loop system is asymptotically stable at the origin.

2.5 Conclusion

In this chapter, we introduced three MPC algorithms: dynamic matrix control, state space-based MPC, and dual mode MPC. These three algorithms are very important and are the fundamental of the distributed predictive controls which will be introduced in the following chapters.

3

Control Structure of Distributed MPC

3.1 Introduction

As introduced in Chapter 2, MPC is a control strategy typically used in a discrete-time control context, i.e., control actions are determined in discrete control cycles of a particular duration. From the beginning of one control interval until the beginning of the next control interval, the manipulated variables (input actions) stay fixed, i.e., a zero-order hold strategy is employed. The objective of the predictive control is to determine those actions that optimize the behavior of the system and minimize costs as specified through the performance index. In order to find the optimal inputs that could be able to lead to the best performance of the closed-loop system, the predictive controller uses the prediction model to predict the future behavior of the system under various inputs and measurable disturbances over a certain prediction horizon. Once the predictive controller has determined the inputs that minimize the system performance index, it implements the resulting optimal inputs until the beginning of the next time instant. And in the next time instant, the predictive controller determines new inputs using updated information. Hence, the predictive control operates in a receding horizon or rolling horizon fashion to determine its actions, and sometimes is called receding horizon control.

The control structure of MPC is a very general concept. It includes how to schedule the controllers, the inputs/outputs of each controller, and the number of controllers that constitute the MPC control system, e.g., the control system can consist of a single MPC controller or multiple MPC controllers. Some properties in which MPC control structure can differ are

- the number of controllers,
- the sensors and actuators that each MPC controller is able to access to,
- the information feeding to each MPC,
- the authority relations between the MPCs,
- the communication protocol that the controllers have among one another, serial or parallel.

It is difficult to define all types of structure due to the large amount of properties that they can have. Fortunately, some mostly accepted types of MPC control structure can be identified,

Distributed Model Predictive Control for Plant-Wide Systems, First Edition. Shaoyuan Li and Yi Zheng.

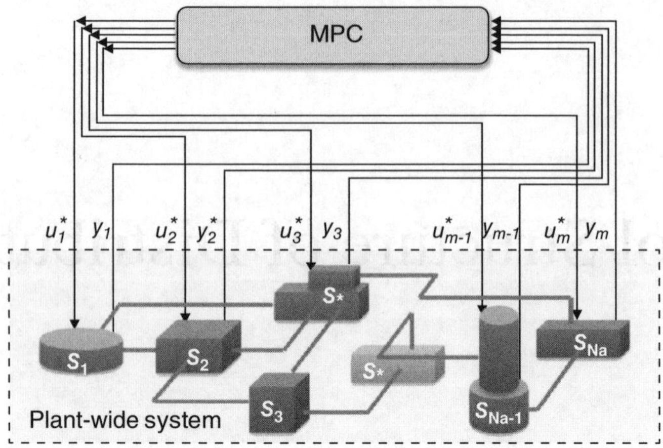

Figure 3.1 The centralized MPC control structure

and that are commonly encountered in theory and practice. Referring to [11], we will introduce the centralized MPC control structure, the single-layer DMPC, and the hierarchical DMPC in this chapter.

3.2 Centralized MPC

As illustrated in Figure 3.1, when there is only one MPC controller that has access to all actuators and sensors of the process network and thus directly controls the physical network, to control the entire system, then the control structure is referred to as a centralized MPC control structure.

The advantage of a centralized MPC control structure in general is that it can obtain the best performance, and it has been extensively studied in the literature. There are a large number of reference materials available for designing the centralized MPC, in particular for small-scale systems. In addition, there are many mature commercial centralized MPC software packages available which make it more convenient for control application engineers to apply centralized MPC to industrial processes. However, there are several issues that obstacle the implementation of a centralized MPC control structure to large-scale plant-wide systems, such as follows:

- Global communication which leads to a severe requirement on communication speed and network safety.
- Heavy computational requirements which make it only suitable for very slow large-scale systems.
- Commercial, legal, and political issues related to unavailability of information and restricted control access.
- Undesirable properties with respect to reliability, scalability, error tolerance, and structure flexibility.

3.3 Single-Layer Distributed MPC

When there are multiple MPC controllers, each of them considering only its own part of subsystem and being able to access only sensors and actuators in that particular part of subsystem, then the control structure is referred to as a single-layer decentralized/DMPC control structure, as illustrated in Figure 3.2. To simplify, we call it decentralized/DMPC. Typically, if the MPCs in the control structure do not communicate with each other, the control structure is called the decentralized MPC control structure. On the contrary, if the MPCs do communicate with each other, the control structure is called the DMPC control structure. In decentralized/DMPC, the MPC controllers first measure the subsystem state, determine the best actions over the predicted subsystem evolution, and then implement the optimal actions.

The reasons that motivate the use of decentralized/DMPC control structures [79, 80] are that these structures are expected to be able to deal or at least relieve the issues presented in the centralized control structure. The decentralized/DMPC structure can:

- improve robustness and reliability, since if one MPC controller fails, another can work normally, or take over and improve responsiveness, since the controllers typically use only local measurements and therefore can react quicker to changing situations;
- reduce communication delays and computational loads, since the MPCs operate locally and therefore solve problems that are smaller, and since communication takes place among nearby control agents;
- deal with unavailability of information and restricted control access, since the MPCs could determine actions of their own part of subsystem only according to the information of their own part of subsystem.

However, since the MPCs in a decentralized/distributed control structure usually have no global overview and can only access a relatively small number of sensors and actuators,

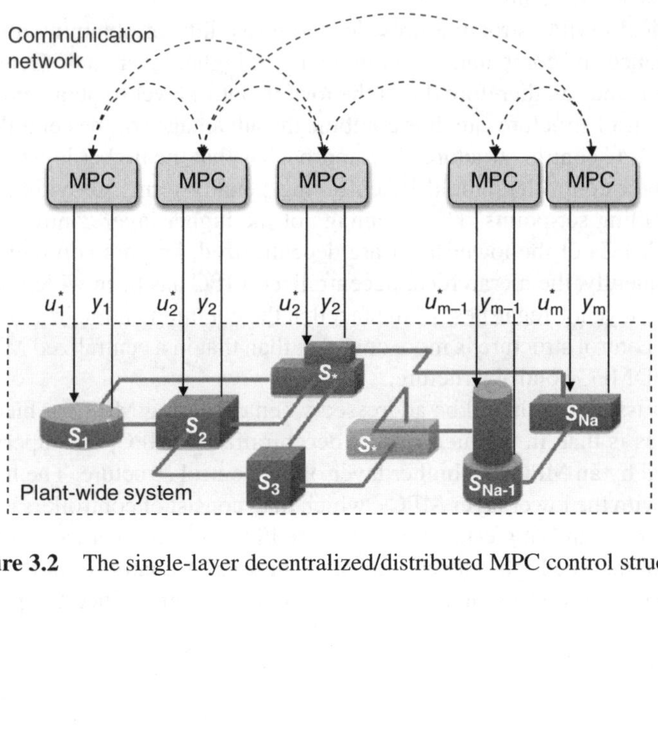

Figure 3.2 The single-layer decentralized/distributed MPC control structure

predicting the evolution of a subsystem over a horizon involves even more uncertainty than when a centralized MPC is employed. This is one of the reasons why typically the decentralized/DMPC control structure has a lower performance than that of the centralized MPC. The challenge in implementing such a decentralized/DMPC strategy comes from ensuring that the actions that the individual MPCs choose result in the performance that is as good as when a hypothetical centralized MPC control structure in which all information is available would be used.

Compared with decentralized MPC and DMPC, the advantage of the decentralized over the single-layer DMPC control structures is that there is no communication between the MPCs, resulting in lower computational requirements and faster control. However, this advantage will typically be at the price of decreased overall performance. The advantage of a DMPC control structure is therefore that improved performance can be obtained, although at the price of increased computation time due to cooperation, communication, and perhaps negotiation among subsystem-based MPCs. Consequently, with the development of network communication, fieldbus, and computer technologies, the study on how to design DMPC coordination strategies for improving the performance of the entire system has become a very popular topic.

3.4 Hierarchical Distributed MPC

When there are multiple MPC controllers, and some of the MPC controllers can force or guide other MPC controllers, we call the control structure a multilayer DMPC structure, as illustrated in Figure 3.3. A typical multilayer MPC control structure is the one in which MPC determines a set-point to a group of other MPCs that work in a decentralized or distributed way. Due to the authority relationship between controllers or groups of controllers, the multilayer MPC control structure is also referred to as a supervisory control structure, or a hierarchical decentralized/DMPC structure.

The hierarchical DMPC structure provides the possibility of obtaining a tradeoff between system performance and computational complexity. A higher layer controller considers a large part of the system and can therefore direct the lower control layer to obtain coordination. Such a hierarchical control structure can thus combine the advantages of the centralized MPC structure with the DMPC control structure. The approaches that the higher layer MPC coordinates with the low-layer controllers could be enforcing penalty terms, providing additional constraints, or providing set-points. The advantage of the higher layer controller is in particular clear when the MPCs of the lower layer are decentralized, i.e., not communicating with one another. Consequently, the hierarchical decentralized MPC has been widely used in chemical processes in recent years, regardless of the fact that the communication in a hierarchical decentralized/DMPC control structure is more complex than that in a centralized MPC structure and a decentralized/DMPC control structure.

An important issue that should be addressed when designing MPC for hierarchical DMPC control structures is that, in the hierarchical decentralized/DMPC, the supervisory MPC can also be regulated by an MPC in a higher layer of the control structure. The higher layer MPC can coordinate with the lower layer MPCs, which may consist of controllers using hierarchical decentralized/DMPC and single-layer decentralized/DMPC. Another issue is the choice of the prediction model that the higher layer control agent uses. A controller in the higher layer has to be able to make relevant predictions of the physical system. Since the physical system is

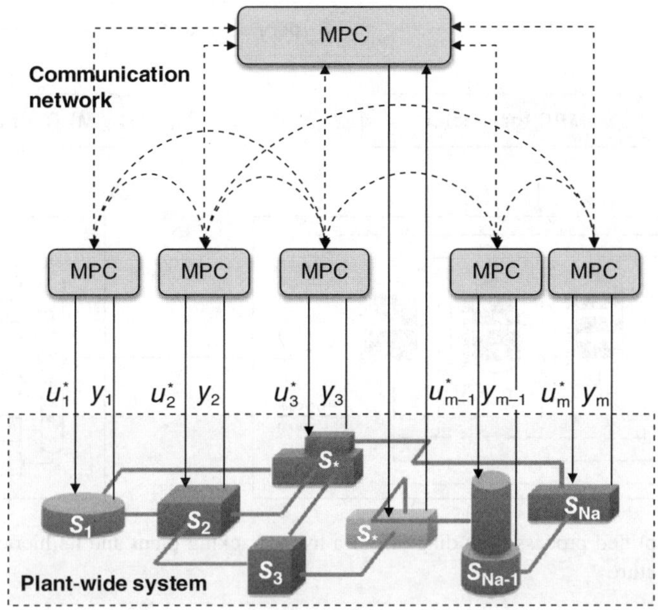

Figure 3.3 The hierarchical decentralized/distributed MPC control structure

under control of the controllers in the lower layer, the controllers of the lower layer have to be taken into account in the predictive model of the higher layer MPC controller. In addition, the higher layer MPCs may involve both nonlinear and linear, continuous and discrete elements, and fast rate and slow rate problems, since they have to consider a larger part of the system than lower-layer controllers. This makes the resulting MPC control problem more complex.

3.5 Example of the Hierarchical DMPC Structure

Example 3.1 The hierarchical DMPC control structure for a hydrocracking process (this example mainly refers to [81]).

Hydrocracking is a catalytic chemical process which converts high-boiling hydrocarbons to more valuable lighter products such as diesel, kerosene, naphtha, and light ends. Hydrocracking is carried out in multiple catalytic beds in the presence of excess hydrogen at high temperatures and pressures. Due to the highly exothermic nature of the process, hydrogen quench is used for interstage cooling. Sulfur and nitrogen compounds are removed in the hydrotreater to yield low-impurity reactor effluent. A significant amount of hydrogen is consumed in cracking reactions. To keep hydrogen circulation constant, make-up hydrogen is added in the process as shown in the middle of Figure 3.4. In the plant fractionation part, the reactor effluent is separated into the final valuable products. Figure 3.4 shows the flow sheet of this industrial plant.

The feed is HVGO (heavy vacuum gas oil) and the products are light ends (LE), light naphtha (LN), heavy naphtha (HN), kerosene (Krs), diesel (Dsl), and bottom (Btm).

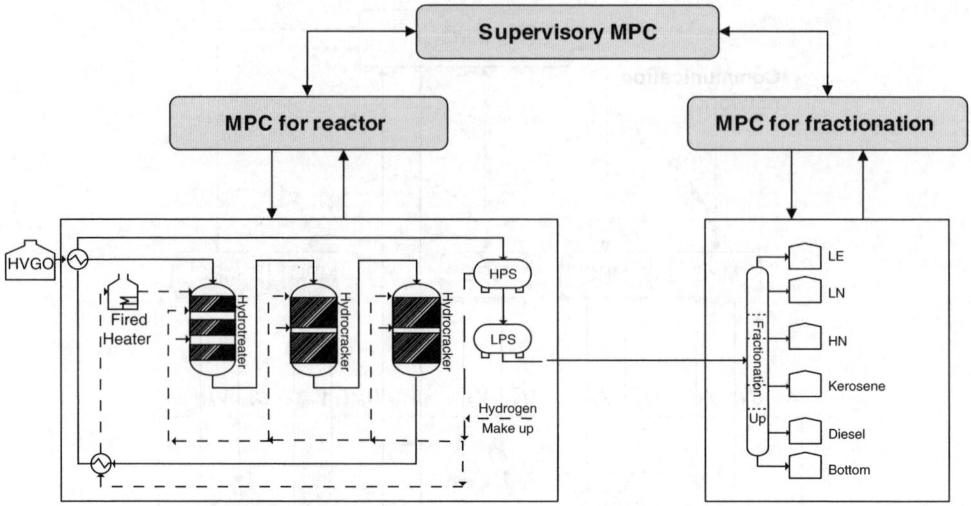

Figure 3.4 Simplified process flow diagram of a hydrocracking plant and its hierarchical distributed MPC control structure

In refineries, hydrocracking plants operate at large production quantity rate, the quality of feeding materials is different and time-varying, and there are many different working conditions corresponding to different product specifications dictated by the market. Such operating characteristics make the real-time economic optimization has quite high potential to create benefits. To fully realize these benefits, interactions among different plant subsystems have to be explored and the lower layer controller must be carefully coordinated.

For these reasons, a hierarchical control structure which is based on the spatial decomposition of the plant into the reactor and fractionation subsystems is used in [81].

The lower layer controllers are the decentralized reactor and fractionator MPC controllers. The higher layer is a supervisory MPC to control the plant-wide economic output variables and coordinate the two lower-layer MPCs.

- Reactor MPC controls the bed exit temperatures by manipulating the set-points of bed inlet temperatures. PID controllers regulate the bed inlet temperatures at the set-points calculated by reactor MPC by adjusting the hydrogen quench flow rates.
- Fractionator MPC controls the production quality by adjusting the set-points of the fractionation PID controllers.
- The supervisory MPC controls the plant-wide economic output variables. It keeps these outputs at their optimal set-points by adjusting the set-points of the reactor exit temperatures and the production parameters as shown in Figure 3.4.

In this fashion, the supervisory MPC coordinates with the two local MPCs to make the hydrocracking plants economically operating. In addition, since the supervisory MPC responds to the effects of slower disturbances, it executes its coordinating control action much less frequently than the decentralized local MPCs.

3.6 Conclusion

In this chapter, a typical MPC control structure was introduced, including the centralized MPC, the single-layer multicontroller decentralized/distributed MPC, and the multilayer multicontroller decentralized/distributed MPC control structures. The properties of the MPC control structures were introduced. Finally, an industrial example was depicted to illustrate the usage of one of the DMPC control structures.

4

Structure Model and System Decomposition

4.1 Introduction

In design distributed control, it is usual to represent the whole system as a collection of interconnected subsystems. The types or forms of the subsystem model are very important for the design of distributed predictive control. Different presentations of subsystem models may cause different designs of MPC and, consequently, may cause different characteristics of designed DMPC. Thus, the existing model formulation for the subsystem, including the state evolution models, the interacting models (input interacting model and state interaction model), etc., is introduced first. We also give some interpretation of how to transfer the presentation of these models among each other.

In addition, how to divide the large-scale system into many subsystems is another very important problem. In certain cases, the decomposition of a large-scale system can be derived directly from the physical description of the problem, which suggests a "natural" grouping of the state variables. More often than not, however, the only information that we have about the system dynamics comes from a mathematical model whose properties provide little or no insight into how the subsystems should be chosen. In order to deal with such problems in a systematic manner, one obviously needs to introduce the decomposition algorithms that are based on the structure of the underlying equations. Thus, in addition to the mathematic models, the structure models are also introduced here to analyze the logical relationship between each subsystem and to help modeling the mathematic model. The structure is an important supplementary of mathematic model, and it is also an important approach for directly looking into the mechanisms of large-scale distributed systems. Here, the function of the structure model in system decomposition, the adjacent matrices, input–output accessibility, and the structure controllability are introduced to investigate the characteristics of the system.

Finally, we refer to the traditional methods used for the MIMO system. Here, we briefly introduce the classic RGA method. The RGA is widely used in designing and analyzing a MIMO control scheme for a process in steady state. The RGA provides a quantitative approach to the analysis of the interactions between the controls and the output, and thus provides a method of pairing manipulated and controlled variables to generate a control scheme.

Distributed Model Predictive Control for Plant-Wide Systems, First Edition. Shaoyuan Li and Yi Zheng.

4.2 System Mathematic Model

Consider the distributed system as shown in Figure 4.1. A distributed system, as illustrated in Figure 4.1, is composed of many interacting subsystems, each of which is controlled by a subsystem-based controller, which in turn is able to exchange information with other subsystem-based controllers.

Suppose that the distributed system $\mathcal{S}$ is composed of m discrete-time linear subsystems $\mathcal{S}_i$, $i \in P = \{1, 2, \ldots, m\}$, and m controllers $\mathcal{C}_i$, $i \in P = \{1, 2, \ldots, m\}$. Let the subsystems interact with each other through their states. If subsystem $\mathcal{S}_i$ is affected by $\mathcal{S}_j$, for any $i \in \mathcal{P}$ and $j \in \mathcal{P}$, subsystem $\mathcal{S}_i$ is said to be a downstream subsystem of subsystem $\mathcal{S}_j$, and subsystem $\mathcal{S}_j$ is an upstream system of $\mathcal{S}_i$. Let $\mathcal{P}_{+i}$ denote the set of the subscripts of the upstream systems of $\mathcal{S}_i$, and $\mathcal{P}_{-i}$ the set of the subscripts of the downstream systems of $\mathcal{S}_i$. Then, subsystem $\mathcal{S}_i$ can be expressed as

$$\begin{cases} \mathbf{x}_i(k+1) = \mathbf{A}_{ii}\mathbf{x}_i(k) + \mathbf{B}_{ii}\mathbf{u}_i(k) + \sum_{j \in \mathcal{P}_i^u} (\mathbf{A}_{ij}\mathbf{x}_j(k) + \mathbf{B}_{ij}\mathbf{u}_j(k)) \\ \mathbf{y}_i(k+1) = \mathbf{C}_{ii}\mathbf{x}_i(k) + \mathbf{C}_{ij}\mathbf{x}_j(k) \end{cases} \tag{4.1}$$

where $\mathbf{x}_i(k) \in \mathbb{R}^{n_{xi}}$, $\mathbf{u}_i(k) \in \mathcal{U}_i \subset \mathbb{R}^{n_{ui}}$, and $\mathbf{y}_i(k) \in \mathbb{R}^{n_{yi}}$ are, respectively, the local state, input, and output vectors, and $\mathcal{U}_i$ is the feasible set of the input $\mathbf{u}_i(k)$, which is used to bound the input according to the physical constraints on the actuators, the control requirements, or the characteristics of the plant. A nonzero matrix $\mathbf{A}_{ij}$, that is, $j \in \mathcal{P}_{+i}$, indicates that $\mathcal{S}_i$ is affected by $\mathcal{S}_j$. In the concatenated vector form, the system dynamics can be written as

$$\begin{cases} \mathbf{x}(k+1) = \mathbf{A}\mathbf{x}(k) + \mathbf{B}\mathbf{u}(k) \\ \mathbf{y}(k+1) = \mathbf{C}\mathbf{x}(k) \end{cases} \tag{4.2}$$

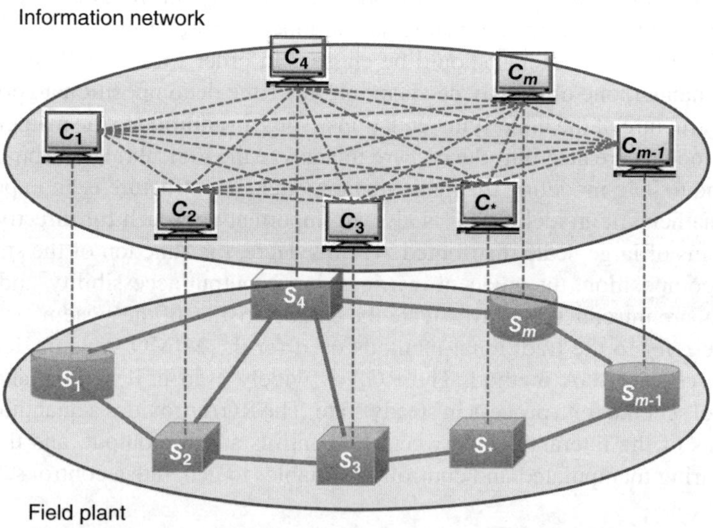

Figure 4.1 The schematic of the distributed system

where

$$\mathbf{x}(k) = \left[\mathbf{x}_1^{\mathrm{T}}(k)\ \mathbf{x}_2^{\mathrm{T}}(k)\ \cdots\ \mathbf{x}_m^{\mathrm{T}}(k)\right]^{\mathrm{T}} \in \mathbb{R}^{n_x}$$
$$\mathbf{u}(k) = \left[\mathbf{u}_1^{\mathrm{T}}(k)\ \mathbf{u}_2^{\mathrm{T}}(k)\ \cdots\ \mathbf{u}_m^{\mathrm{T}}(k)\right]^{\mathrm{T}} \in \mathcal{U} \subset \mathbb{R}^{n_u}$$
$$\mathbf{y}(k) = \left[\mathbf{y}_1^{\mathrm{T}}(k)\ \mathbf{y}_2^{\mathrm{T}}(k)\ \cdots\ \mathbf{y}_m^{\mathrm{T}}(k)\right]^{\mathrm{T}} \in \mathbb{R}^{n_y}$$

are, respectively, the concatenated state, control input, and output vectors of the overall system $\mathcal{S}$. Also, $\mathbf{u}(k) \in \mathcal{U} = \mathcal{U}_1 \times \mathcal{U}_2 \times \cdots \times \mathcal{U}_m$. **A**, **B**, and **C** are constant matrices of appropriate dimensions and are defined as follows:

$$\mathbf{A} = \begin{bmatrix} \mathbf{A}_{11} & \mathbf{A}_{12} & \cdots & \mathbf{A}_{1m} \\ \mathbf{A}_{21} & \mathbf{A}_{22} & \cdots & \mathbf{A}_{2m} \\ \vdots & \vdots & \ddots & \vdots \\ \mathbf{A}_{m1} & \mathbf{A}_{m2} & \cdots & \mathbf{A}_{mm} \end{bmatrix}^{\mathrm{T}}$$

$$\mathbf{B} = \begin{bmatrix} \mathbf{B}_{11} & \mathbf{B}_{12} & \cdots & \mathbf{B}_{1m} \\ \mathbf{B}_{21} & \mathbf{B}_{22} & \cdots & \mathbf{B}_{2m} \\ \vdots & \vdots & \ddots & \vdots \\ \mathbf{B}_{m1} & \mathbf{B}_{m2} & \cdots & \mathbf{B}_{mm} \end{bmatrix}^{\mathrm{T}}$$

$$\mathbf{C} = \begin{bmatrix} \mathbf{C}_{11} & \mathbf{C}_{12} & \cdots & \mathbf{C}_{1m} \\ \mathbf{C}_{21} & \mathbf{C}_{22} & \cdots & \mathbf{C}_{2m} \\ \vdots & \vdots & \ddots & \vdots \\ \mathbf{C}_{m1} & \mathbf{C}_{m2} & \cdots & \mathbf{C}_{mm} \end{bmatrix}^{\mathrm{T}}$$

If there is only the state interacting term, we call the model the *state interacted model* and it can be expressed as

$$\mathbf{x}_i(k+1) = \mathbf{A}_{ii}\mathbf{x}_i(k) + \mathbf{B}_{ii}\mathbf{u}_i(k) + \sum_{j\in\mathcal{P}_{+i}} \mathbf{A}_{ij}\mathbf{x}_j(k) \tag{4.3}$$

Similarly, if there is only the input interacting term of input, we call the model the *input interacted model* and it can be expressed as

$$\mathbf{x}_i(k+1) = \mathbf{A}_{ii}\mathbf{x}_i(k) + \mathbf{B}_{ii}\mathbf{u}_i(k) + \sum_{j\in\mathcal{P}_i^u} \mathbf{B}_{ij}\mathbf{u}_j(k) \tag{4.4}$$

In fact, the state interacted model can be transformed into the input interacted model by selecting a suitable virtue states. Let us take the following system which is composed of two state interacted subsystems as an example.

1. Model of subsystem $\mathcal{S}_1$

$$\begin{bmatrix} \mathbf{x}_1^{(1)}(k+1) \\ \mathbf{x}_1^{(2)}(k+1) \end{bmatrix} = \begin{bmatrix} \mathbf{A}_{11}^{(1)} & 0 \\ 0 & \mathbf{A}_{11}^{(2)} \end{bmatrix} \begin{bmatrix} \mathbf{x}_1^{(1)}(k) \\ \mathbf{x}_1^{(2)}(k) \end{bmatrix} + \begin{bmatrix} \mathbf{B}_{11}^{(1)} \\ \mathbf{B}_{11}^{(2)} \end{bmatrix} \mathbf{u}_1(k) \tag{4.5}$$

2. Model of subsystem $\mathcal{S}_2$:

$$\mathbf{x}_2(k+1) = \mathbf{A}_{22}\mathbf{x}_2(k) + \mathbf{B}_{22}\mathbf{u}_2(k) + \mathbf{A}_{21}^{(2)}\mathbf{x}_1^{(2)}(k) \tag{4.6}$$

where $\mathbf{x}_1^{(l)}(k)$, $l = 1, 2$, expresses the lth state block of subsystem $\mathcal{S}_1$, and $\mathbf{A}_{11}^{(1)}$, $\mathbf{B}_{11}^{(1)}$, and $\mathbf{A}_{21}^{(2)}$ are the coefficients corresponding to the lth state block.

In the above system, subsystem $\mathcal{S}_1$ impacts subsystem $\mathcal{S}_2$ by state $\mathbf{x}_1^{(2)}(k)$. If we append an additional state $\mathbf{x'}_1^{(2)}(k)$ as a virtue state into $\mathcal{S}_2$, and let the dynamics of $\mathbf{x'}_1^{(2)}(k)$ be the same as $\mathbf{x}_1^{(2)}(k)$, then the model of subsystem $\mathcal{S}_2$ becomes

$$\begin{bmatrix} \mathbf{x'}_1^{(2)}(k+1) \\ \mathbf{x}_2(k+1) \end{bmatrix} = \begin{bmatrix} \mathbf{A}_{11}^{(1)} & 0 \\ 0 & \mathbf{A}_{22} \end{bmatrix} \begin{bmatrix} \mathbf{x'}_1^{(1)}(k) \\ \mathbf{x}_2(k) \end{bmatrix} + \begin{bmatrix} \mathbf{B}_{11}^{(2)} & 0 \\ 0 & \mathbf{B}_{22} \end{bmatrix} \begin{bmatrix} \mathbf{u}_1(k) \\ \mathbf{u}_2(k) \end{bmatrix} \tag{4.7}$$

It can be seen from Equations (4.5)–(4.7) that the state interacted model representations of subsystems $\mathcal{S}_1$ and $\mathcal{S}_2$, see (4.5) and (4.6), are expressed by two input interacting models, see (4.5) and (4.7). The resulting input interacting models can be directly used in the design of DMPC. Similarly, we can transform the input interacted model into the state interacted model if we take the inputs as states. These transformations provide more freedom to the control engineers in selecting appropriate controller design methods according to the requirement.

4.3 Structure Model and Structure Controllability

4.3.1 Structure Model

Here, we induce the concept of a structure model, which describes whether there are interacted relationships between each subsystem and helps the system to decompose. This section mainly refers to [82].

There are many different methods to express the structure of distributed systems, e.g., graphs, tables, and matrices. The most convenient one is the directed graph, which uses a graph to express the system structure. In the directed graph, one node expresses one unit, which can be denoted as $\mathcal{S}_i$. If unit $\mathcal{S}_i$ directly affects unit $\mathcal{S}_j$, then we can induce an arrow to describe this relationship. This directed graph could provide the most intuitive imagination on the relationship among units. However, when the number of subsystems increases, the directed graph will become more and more complicated, and will not be suitable for computation. Thus, here we use matrices to describe the system structure. Among them the most direct one is the *adjacency matrix*, which is completely equivalent to the *directed graph*.

To interpret the structure model, let us take the multizone building temperature regulation systems as an example, which is a class of typical spatially distributed systems, as shown in Figure 4.2, which are composed of six physically interacted subsystems (rooms or zones) labeled with $\mathcal{S}_1, \mathcal{S}_2, \ldots, \mathcal{S}_6$, respectively. The thermal influences between rooms of the same building occur through internal walls (the internal walls' isolation is weak) and/or door openings. A thermal meter is installed in each room, which is used to measure the temperature of the multizone building.

According to the heat transfer mechanism, we can write the state space mathematical model of the whole system as

$$\mathbf{x}(k+1) = \mathbf{A}\mathbf{x}(k) \tag{4.8}$$

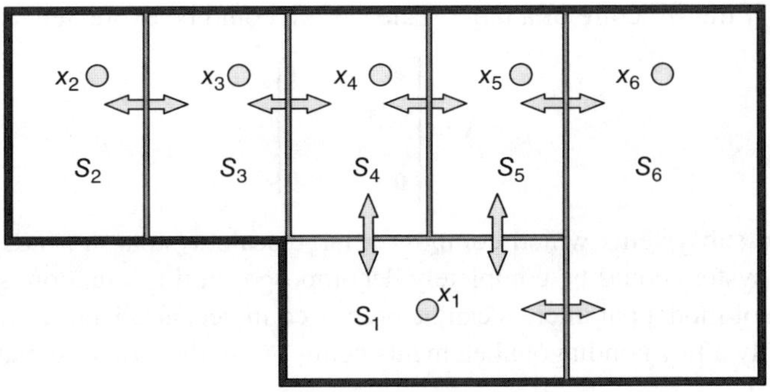

Figure 4.2 A multizone building temperature regulation system

where $\mathbf{x}(k) = \left[x_1\,(k)\; x_2(k)\; \cdots\; x_6(k)\right]^{\mathrm{T}}$ is the temperature of each room, and $\mathbf{A}$ is the system matrix

$$\mathbf{A} = \begin{bmatrix} 0.6 & 0 & 0 & 0.09 & 0.09 & 0.09 \\ 0 & 0.6 & 0.18 & 0 & 0 & 0 \\ 0 & 0.18 & 0.6 & 0.18 & 0 & 0 \\ 0.09 & 0 & 0.18 & 0.6 & 0.18 & 0 \\ 0.09 & 0 & 0 & 0.18 & 0.6 & 0.18 \\ 0.06 & 0 & 0 & 0 & 0.12 & 0.7 \end{bmatrix}$$

For the same system, if only the interaction relationship is concerned, the structure model can be adopted. If we consider each room as a unit, the system adjacent matrix $\overline{\mathbf{A}}$ can be expressed as

$$\overline{\mathbf{A}} = \begin{bmatrix} * & 0 & 0 & * & * & * \\ 0 & * & * & 0 & 0 & 0 \\ 0 & * & * & * & 0 & 0 \\ * & 0 & * & * & * & 0 \\ * & 0 & 0 & * & * & * \\ * & 0 & 0 & 0 & * & * \end{bmatrix}$$

The notation "*" in the ith row and jth column expresses that the unit S_i is impacted by S_j. Since the structure model focuses on the logical relationship among subsystems, the included information in which is less than that in the mathematical model. However, it could provide more intuitive comprehensive to the logical relationship among subsystems. In addition, the computation of the structure model is more convenient than the mathematical model.

4.3.2 Function of the Structure Model in System Decomposition

The structure model is more general than the mathematic model in the decomposition of the system structure since it could directly express the characteristics of the system structure and gives suggestion for system decomposition.

For example, if the structure of a large-scale system could be expressed as

$$\overline{\mathbf{A}} = \begin{bmatrix} * & & & \mathbf{0} \\ & * & & \\ & & \ddots & \\ \mathbf{0} & & & * \end{bmatrix}$$

it refers that the subsystems which compose a large-scale system are independent of each other. Then the system could be completely decomposed. In this situation, system modeling or identification of model parameters can be processed in decentralized formulation.

If there are only a few nondiagonal elements being "*" in the structure matrix, for example

$$\overline{\mathbf{A}} = \begin{array}{c} \\ 1 \\ 2 \\ 3 \\ 4 \end{array} \begin{array}{c} \begin{array}{cccc} 1 & 2 & 3 & 4 \end{array} \\ \begin{bmatrix} * & 0 & 0 & * \\ 0 & * & 0 & 0 \\ 0 & 0 & * & 0 \\ 0 & 0 & 0 & * \end{bmatrix} \end{array}$$

it means that each subsystem is weakly interacting. After some appropriate transformation, the structure matrix may be rewritten in the following form:

$$\overline{\mathbf{A}} = \begin{array}{c} \\ 1 \\ 4 \\ 2 \\ 3 \end{array} \begin{array}{c} \begin{array}{cccc} 1 & 4 & 2 & 3 \end{array} \\ \left[\begin{array}{cc:c:c} * & * & & 0 \\ & * & & \\ \hdashline & & * & \\ \hdashline & & & * \end{array}\right] \end{array}$$

Here, unit 1st and unit 4th compose a new subsystem; unit 2nd and unit 3rd are completely decomposable.

If the interaction among subsystems is hierarchical and the links go only one way, then we could use the upper (or down) triangular matrix to describe the structure, e.g.,

$$\overline{\mathbf{A}} = \begin{bmatrix} * & 0 & \cdots & 0 \\ * & * & \ddots & \vdots \\ \vdots & \ddots & \ddots & 0 \\ * & \cdots & * & * \end{bmatrix}$$

where unit 1st is independent of all other subsystems. If the states of unit 1st are fixed, the states of unit 2nd will be independent of all other subsystems. And the rest may be deduced similarly. These types of mathematical models could be solved in cascade.

If there are many nondiagonal elements being "*" in the structure matrix, we call this type of system the strong interacted system, e.g.,

$$\overline{\mathbf{A}} = \begin{array}{c} \\ 1 \\ 2 \\ 3 \\ 4 \end{array} \begin{array}{c} \begin{array}{cccc} 1 & 2 & 3 & 4 \end{array} \\ \begin{bmatrix} * & * & 0 & 0 \\ 0 & * & 0 & * \\ * & 0 & * & 0 \\ 0 & 0 & * & * \end{bmatrix} \end{array}$$

where unit 1st impacts unit 2nd directly ($\overline{A}_{12}$ = "*"), and unit 2nd indirectly impacts unit 1st through unit 3rd and unit 4th ($\overline{A}_{24} = \overline{A}_{43} = \overline{A}_{31}$ ="*"). In general, there exist interacting loops in the strong interacted structure, which means all the elements of $\overline{a}_{i_1i_2}, \overline{a}_{i_2i_3}, \dots, \overline{a}_{i_pi_1}$ are "*" in the structure matrix. This structure cannot transform into a diagonal form or triangular form. In this case, the system model cannot be completely divided.

In many large-scale systems, the structure models can be directly deduced by the physical composition of systems or the product processing. However, the decomposability of system may not be obtained directly in this structure model. In this case, we can transform the system model into a triangular or diagonal form to divide the system. If some subsystems consist of many units after partitioned by the triangular form, we can continue to, if possible, transform the subsystem to the diagonal form for further decomposition.

4.3.3 *Input–Output Accessibility*

By the structure model, we are not only able to analyze the static characteristics of a system for providing the important information for system decomposition but also able to analyze the dynamic characteristics of a system. In this section, we induce the concept of *structure accessibility* and *general rank* of the structure matrix, which are very important for discussing the *structure controllability* of a system.

Consider the following LTI system:

$$\begin{cases} \mathbf{x}(k+1) = \mathbf{A}\mathbf{x}(k) + \mathbf{B}\mathbf{u}(k) \\ \mathbf{y}(k+1) = \mathbf{C}\mathbf{x}(k) \end{cases} \tag{4.9}$$

where **A**, **B**, and **C** are coefficient matrices. There is a structure description $(\overline{\mathbf{A}}, \overline{\mathbf{B}}, \overline{\mathbf{C}})$ corresponding to the mathematical model (**A**, **B**, **C**), where zero in $\overline{\mathbf{A}}, \overline{\mathbf{B}}, \overline{\mathbf{C}}$ expresses the "0" element in **A**, **B**, and **C** and "*" expresses nonzero elements in **A**, **B**, and **C**. It can be seen that when two mathematical models $(\mathbf{A}_1, \mathbf{B}_1, \mathbf{C}_1)$ and $(\mathbf{A}_2, \mathbf{B}_2, \mathbf{C}_2)$ are not equal to each other, their structure model may be the same and expressed as $(\overline{\mathbf{A}}, \overline{\mathbf{B}}, \overline{\mathbf{C}})$. Here, we think that these two systems are *structure equivalence.*

Above, we have induced the neighbor matrix to describe the relationship of each subsystem. For dynamic system (4.9), the inputs, states, and outputs are the basic elements of the system; thus, we can express the neighboring connection matrix as

$$\mathbf{A}_a = \begin{array}{c} \\ \mathbf{x} \\ \mathbf{u} \\ \mathbf{y} \end{array} \begin{array}{c} \begin{array}{ccc} \mathbf{x} & \mathbf{u} & \mathbf{y} \end{array} \\ \begin{bmatrix} \overline{\mathbf{A}}^{\mathrm{T}} & \mathbf{0} & \overline{\mathbf{C}}^{\mathrm{T}} \\ \overline{\mathbf{B}}^{T} & \mathbf{0} & \mathbf{0} \\ \mathbf{0} & \mathbf{0} & \mathbf{0} \end{bmatrix} \end{array}$$

$$\begin{bmatrix} \mathbf{x}_{a,k+1} \\ \mathbf{u}_{a,k+1} \\ \mathbf{y}_{a,k} \end{bmatrix} = \begin{bmatrix} \mathbf{A} & \mathbf{B} & \mathbf{0} \\ \mathbf{0} & \mathbf{0} & \mathbf{0} \\ \mathbf{C} & \mathbf{0} & \mathbf{0} \end{bmatrix} \begin{bmatrix} \mathbf{x}_{a,k} \\ \mathbf{u}_{a,k} \\ \mathbf{y}_{a,k-1} \end{bmatrix}$$

$$\overline{\mathbf{A}}_d = \begin{bmatrix} \overline{\mathbf{A}} & \overline{\mathbf{B}} & \mathbf{0} \\ \mathbf{0} & \mathbf{0} & \mathbf{0} \\ \overline{\mathbf{C}} & \mathbf{0} & \mathbf{0} \end{bmatrix}$$

For example, consider the dynamic system

$$\begin{cases}\begin{bmatrix}x_1(k+1)\\x_2(k+1)\\x_3(k+1)\end{bmatrix}=\begin{bmatrix}0.1&0.2&0\\0.2&0.5&0.2\\0&0.2&0.1\end{bmatrix}\begin{bmatrix}x_1(k)\\x_2(k)\\x_3(k)\end{bmatrix}+\begin{bmatrix}1&0\\0&1\\0&0\end{bmatrix}\begin{bmatrix}u_1(k)\\u_2(k)\end{bmatrix}\\[2ex]\begin{bmatrix}y_1(k)\\y_2(k)\end{bmatrix}=\begin{bmatrix}1&0&0\\0&0&1\end{bmatrix}\begin{bmatrix}x_1(k)\\x_2(k)\\x_3(k)\end{bmatrix}\end{cases}$$

We have

$$\overline{\mathbf{A}}=\begin{bmatrix}*&*&0\\*&*&*\\0&*&*\end{bmatrix}\quad\overline{\mathbf{B}}=\begin{bmatrix}*&0\\0&*\\0&0\end{bmatrix}\quad\overline{\mathbf{C}}=\begin{bmatrix}*&0&0\\0&0&*\end{bmatrix}$$

Then the adjacent matrix of (4.9) can be expressed as

$$\mathbf{A}_a=\begin{matrix}\\x_1\\x_2\\x_3\\u_1\\u_2\\y_1\\y_2\end{matrix}\begin{matrix}\begin{matrix}x_1&x_2&x_3&u_1&u_2&y_1&y_2\end{matrix}\\\begin{bmatrix}*&*&0&0&0&*&0\\*&*&*&0&0&0&0\\0&*&*&0&0&0&*\\*&0&0&0&0&0&0\\0&*&0&0&0&0&0\\0&0&0&0&0&0&0\\0&0&0&0&0&0&0\end{bmatrix}\end{matrix}\tag{4.10}$$

It should be noticed that the structure matrices $\overline{\mathbf{A}}$, $\overline{\mathbf{B}}$, and $\overline{\mathbf{C}}$ appeared in $\mathbf{A}_a$ are the transposes of them. It should also be noticed that the columns which consist of zeros are corresponding to the inputs in the adjacent matrix, and the rows which consist of zeros are corresponding to the outputs in the adjacent matrix.

It can be seen from (4.10) that the adjacent matrix only reflects the direct impaction among each subsystem. In fact, we are usually concern about if one unit indirectly impacts another unit through some intermediate units. In this case, the k power of $\mathbf{A}_a$ can be used to express the interaction during k step ahead. Here, the power of the structure matrix can be calculated according to the rule of matrix power, where the operation of "plus" and "times" are substituted by logic "plus" and logic "times" (refer to the maximizing-minimum operation in fuzzy logic).

The element $a_{ij}^{(k)}$ = " $*$ " in $\mathbf{A}_a^k$ expresses that the unit $\mathcal{S}_i$ will impact unit $\mathcal{S}_j$ at k step ahead. If $\mathcal{S}_i$ could directly or indirectly impact unit $\mathcal{S}_j$, we say that $\mathcal{S}_i$ is accessible to $\mathcal{S}_j$.

In a system which consists of N units, for any two units $\mathcal{S}_i$ and $\mathcal{S}_j$, if one is accessible to the other, then the relationship must be reflected in the matrix $\mathbf{A}_a, \mathbf{A}_a^2, \ldots, \mathbf{A}_a^{N-1}$. If each unit is accessible to itself, then all accessible relationships can be described by the following matrix:

$$\mathbf{R}=\mathbf{I}\cup\mathbf{A}_a\cup\cdots\cup\mathbf{A}_a^{N-1}\tag{4.11}$$

The matrix $\mathbf{R}$ is so-called the *accessible matrix*. It is also a kind of structure matrix, where the ith row and jth element r_{ij} equal to zero expresses that $\mathcal{S}_i$ is unaccessible to $\mathcal{S}_j$. By logic operating rules, the above equation can be simply rewritten as

$$\mathbf{R} = (\mathbf{I} \cup \mathbf{A}_a)^{N-1} \tag{4.12}$$

In dynamic systems, there are three basic facts: system state, inputs, and outputs. To find out the interactions between subsystems, to find out whether the states are able to be accessed by the input, to find out whether the states of the system impact the output, the system dynamic structure analysis is very important. If every state x_i can be accessed by at least one input u_j, then we say that this system is input accessible. If every state x_i can be accessed to at least one output y_l, we say that this system is output accessible. The input accessibility and output accessibility reflect whether the system states can be adjusted by inputs and whether all the system states can be reflected by outputs.

The accessible matrix (4.12) can be rewritten in the following bock form:

$$\mathbf{R} = (\mathbf{I} \cup \mathbf{A}_a)^{N-1} = \begin{array}{c} \\ \mathbf{x} \\ \mathbf{u} \\ \mathbf{y} \end{array} \begin{array}{c} \begin{array}{ccc} \mathbf{x} & \mathbf{u} & \mathbf{y} \end{array} \\ \begin{bmatrix} \mathbf{R}_{xx} & \mathbf{R}_{xu} & \mathbf{R}_{xy} \\ \mathbf{R}_{ux} & \mathbf{R}_{uu} & \mathbf{R}_{uy} \\ \mathbf{R}_{yx} & \mathbf{R}_{yu} & \mathbf{R}_{yy} \end{bmatrix} \end{array} \tag{4.13}$$

Obviously, the input accessibility and output accessibility can be described by the structure matrices $\mathbf{R}_{ux}$ and $\mathbf{R}_{xy}$, respectively. Consider that

$$\mathbf{R} = (\mathbf{I} \cup \mathbf{A}_a)^{N-1} = \begin{bmatrix} \left(\overline{\mathbf{A}}^{\mathrm{T}} + \mathbf{I}\right)^{N-1} & \mathbf{0} & (\overline{\mathbf{A}}^{\mathrm{T}} + \mathbf{I})^{N-2}\overline{\mathbf{C}}^{\mathrm{T}} \\ \overline{\mathbf{B}}^{\mathrm{T}}(\overline{\mathbf{A}}^{\mathrm{T}} + \mathbf{I})^{N-2} & \mathbf{I} & \overline{\mathbf{B}}^{\mathrm{T}}(\overline{\mathbf{A}}^{\mathrm{T}} + \mathbf{I})^{N-3}\overline{\mathbf{C}}^{\mathrm{T}} \\ \mathbf{0} & \mathbf{0} & \mathbf{I} \end{bmatrix} \tag{4.14}$$

and notice that the interaction relationship can at most go through n units in $\mathbf{R}$, just like $u_j \to x_{i_1} \to x_{i_2} \to \cdots \to x_{i_n}$ and $x_{i_1} \to x_{i_2} \to \cdots \to x_{i_n} \to y_k$, where n is the dimension of state $\mathbf{x}$. We have the *input accessibility matrix*

$$\mathbf{R}_{ux} = \overline{\mathbf{B}}^{\mathrm{T}}(\overline{\mathbf{A}}^{\mathrm{T}} + \mathbf{I})^{n-1},$$

and the *output accessibility matrix*

$$\mathbf{R}_{xy} = (\overline{\mathbf{A}}^{\mathrm{T}} + \mathbf{I})^{n-1}\overline{\mathbf{C}}^{\mathrm{T}}.$$

If there are no columns where all elements are zeros in $\mathbf{R}_{ux}$, this system is input accessible. If there are no rows where all elements are zeros in $\mathbf{R}_{xy}$, this system is output accessible.

Example 4.1 Consider the following system:

$$\overline{\mathbf{A}} = \begin{bmatrix} 0 & 0 & * & * & 0 \\ 0 & 0 & * & 0 & * \\ 0 & 0 & * & * & 0 \\ 0 & 0 & * & * & 0 \\ 0 & 0 & * & 0 & 0 \end{bmatrix} \tag{4.15}$$

$$\overline{\mathbf{B}} = \begin{bmatrix} * & 0 & 0 & 0 \\ 0 & 0 & * & * \\ 0 & 0 & 0 & * \\ * & * & 0 & 0 \\ * & 0 & 0 & * \end{bmatrix} \tag{4.16}$$

$$\overline{\mathbf{C}} = \begin{bmatrix} 0 & 0 & * & * & 0 \\ 0 & 0 & 0 & 0 & * \end{bmatrix} \tag{4.17}$$

Judge its input accessibility and output accessibility.
Here $n = 5$; according to

$$\mathbf{R}_{ux} = \overline{\mathbf{B}}^T (\overline{\mathbf{A}}^T + \mathbf{I})^4 = \begin{bmatrix} * & * & * & * & * \\ * & * & * & * & * \\ 0 & * & 0 & 0 & 0 \\ * & * & * & * & * \end{bmatrix}$$

the system is input accessible. In addition, since

$$\mathbf{R}_{ux} = (\overline{\mathbf{A}}^T + \mathbf{I})^4 \overline{\mathbf{C}}^T = \begin{bmatrix} 0 & 0 \\ 0 & 0 \\ * & * \\ * & * \\ 0 & * \end{bmatrix}$$

the system is output inaccessible, where x_1 and x_2 cannot be reflected by any output.

Remark 4.1 *In some literature works, the accessibility is called connectability. They all reflect the characteristic of whether one unit could impact another one.*

4.3.4 General Rank of the Structure Matrix

Considering that there is no quantity description of the structure matrix, and the rank of structure matrix cannot be defined as the numeric matrix; thus we define the maximum rank of its corresponding numeric matrix rank as the rank of the structure matrix. To identify from the rank of the numeric matrix, we call it the "generic rank" that is

$$\mathrm{gr}(\overline{\mathbf{S}}) = \max(\mathrm{rank}(\mathbf{S}))$$

The generic rank of the structure matrix can be calculated by a different method. See reference [82].

4.3.5 Structure Controllability

In the control of a dynamic large-scale system, to get a feasible control solution, the system must be controllable. Thus, it is important to test the controllability of a system. Here, we will introduce how to analyze the controllability of the dynamic system by the method of structure analysis. The analysis of structure controllability is not quantified and cannot give the results of whether the controllability is weak or strong. However, it is very convenient since it only involves some logical relationship.

Consider the structured linear system

$$\mathbf{x}(k+1) = \overline{\mathbf{A}}\mathbf{x}(k) + \overline{\mathbf{B}}\mathbf{u}(k) \tag{4.18}$$

where $\mathbf{x} \in \mathbb{R}^n$, $\mathbf{u} \in \mathbb{R}^{n_u}$, and $\overline{\mathbf{A}}, \overline{\mathbf{B}}$ are structure matrices. If there is a controllable general system whose structure is equivalent to $(\overline{\mathbf{A}}, \overline{\mathbf{B}})$, then we say $(\overline{\mathbf{A}}, \overline{\mathbf{B}})$ is structure controllable.

It can be seen from the definition that if $(\overline{\mathbf{A}}, \overline{\mathbf{B}})$ is structure uncontrollable, then the numeric system is definitely uncontrollable in spite of any values of the matrix's elements.

For a general numeric system $(\mathbf{A}, \mathbf{B})$, Rosenbrock provided a condition related to the rank of an extent matrix,

$$rank(k\mathbf{G}_c) = n^2 \tag{4.19}$$

to judge the controllability of $(\mathbf{A}, \mathbf{B})$, where $n^2 \times n(n+m-1)$-dimensional matrix $\mathbf{G}_c$ is defined as

$$\mathbf{G}_c = \begin{bmatrix} \mathbf{B} & \mathbf{I} & 0 & 0 & 0 & \cdots & 0 & 0 \\ 0 & -\mathbf{A} & \mathbf{B} & -\mathbf{I} & 0 & \cdots & 0 & 0 \\ 0 & 0 & 0 & -\mathbf{A} & 0 & \cdots & 0 & 0 \\ \cdots & \cdots & \cdots & \cdots & \ddots & \cdots & \cdots & \cdots \\ 0 & \cdots & \cdots & 0 & 0 & \mathbf{B} & \mathbf{I} & 0 \\ 0 & \cdots & \cdots & 0 & 0 & 0 & -\mathbf{A} & \mathbf{B} \end{bmatrix} \tag{4.20}$$

Then, this result is introduced to the study of structure controllability by Shields and Pearson. They provided the following criterion.

Theorem 4.1 *System (4.18) is controllable if and only if*

$$\mathrm{gr}(\overline{\mathbf{G}}_c) = n^2,$$

where

$$\overline{\mathbf{G}}_c = \begin{bmatrix} \overline{\mathbf{B}} & \overline{\mathbf{I}} & 0 & 0 & 0 & \cdots & 0 & 0 \\ 0 & -\overline{\mathbf{A}} & \overline{\mathbf{B}} & -\overline{\mathbf{I}} & 0 & \cdots & 0 & 0 \\ 0 & 0 & 0 & -\overline{\mathbf{A}} & 0 & \cdots & 0 & 0 \\ \cdots & \cdots & \cdots & \cdots & \ddots & \cdots & \cdots & \cdots \\ 0 & \cdots & \cdots & 0 & 0 & \overline{\mathbf{B}} & \overline{\mathbf{I}} & 0 \\ 0 & \cdots & \cdots & 0 & 0 & 0 & -\overline{\mathbf{A}} & \overline{\mathbf{B}} \end{bmatrix} \tag{4.21}$$

In addition, combined with the input connectivity (accessibility) and the general rank criterion, Division and Morari give another necessary and sufficient condition.

Theorem 4.2 *System (4.18) is controllable if and only if the following two conditions are satisfied:*

1. *System (4.18) is input accessible;*
2. $\mathrm{gr}\left(\left[\overline{\mathbf{A}}\ \overline{\mathbf{B}}\right]\right) = n.$

In this theorem, the first condition points out that the inputs could be able to affect all the states. The second condition points out that the degree of freedom of inputs is large enough to regulate the state to the desired value.

Example 4.2 Judge the structure controllability of the structured system $(\overline{\mathbf{A}}, \overline{\mathbf{B}})$ expressed by (4.15) and (4.16).

First, let us check the input accessibility according to condition (1) in Theorem 4.2. From Example 4.1, the system $(\overline{\mathbf{A}}, \overline{\mathbf{B}})$ is input accessible.

Then what we need to do is to check the generic rank of $[\overline{\mathbf{A}}, \overline{\mathbf{B}}]$, according to condition (2) in Theorem 4.2

$$\begin{bmatrix} \overline{\mathbf{A}} & \overline{\mathbf{B}} \end{bmatrix} = \begin{bmatrix} 0 & 0 & * & * & 0 & * & 0 & 0 & 0 \\ 0 & 0 & * & 0 & * & 0 & 0 & * & * \\ 0 & 0 & * & * & 0 & 0 & 0 & 0 & * \\ 0 & 0 & * & * & 0 & * & * & 0 & 0 \\ 0 & 0 & * & 0 & 0 & * & 0 & 0 & * \end{bmatrix} \tag{4.22}$$

Delete the first and second columns and adjust the matrix; the following normalized form can be deduced:

$$\mathbf{S} = \begin{matrix} \\ x_1 \\ x_3 \\ x_5 \\ x_4 \\ x_2 \end{matrix} \begin{matrix} \begin{matrix} x_3 & x_4 & x_1 & x_4 & u_2 & u_5 & u_3 \end{matrix} \\ \begin{bmatrix} * & * & * & 0 & 0 & 0 & 0 \\ * & * & 0 & * & 0 & 0 & 0 \\ * & 0 & * & * & 0 & 0 & 0 \\ * & * & * & 0 & * & 0 & 0 \\ * & 0 & 0 & * & 0 & * & * \end{bmatrix} \end{matrix} \tag{4.23}$$

Obviously, $\text{gr}\left(\begin{bmatrix} \overline{\mathbf{A}} & \overline{\mathbf{B}} \end{bmatrix}\right) = \text{gr}(\mathbf{S})) = 5 = n$ satisfies the second condition; consequently, the system $(\overline{\mathbf{A}}, \overline{\mathbf{B}})$ is structure controllable.

4.4 Related Gain Array Decomposition

For the distributed system, one or more manipulated variables can affect the interactions of controlled variables in a specific loop or all other control loops. Thus, understanding the dependence of different manipulated and controlled variables in a large-scale distributed control scheme could be extremely helpful in designing and implementing a control scheme for a process.

To design and analyze the control scheme for the distributed system, we can refer to the traditional methods used for the MIMO system. Here, we will briefly introduce the classic RGA method by mainly referring to [83–85]. The RGA provides a quantitative approach to the analysis of the interactions between the manipulated variables and the controlled variables, and then provides a method of pairing manipulated and controlled variables to generate a control scheme. The RGA has been widely used in designing and analyzing a MIMO control scheme for a process in steady state.

4.4.1 RGA Definition

We follow [86] here and show that the RGA provides a measure of interactions. Let u_j and y_i denote a particular manipulated variable-controlled variable pair for the MIMO plant $\mathbf{G}(s)$, and assume that our task is to use u_j to control y_i.

In general, the process gain $\partial y_i/\partial u_j$ can be evaluated in two extreme cases [86]:

- Other loop is open

$$\left(\frac{\partial y_i}{\partial u_j}\right)_{u_k=0,k\neq j} = g_{ij} \tag{4.24}$$

- Other loop is closed

$$\left(\frac{\partial y_i}{\partial u_j}\right)_{y_k=0,k\neq j} = \hat{g}_{ij} \tag{4.25}$$

where $g_{ij} = \text{Element}(\mathbf{G}, i, j)$ is the ith row and jth column element of $\mathbf{G}$, whereas $\hat{g}_{ij} = 1/\text{Element}(\mathbf{G}^{-1}, j, i)$ is the inverse of the jth row and ith column element of $\mathbf{G}^{-1}$.

Example 4.3 To make the concept clear, the TITO system shown in Figure 4.3 is taken as an example.

Considering the control of y_2 using u_2, the process gain is either

$$\left(\frac{\partial y_2}{\partial u_2}\right)_{u_1=0} = g_{22}$$

or

$$\left(\frac{\partial y_2}{\partial u_2}\right)_{y_1=0} = g_{22} - \frac{g_{12}g_{21}}{g_{11}}$$

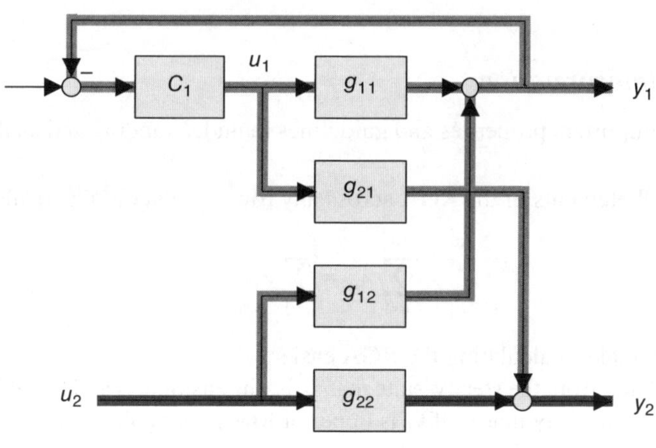

Figure 4.3 The two-input–two-output system (TITO)

To derive (4.25), we note that

$$\mathbf{y} = \mathbf{G}\mathbf{u} \quad \Rightarrow \quad \left(\frac{\partial y_i}{\partial u_j}\right)_{u_k=0, k\neq j} = \text{Element}(\mathbf{G}, i, j)$$

Interchange the roles of **G** and $\mathbf{G}^{-1}$, the roles of **u** and **y**, and the roles of subscribe i and j, it yields [83]

$$\mathbf{u} = \mathbf{G}^{-1}\mathbf{y} \quad \Rightarrow \quad \left(\frac{\partial u_j}{\partial y_i}\right)_{y_k=0, k\neq i} = \text{Element}(\mathbf{G}^{-1}, j, i)$$

and (4.25) follows. Reference [86] points out that the ratio between the gains in (4.24) and (4.25) is a useful measure of interactions between the jth input and the ith output, and defines the relative gain of the jth input to the ith output λ_{ij} as

$$\lambda_{ij} \triangleq \frac{g_{ij}}{\hat{g}_{ij}} = \text{Element}(\mathbf{G}, i, j) \cdot \text{Element}(\mathbf{G}^{-1}, j, i) \tag{4.26}$$

The RGA is the corresponding matrix of relative gains. From (4.26), the RGA can be expressed as

$$\Lambda(\mathbf{G}) = \mathbf{G} \otimes (\mathbf{G}^{-1})^{\mathrm{T}} = \begin{bmatrix} \lambda_{11} & \lambda_{12} & \cdots & \lambda_{1n} \\ \lambda_{21} & \lambda_{22} & \cdots & \lambda_{2n} \\ \vdots & \vdots & \ddots & \vdots \\ \lambda_{n1} & \lambda_{n2} & \cdots & \lambda_{nn} \end{bmatrix} \tag{4.27}$$

where operation $\otimes$ denotes element-by-element multiplication (the *Schur* product).

The array is a matrix with one column for each input variable and one row for each output variable in the MIMO system. This format allows a process engineer to easily compare the relative gains associated with each input–output variable pair. It helps engineers to match the input and output variables that have the biggest effect on each other, and to minimize the undesired side effects at the same time.

4.4.2 RGA Interpretation

There are some important properties and guidelines in understanding and analyzing the RGA:

1. The sum of all elements of the RGA across any row or any column equals 1:

$$\sum_{i=1}^{n} \lambda_{ij} = \sum_{j=1}^{n} \lambda_{ij} = 1$$

2. This property makes calculating the RGA easier.
3. The λ_{ij} calculated from the steady-state matrix is dimensionless and not affected by scaling.
4. The RGA is the identity matrix if **G** is upper or lower triangular.
5. A change in an element of **G** equals the negative inverse of its corresponding RGA element $g_{ij}(1 - 1/\lambda_{ij})$.
6. Plants with large RGA elements are always ill-conditioned, but the reverse may not hold.

Table 4.1 The meaning of the value for RGA elements

Value	Interpretation
$\lambda_{ij} = 1$	Open-loop gain and closed-loop gain are identical Interaction does not affect the pairing of (y_i, u_j)
$\lambda_{ij} = 0$	Open-loop gain is zero The *j*th input does not affect the *j*th output
$0 < \lambda_{ij} < 1$	Closed-loop interaction increases gain Interaction is most severe when $\lambda_{ij} = 0.5$
$\lambda_{ij} > 1$	Closed-loop interaction reduces gain Higher values indicate more severe interaction
$\lambda_{ij} < 0$	Closed-loop gain is in the opposite direction from the open-loop gain

In order to better comprehend the meaning of the RGA, the meaning of the values of the elements of the RGA is shown in Table 4.1.

Therefore, considering all of the above shown in Table 4.1, the pairing strategy should match the variables where λ_{ij} is nearest to 1 while avoiding the variables where λ_{ij} is zero or negative.

4.4.3 Pairing Rules

Before introducing the paring rules, let us first introduce the *Niederlinksi index (NI)* which is used to analyze the stability of the control loop pairings using the result of the RGA [86]:

$$\text{NI} = \frac{\det(\mathbf{G})}{\prod_{i=1}^{n} g_{ii}} \tag{4.28}$$

The goal of the RGA and NI analysis is to quantitatively determine the optimal variable pairing for a given process. A negative NI value indicates instability in the control loop. Intuitively, we prefer to pair variables u_j and y_j so that λ_{ij} is close to 1 at all frequencies for decentralized control; the reason is that the gain from u_j to y_j is not affected by closing the other loops. However, one should avoid pairings where the sign of the steady-state gain from u_j to y_j may change depending on the control of the other outputs, because this will yield instability with integral action in the loop. More precisely, the basic rules that should be remembered when someone attempts to obtain an optimal pairing of control loops in for decentralized control are as follows:

- **Rule 1:** Select pairing with positive RGA coefficients close to unity in the frequency range where performance is sought in the frequency range of interest; the control loops will be effectively decoupled.
- **Rule 2:** If the NI value is negative, the loop pairing for that control system configuration is unacceptable.

In addition, it should be noted that NI should not be used for systems that have time delays (dead time), as the stability predictions assume immediate feedback to the controller.

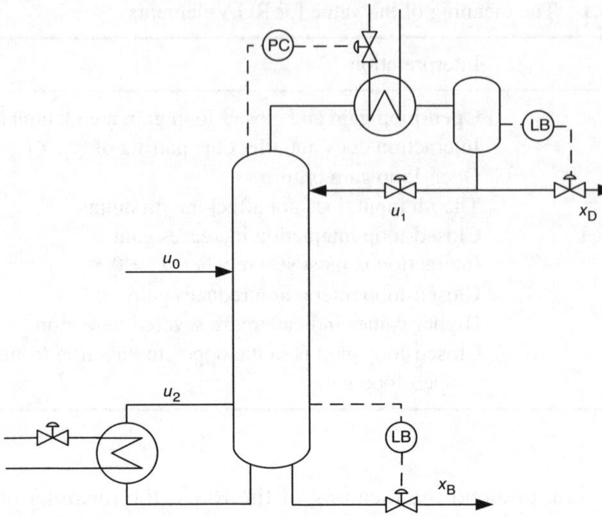

Figure 4.4 Distillation column controlled with the LV-configuration

Example 4.4 A binary distillation column shown in Figure 4.4 is considered, where the three basic control loops for liquid holdup M_D,M_B and pressure P are closed, F is the feed material, and the light product composition x_D and the heavy product composition x_B are controlled by the reflux L and boilup V.

The system transfer function matrix is [87]

$$\begin{bmatrix} x_D \\ x_B \end{bmatrix} = \begin{bmatrix} \dfrac{12.8}{16.7s+1}e^{-1} & \dfrac{-18.9}{21s+1}e^{-3} \\ \dfrac{6.6}{10.9s+1}e^{-7} & \dfrac{-19.4}{14.4+1}e^{-3} \end{bmatrix} \begin{bmatrix} u_1 \\ u_2 \end{bmatrix} + \begin{bmatrix} \dfrac{3.8}{14.9s+1}e^{-8.1} \\ \dfrac{4.9}{13.2s+1}e^{-3.4} \end{bmatrix} u_0 \tag{4.29}$$

The problem is how to pair the inputs u_1, u_2 and the outputs x_B and x_D.
From Equation (4.29), the column steady-state model can be derived and expressed as

$$\begin{bmatrix} x_D \\ x_B \end{bmatrix} = \begin{bmatrix} 12.8 & -18.9 \\ 6.6 & -19.4 \end{bmatrix} \begin{bmatrix} u_1 \\ u_2 \end{bmatrix} + \begin{bmatrix} 3.8 \\ 4.9 \end{bmatrix} u_0 \tag{4.30}$$

By (4.27), we have the RGA:

$$\Lambda = \begin{bmatrix} 2 & -1 \\ -1 & 2 \end{bmatrix}$$

To ensure stability, the column must be paired diagonally (using u_1 to control x_D, and using u_2 to control x_D).

4.5 Conclusion

In this chapter, the existing model formulation for a subsystem, including the state evolution models, the interacting models (input interacting model and state interaction model), etc., was introduced first. In addition to the mathematical models, the structure models, important supplementary of mathematical models, are also introduced here to analyze the logical relationship between each subsystem and to help modeling the mathematical model. The function of the structure model in system decomposition, the adjacent matrices, input–output accessibility, and the structure controllability are introduced to investigate the characteristics of the system. Finally, the classic RGA method is briefly introduced, which provides a quantitative approach to the analysis of the interactions between the controls and the output.

Part Two

Unconstrained Distributed Predictive Control

5

Local Cost Optimization-based Distributed Model Predictive Control

5.1 Introduction

Consider a distributed system as shown in Figure 4.1, which is composed of many interacting subsystems, each of which is controlled by a subsystem-based controller, which in turn is able to exchange information with other subsystem-based controllers. The control objective is to achieve a specific global performance of the entire system (or a common goal of all subsystems).

For controlling such a system, the distributed (or decentralized) framework, where each subsystem is controlled by an independent controller, is usually adopted despite the resulting global performance is not as good as a centralized solution. The reasons are as follows: (1) the classical centralized control solution is often impractical for its lack of tolerance to control faults and the large computational cost. The whole system is out of control when the centralized controller fails, and the control integrity cannot be guaranteed when a control component fails. (2) The distributed framework, in contrast, has the advantages of fault tolerance, less computation, and being flexible to the system structure. (3) The development of communication network technologies in process industries, which allows the distributed control technologies and methodologies to exchange information for improving control, promotes the development of distributed control solutions [27, 47, 57].

Among the distributed solutions, the distributed model predictive control (DMPC), which controls each subsystem by a separate local model predictive control (MPC), has become more and more prosperous [27, 47, 57] since it not only inherits the MPC's ability to explicitly accommodate constraints [22, 66, 67, 71, 88] but also possesses the advantages of the distributed framework as previously mentioned. However, as pointed out by the authors of [32, 41, 42], the performance of the DMPC is, in most cases, worse than that of the centralized MPC.

To improve the global performance of the DMPC, several coordination strategies have appeared in the literature that accommodate different cost functions for the subsystem-based MPC. The simplest and most adopted strategy is that each local controller minimizes its own

Distributed Model Predictive Control for Plant-Wide Systems, First Edition. Shaoyuan Li and Yi Zheng.

subsystem's cost and uses the state prediction of the previous time instant to approximate the state sequence at the current time instant in the procedure of computing an optimal solution. Another commonly used coordinating strategy is that each subsystem-based MPC optimizes the cost of the overall system to improve the global performance [89]. This strategy could achieve a good global performance in some cases, but it reduces the flexibility and increases the communication load. In an effort to achieve a tradeoff between the global performance of the entire system and the computational burden, the authors of [54] proposed the idea that each subsystem-based MPC optimizes the costs of all subsystems that it interacts with. Recently, an intuitively reasonable strategy was proposed in [19, 47, 56, 57], where each subsystem-based MPC only considers the cost of the corresponding subsystem and the subsystems it directly impacts on. Such a design can be referred to as networked distributed MPC (N-DMPC).

In this chapter, the most used and very practiced strategy local cost optimization-based DMPC (LCO-DMPC) is introduced; both the unconstrained iterative and noniterative methods are presented. In Section 5.2, the noniterative LCO-DMPC using the input- and state-interacted model is introduced; the majority of this part of content comes from [46]. The closed-loop solution and stable condition are deduced. Section 5.3 gives the iterative LCO-DMPC which could achieve Nash optimality when the convergence conditions are satisfied. The computational convergence and nominal stability of the distributed MPC as well as the performance analysis with single-step horizon control under communication failure are given in detail that characterizes the Nash optimization-based DMPC.

5.2 Local Cost Optimization-based Distributed Predictive Control

5.2.1 Problem Description

Without losing generality, suppose that the whole system is composed of n linear, discrete-time subsystems $\mathcal{S}_i, i = 1, 2, \dots, m$, and each subsystem interacts with each other by both inputs and states; then the state–space model of subsystem $\mathcal{S}_i$ can be expressed as $\mathbf{A}_j$:

$$\begin{cases} \mathbf{x}_i(k+1) = \mathbf{A}_{ii}\mathbf{x}_i(k) + \mathbf{B}_{ii}\mathbf{u}_i(k) + \sum\limits_{j=1(j\neq i)}^{m} \mathbf{A}_{ij}\mathbf{x}_j(k) + \sum\limits_{j=1(j\neq i)}^{m} \mathbf{B}_{ij}\mathbf{u}_j(k) \\ \mathbf{y}_i(k) = \mathbf{C}_{ii}\mathbf{x}_i(k) + \sum\limits_{j=1(j\neq i)}^{n} \mathbf{C}_{ij}\mathbf{x}_j(k) \end{cases} \tag{5.1}$$

where vectors $\mathbf{x}_i \in \mathbb{R}^{n_{xi}}, \mathbf{u}_i \in \mathbb{R}^{n_u}$, and $\mathbf{y}_i \in \mathbb{R}^{n_{yi}}$ are the local state, control input, and output vectors, respectively. When at least one of the matrices $\mathbf{A}_{ij}$, $\mathbf{B}_{ij}$, or $\mathbf{C}_{ij}$ is not null, it means that $\mathcal{S}_j$ interacts with $\mathcal{S}_i$. In this case, $\mathcal{S}_j$ is called the upstream neighboring subsystem of $\mathcal{S}_i$ and $\mathcal{S}_i$ is called the downstream neighboring subsystem of $\mathcal{S}_j$. $\mathcal{S}_i$ and $\mathcal{S}_j$ are called the neighboring subsystems or neighbors. To proceed, we define that the upstream (downstream) neighborhood $\mathcal{P}_{+i}(\mathcal{P}_{-i})$ of subsystem $\mathcal{S}_i$ is the set of the subscripts of all its upstream (downstream) neighbors:

$$\mathcal{P}_{+i} = \{\mathcal{S}_i, \mathcal{S}_j | \mathcal{S}_j \text{ is an upstream neighbor of } \mathcal{S}_i\}$$

$$\mathcal{P}_{-i} = \{\mathcal{S}_i, \mathcal{S}_j | \mathcal{S}_j \text{ is a downstream neighbor of } \mathcal{S}_i\}$$

The neighborhood $\mathcal{P}_i$ of subsystem $\mathcal{S}_i$ is the set of all its neighbors:

$$\mathcal{P}_i = \mathcal{P}_{+i} \cup \mathcal{P}_{-i} \cup \mathcal{S}_i$$

and defines the state and output interaction vectors $\mathbf{w}_i$ and $\mathbf{v}_i$:

$$\begin{cases} \mathbf{w}_i(k) \triangleq \sum_{j=1(j\neq 1)}^{n} \mathbf{A}_{ij}\mathbf{x}_j(k) + \sum_{j=1(j\neq 1)}^{m} \mathbf{B}_{ij}\mathbf{u}_j(k) \\ \mathbf{v}_i(k) \triangleq \sum_{j=1(j\neq 1)}^{m} \mathbf{C}_{ij}\mathbf{x}_j(k) \end{cases} \tag{5.2}$$

The whole system can be expressed as

$$\begin{cases} \mathbf{x}(k+1) = \mathbf{A}\mathbf{x}(k) + \mathbf{B}\mathbf{u}(k) \\ \mathbf{y}(k) = \mathbf{C}\mathbf{x}(k) \end{cases} \tag{5.3}$$

where $\mathbf{x} \in \mathbb{R}^{n_x}, \mathbf{u} \in \mathbb{R}^{n_u}$, and $\mathbf{y} \in \mathbb{R}^{n_y}$ are the state vector, control input vector, and control output vector, respectively.

The control objective of this system is to minimize a global performance index at time k under the distributed control framework, and

$$J(k) = \sum_{i=1}^{m} \left[\sum_{l=1}^{P} \left\| \mathbf{y}_i(k+l) - \mathbf{y}_i^d(k+l) \right\|_{\mathbf{Q}_i}^2 + \sum_{l=1}^{M} \| \Delta\mathbf{u}_i(k+l-1) \|_{\mathbf{R}_i}^2 \right] \tag{5.4}$$

where $\mathbf{Q}_i$ and $\mathbf{R}_i$ are the weight matrices, P and $M \in \mathbb{N}$ are the predictive horizon and control horizon, respectively, and $P \geq M$, $\mathbf{y}_i^d$ is the set-point of subsystem $\mathcal{S}_i$; $\Delta\mathbf{u}_i(k) = \mathbf{u}_i(k) - \Delta\mathbf{u}_i(k-1)$ is the input increment vector of subsystem $\mathcal{S}_i$.

5.2.2 DMPC Formulation

In the DMPC, the control framework introduced is based on a set of an independent subsystem-based controller $\mathcal{C}_i$; $i = 1, \ldots, n$, implementing an MPC algorithm for subsystem $\mathcal{S}_i$ using both local information acquired on $\mathcal{S}_i$ and the estimate of the interactions among $\mathcal{S}_i$ and is upstream neighbors $\mathcal{P}_{+i}$. The resulting optimal sequence and the future prediction of the state over the prediction horizon have to be exchanged among subsystems through a local area network.

The simplifying hypothesis of accessible local states $\mathbf{x}_i(k)$ is considered in this chapter. Moreover, the sets of the prediction and control horizons are the same to each MPC controller $\mathcal{C}_i$; $i = 1, \ldots, n$, and are considered as P and M, respectively. Let all subsystem-based MPCs be synchronous. In addition, the following assumption is required.

Assumption 5.1

1. Control agents communicate only once within a sampling time interval.
2. The communication channel introduces a delay of a single sampling time interval.

Table 5.1 Notations used in this chapter

5 Notations	6 Explanations
$\mathbf{0}_{a\times b}$	The $a \times b$ null matrix
$\mathbf{0}_a$	The $a \times a$ null matrix
$\mathbf{I}_a$	The $a \times a$ identity matrix
$\text{diag}_a\{\mathbf{A}\}$	A diagonal block matrix made by a blocks equal to $\mathbf{A}$
$\|\mathbf{v}\|^2 \triangleq \mathbf{v}^\mathrm{T}\mathbf{v}$	The square of the norm induced by matrix $\mathbf{v} \geq 0$.
$\hat{\mathbf{x}}(l\|h)$	The prediction of $\mathbf{x}(l)$ computed at time h; $l, h \in \mathcal{P}$, $h < l$
$\mathbf{u}(l\|h)$	The value of $\mathbf{u}(l)$ computed at time h; $l, h \in \mathcal{P}$, $h < l$
$\Delta\mathbf{u}_i \in \mathbb{R}^{n_{u_i}}$	The control increment $\Delta\mathbf{u}_i \triangleq \mathbf{u}_i(k) - \mathbf{u}_i(k-1)$
$\hat{\mathbf{w}}_i(l\|h)$, $\hat{\mathbf{v}}_i(l\|h)$	The estimations of $\mathbf{w}_i(l)$ and $\mathbf{v}_i(l)$ computed at time h, $l, h \in \mathcal{P}$ and $h < l$
$\mathbf{Y}_i^d \in \mathbb{R}^{n_{y_i}}$	The desired output that can be either known or predicted
	A stacked vector made by the vectors $\Delta\mathbf{u}_i(k\|k)$, …, $\Delta\mathbf{u}_i(k+p-1\|k)$ and $\Delta\mathbf{U}_i(k,p\|k) \triangleq \{\Delta\mathbf{u}_i(k\|k), \ldots, \Delta\mathbf{u}_i(k+p-1\|k)\}$
$\mathbf{X}_i(l,p\|h)$	A stacked vector made by the vectors $\mathbf{x}_i(l\|h)$, …, $\mathbf{x}_i(l+p-1\|h)$, and $\mathbf{X}_i(l,p\|h) \triangleq [\mathbf{x}_i^\mathrm{T}(l\|h), \ldots, \mathbf{x}_i^\mathrm{T}(l+p-1\|h)]$, $p, l, h \in \mathbb{N}$ and $h < l$
$\mathbf{X}(l,p\|h)$	A complete stacked vector made by the vectors $\mathbf{X}_1(k,m\|h)$, …, $\mathbf{X}_n(k,m\|h)$, and $\mathbf{X}(l,p\|h) \triangleq [\mathbf{X}_1^\mathrm{T}(l,p\|h), \ldots, \mathbf{X}_i^\mathrm{T}(l,p\|h)]$, $p, l, h \in \mathbb{N}$ and $h < l$

These two assumptions are not restrictive. The first assumption of a single information exchange within a sampling time interval is due to the necessity of minimizing the amount of data exchange through the network. In real situations, an instantaneous data transfer is not possible. Therefore, the unit delay is required. This hypothesis is equivalent to assume a simple deterministic model of the network, where the transmission time delay is constant and equal to a sampling time interval. We also define the necessary notation in Table 5.1.

5.2.2.1 Performance Index

For the large-scale system, considered here, the global performance index (5.4) can be decomposed in terms of the local performance index J_i for each subsystem $\mathcal{S}_i$, $i = 1, 2, \ldots, m$ [90]

$$J_i(k) = \sum_{l=1}^{P} \|\hat{\mathbf{y}}_i(k+l|k) - \mathbf{y}_i^d(k+l|k)\|_{\mathbf{Q}_i}^2 + \sum_{l=1}^{M} \|\Delta\mathbf{u}_i(k+l-1|k)\|_{\mathbf{R}_i}^2 \tag{5.5}$$

The local control decision of $\mathcal{S}_i$ is computed by solving the optimization problem $\min_{\Delta\mathbf{U}(k,M|k)} J_i(k)$ with local input/output variables and constraints in the distributed MPC based on the state (or input) estimations of neighbors at time $k-1$.

5.2.2.2 Predictive Model

To predict the future state of the current subsystem $\mathcal{S}_i$, the only information that each controller $\mathcal{C}_i, i = 1, \ldots, n$, needs is the future behavior of subsystems $\mathcal{S}_j$ controlled by the agents $\mathcal{C}_j \in \mathcal{P}_{+i}$. Similarly, $\mathcal{C}_i$ should broadcast the future behavior of the local variables only to the agents $\mathcal{C}_j \in \mathcal{P}_{-i}$. Then the states and outputs of the downstream neighbors in l-step ahead can be predicted by

$$\begin{cases} \hat{\mathbf{x}}_i(k+l|k) = \mathbf{A}_{ii}^l \hat{\mathbf{x}}_i(k|k) + \sum_{s=1}^{l} \mathbf{A}_{ii}^{s-1} \mathbf{B}_{ii} \mathbf{u}_i(k+l-s|k) \\ \quad + \sum_{s=1}^{l} \mathbf{A}_{ii}^{s-1} \hat{\mathbf{w}}_i(k+l-s|k-1) \\ \hat{\mathbf{y}}_i(k+l|k) = \mathbf{C}_{ii} \hat{\mathbf{x}}_i(k+l|k) + \hat{\mathbf{v}}_i(k+l|k-1) \end{cases} \tag{5.6}$$

5.2.2.3 Optimization Problem

From above, the optimization problem for each subsystem-based local cost optimization-based MPC (LCO-MPC) in each control cycle can be concluded as follows.

Problem 5.1 For each independent controller $\mathcal{C}_i$, $i = 1, \ldots, m$, the unconstrained LCO-DMPC problem with the prediction horizon P and control horizon M, $M < P$, at time k solves the following optimization problem:

$$\min_{\Delta \mathbf{U}_i(k,M|k)} \overline{J}_i(k) = \sum_{l=1}^{P} \left\| \hat{\mathbf{y}}_i(k+l|k) - \mathbf{y}_i^{\mathrm{d}}(k+l|k) \right\|_{\mathbf{Q}_i}^2 + \sum_{l=1}^{M} \left\| \Delta \mathbf{u}_i(k+l-1|k) \right\|_{\mathbf{R}_i}^2 \tag{5.7}$$

for $i = 1, \ldots, m$, subject to constraints

$$\hat{\mathbf{x}}_i(k+l|k) = \mathbf{A}_{ii}^l \hat{\mathbf{x}}_i(k|k) + \sum_{s=1}^{l} \mathbf{A}_{ii}^{s-1} \mathbf{B}_{ii} \mathbf{u}_i(k+l-s|k) + \sum_{s=1}^{l} \mathbf{A}_{ii}^{s-1} \hat{\mathbf{w}}_i(k+l-s|k-1) \tag{5.8}$$

$$\hat{\mathbf{y}}_i(k+l|k) = \mathbf{C}_{ii} \hat{\mathbf{x}}_i(k+l|k) + \hat{\mathbf{v}}_i(k+l|k-l) \tag{5.9}$$

where $\Delta \mathbf{U}_i(k, M|k) = \{\Delta \mathbf{u}_i(k|k), \ldots, \Delta \mathbf{u}_i(k+M-1|k)\}$.

Each controller $\mathcal{C}_i$ is composed of three parts: an optimizer, a state predictor, and an interaction predictor. At time k, based on the exchanged information, the interaction predictor of the MPC controller $\mathcal{C}_i$ estimates the future interaction sequence over the prediction horizon $\hat{\mathbf{w}}_i(k+l-1|k-1)$, $l = 1, \ldots, P$. Then, combing with the local state measurement of $x_i(k)$, Problem 5.1 can be solved by the optimizer that computes the optimal manipulated variable increments sequence $\Delta\mathbf{U}_i^*(k, M|k)$ over the control horizon. The first element of $\Delta\mathbf{U}_i(k, M|k)$, $\Delta u_i(k|k)$, is selected and $u_i(k) = u_i(k-1) + \Delta u_i(k|k)$ is applied as a control input to subsystem $\mathcal{S}_i$. Finally, the state predictor computes an estimate of the future state trajectory over the prediction horizon by (5.8) and broadcasts it and the optimal control sequence $\Delta\mathbf{U}_i(k, M|k)$ over the control horizon to its output neighbors $\mathcal{S}_j$, $j \in \mathcal{P}_{-i}$. At time $k+1$, the interaction prediction part of each controller uses this information to estimate the interaction predictions $\hat{\mathbf{w}}_i(k+l-1|k)$ and the whole procedure is repeated.

It should be noticed that in Problem 5.1, the future interaction sequences are substituted by the estimate of the future interaction sequence $\hat{\mathbf{w}}_i(k+l-s|k-1)$ and $\hat{\mathbf{v}}_i(k+l|k-1)$ based on the information broadcast at time $k-1$ from the agents $\mathcal{C}_j \in \mathcal{P}_{+i}$, because at time k, the predictions $\hat{\mathbf{w}}_i(k+l-s|k)$ and $\hat{\mathbf{v}}_i(k+l|k)$ are unknown for the controller $\mathcal{C}_i$. That is why in Equations (5.8) and (5.9) the controller $\mathcal{C}_i$ has the current formation.

A detailed description of the proposed LCO-DMPC is provided in the next section where a closed-loop solution to the LCO-DMPC problem is introduced by the formalization of the different actions performed by each controller.

5.2.3 Closed-loop Solution

The main result of this section is the computation of an explicit solution to the LCO-DMPC problem. For this purpose, first the expressions for the interaction prediction and the state prediction are provided in an explicit form.

5.2.3.1 Interaction Prediction

According to Equation (5.2), let the last $(P-M+1)$ samples of the control action predictions $\hat{\mathbf{U}}_j(k, P|k-1)$ are equal to the last element of $\mathbf{U}_j(k-1, M|k-1)$. Then, for $i = 1, \ldots, m$, the l-ahead predictions at time k based on the information computed at time $k-1$of the interaction vectors (5.2) can be expressed as

$$\begin{cases} \hat{\mathbf{w}}_i(k+l|k-1) \triangleq \sum_{j=1(j\neq 1)}^{n} \mathbf{A}_{ij}\hat{\mathbf{x}}_j(k+l|k-1) + \sum_{j=1(j\neq 1)}^{m} \mathbf{B}_{ij}\mathbf{u}_j(k+l|k-1) \\ \hat{\mathbf{v}}_i(k) \triangleq \sum_{j=1(j\neq 1)}^{m} \mathbf{C}_{ij}\hat{\mathbf{x}}_j(k+l|k-1) \\ l = 1, 2, \ldots, P \end{cases} \tag{5.10}$$

where

$$\begin{aligned} & u_j(k-1+l|k-1) = u_j(k-1+M|k-1) \\ & l > M \end{aligned} \tag{5.11}$$

Let us define that

$$\begin{aligned} \tilde{\mathbf{A}}_i &\triangleq [\mathrm{diag}_P\{\mathbf{A}_{i,1}\} \cdots \mathrm{diag}_P\{\mathbf{A}_{i,i-1}\}\ \mathbf{0}\ \mathrm{diag}_P\{\mathbf{A}_{i,i+1}\} \cdots \mathrm{diag}_P\{\mathbf{A}_{i,m}\}] \\ \tilde{\tilde{\mathbf{B}}}_i &\triangleq [\mathrm{diag}_P\{\mathbf{B}_{i,1}\} \cdots \mathrm{diag}_P\{\mathbf{B}_{i,i-1}\}\ \mathbf{0}\ \mathrm{diag}_P\{\mathbf{B}_{i,i+1}\} \cdots \mathrm{diag}_P\{\mathbf{B}_{i,m}\}] \\ \tilde{\mathbf{C}}_i &\triangleq [\mathrm{diag}_P\{\mathbf{C}_{i,1}\} \cdots \mathrm{diag}_P\{\mathbf{C}_{i,i-1}\}\ \mathbf{0}\ \mathrm{diag}_P\{\mathbf{C}_{i,i+1}\} \cdots \mathrm{diag}_P\{\mathbf{C}_{i,m}\}] \end{aligned} \tag{5.12}$$

where the diagonal blocks $\text{diag}_P\{\mathbf{A}_{ij}\}$, $\text{diag}_P\{\mathbf{B}_{ij}\}$, and $\text{diag}_P\{\mathbf{C}_{ij}\}$ are the zero blocks of congruent dimensions if $\mathcal{C}_j$ does not belong to the input neighborhood $\mathcal{P}_{+i}$. Moreover, define that

$$\tilde{\mathbf{\Gamma}}_i \triangleq \begin{bmatrix} \mathbf{0}_{(M-1)n_{u_i}\times n_{u_i}} & \mathbf{I}_{(M-1)n_{u_i}} \\ \mathbf{0}_{n_{u_i}\times(M-1)n_{u_i}} & \mathbf{I}_{n_{u_i}} \\ \vdots & \vdots \\ \mathbf{0}_{n_{u_i}\times(M-1)n_{u_i}} & \mathbf{I}_{n_{u_i}} \end{bmatrix}$$

$$\begin{aligned} \tilde{\mathbf{\Gamma}} &\triangleq \text{diag}\{\tilde{\mathbf{\Gamma}}_1, \ldots, \tilde{\mathbf{\Gamma}}_m\} \\ \tilde{\mathbf{B}}_i &\triangleq \tilde{\tilde{\mathbf{B}}}_i\tilde{\mathbf{\Gamma}}, \end{aligned} \tag{5.13}$$

$$\begin{aligned} \tilde{\mathbf{A}} &\triangleq \begin{bmatrix}\tilde{\mathbf{A}}_1^{\mathrm{T}} & \ldots & \tilde{\mathbf{A}}_m^{\mathrm{T}}\end{bmatrix}^{\mathrm{T}} \\ \tilde{\mathbf{B}} &\triangleq \begin{bmatrix}\tilde{\mathbf{B}}_1^{\mathrm{T}} & \ldots & \tilde{\mathbf{B}}_m^{\mathrm{T}}\end{bmatrix}^{\mathrm{T}} \\ \tilde{\mathbf{C}} &\triangleq \begin{bmatrix}\tilde{\mathbf{C}}_1^{\mathrm{T}} & \ldots & \tilde{\mathbf{C}}_m^{\mathrm{T}}\end{bmatrix}^{\mathrm{T}} \end{aligned} \tag{5.14}$$

By representing Equation (5.10) in a stacked form for $l = 1, \ldots, P$, under Assumption 5.1, for each controller $\mathcal{C}_i$, $i = 1, \ldots, m$, the stacked predictions of the interaction vectors at time k based on the information computed at time $k-1$ are given by

$$\begin{cases} \hat{\mathbf{W}}_i(k, P|k-1) = \tilde{\mathbf{A}}_i\hat{\mathbf{X}}(k, P|k-1) + \tilde{\mathbf{B}}_i\mathbf{U}(k-1, M|k-1) \\ \hat{\mathbf{V}}_i(k, P|k-1) = \hat{\mathbf{C}}_i\hat{\mathbf{X}}(k, P|k-1) \end{cases} \tag{5.15}$$

and the complete stacked predictions have the following forms:

$$\begin{cases} \hat{\mathbf{W}}(k, P|k-1) = \tilde{\mathbf{A}}\hat{\mathbf{X}}(k, P|k-1) + \tilde{\mathbf{B}}\mathbf{U}(k-1, M|k-1) \\ \hat{\mathbf{V}}(k, P|k-1) = \hat{\mathbf{C}}\hat{\mathbf{X}}(k, P|k-1) \end{cases} \tag{5.16}$$

Remark 5.1 *In Equation (5.15), the complete stacked vectors* $\hat{\mathbf{X}}(k, M\,|\,k-1)$ *and* $\mathbf{U}_j(k-1, M\,|\,k-1)$ *are defined with the predicted state trajectories of all subsystems and with the open-loop control sequences of all subsystem-based MPC. In the real distributed implementation of LCO-DMPC, the zero blocks on the matrices* $\tilde{\mathbf{A}}_i$, $\tilde{\mathbf{B}}_i$, *and* $\tilde{\mathbf{C}}_i$ *allow us to build the complete stacked vectors* $\hat{\mathbf{X}}(k, P|k-1)$ *and* $\mathbf{U}(k-1, M\,|\,k-1)$. *We can just keep the value of nonzero blocks* $\hat{\mathbf{X}}_j(k, P|k-1)$ *and* $\mathbf{U}_j(k-1, M\,|\,k-1)$ *produced by the controller* $\mathcal{C}_j$, $j \in \mathcal{P}_{+i}$, *and set the other parts of* $\hat{\mathbf{X}}(k, P|k-1)$ *and* $\mathbf{U}(k-1, M\,|\,k-1)$ *except* $\hat{\mathbf{X}}_j(k, P|k-1)$ *and* $\mathbf{U}_j(k-1, M\,|\,k-1)$, $j \in \mathcal{P}_{-i}$, *as the real values.*

5.2.3.2 State Prediction

Now, we will introduce how to obtain the state prediction for each *controller* $\mathcal{C}_i$, $i = 1, \ldots, m$.

By Equations (5.8), (5.9), and definitions (5.9), and imposing (5.11) and

$$\hat{\mathbf{v}}_i(k+p|k-1) = \hat{\mathbf{v}}_i(k+p-1|k-1)$$

the stacked state and output predictions for the controller $\mathcal{C}_i$ take the form

$$\begin{cases} \hat{\mathbf{X}}_i(k+1,P|k) = \overline{\mathbf{S}}_i[\overline{\mathbf{A}}_i\hat{x}_i(k|k) + \overline{\mathbf{B}}_i\mathbf{U}_i(k,M|k) + \hat{\mathbf{W}}_i(k,P|k-1)] \\ \hat{\mathbf{Y}}_i(k+1,P|k) = \overline{\mathbf{C}}_i\hat{\mathbf{X}}_i(k+1,P|k) + \mathbf{T}_i\hat{\mathbf{V}}_i(k,P|k-1) \end{cases} \tag{5.17}$$

Define that

$$\overline{\mathbf{S}}_c \triangleq \begin{bmatrix} \mathbf{A}_{ii}^0 & \cdots & 0 \\ \vdots & \ddots & \vdots \\ \mathbf{A}_{ii}^{p-1} & \cdots & \mathbf{A}_{ii}^0 \end{bmatrix} \tag{5.18}$$

$$\mathbf{T}_i \triangleq \begin{bmatrix} \mathbf{0}_{(P-1)n_{y_i}\times n_{y_i}} & \mathbf{I}_{(P-1)n_{y_i}} \\ \mathbf{0}_{n_{y_i}\times(p-1)n_{y_i}} & \mathbf{I}_{n_{y_i}} \end{bmatrix}$$

$$\begin{aligned} \overline{\mathbf{S}} &\triangleq \mathrm{diag}\{\overline{\mathbf{S}}_1,\ldots,\overline{\mathbf{S}}_n\} \\ \mathbf{T} &\triangleq \mathrm{diag}\{\overline{\mathbf{T}}_1,\ldots,\overline{\mathbf{T}}_n\} \end{aligned} \tag{5.19}$$

$$\overline{\mathbf{A}}_i \triangleq \begin{bmatrix} \mathbf{A}_{ii} \\ \mathbf{0}_{Pn_{y_i}\times n_{y_i}} \end{bmatrix}$$

$$\overline{\mathbf{B}} \triangleq \begin{bmatrix} \mathrm{diag}_M\{\mathbf{B}_{ii}\} & & & \\ \mathbf{0}_{n_{u_i}} & \cdots & \mathbf{0}_{n_{u_i}} & \mathbf{B}_{ii} \\ \vdots & \ddots & \vdots & \vdots \\ \mathbf{0}_{n_{u_i}} & \cdots & \mathbf{0}_{n_{u_i}} & \mathbf{B}_{ii} \end{bmatrix} \tag{5.20}$$

$$\overline{\mathbf{C}}_i \triangleq \mathrm{diag}_P\{\mathbf{C}_{ii}\}$$

$$\begin{aligned} \overline{\mathbf{A}} &\triangleq \mathrm{diag}\{\overline{\mathbf{A}}_1,\ldots,\overline{\mathbf{A}}_m\} \\ \overline{\mathbf{B}} &\triangleq \mathrm{diag}\{\overline{\mathbf{B}}_1,\ldots,\overline{\mathbf{B}}_m\} \\ \overline{\mathbf{C}} &\triangleq \mathrm{diag}\{\overline{\mathbf{C}}_1,\ldots,\overline{\mathbf{C}}_m\} \end{aligned} \tag{5.21}$$

Then, by substituting $\mathbf{W}_i(k,P|k-1)$ and $\mathbf{V}_i(k,P|k-1)$ with their explicit expressions (5.15), it results

$$\begin{cases} \hat{\mathbf{X}}_i(k+1,P|k) = \overline{\mathbf{S}}_i[\overline{\mathbf{A}}_i\hat{\mathbf{x}}_i(k|k) + \overline{\mathbf{B}}_i\mathbf{U}_i(k,M|k) + \tilde{\mathbf{A}}_i\hat{\mathbf{X}}(k,P|k-1) \\ \qquad + \tilde{\mathbf{B}}_i\mathbf{U}(k-1,M|k-1)] \\ \hat{\mathbf{Y}}_i(k+1,P|k) = \overline{\mathbf{C}}_i\hat{\mathbf{X}}_i(k+1,P|k) + \mathbf{T}_i\tilde{\mathbf{C}}_i\hat{\mathbf{X}}(k,P|k-1) \end{cases} \tag{5.22}$$

Thus, under Assumption 5.1, for each controller $\mathcal{C}_i$, $i = 1, \ldots, m$, the stacked decentralized predictions of state and output at time k are expressed by Equation (5.22), and the complete stacked decentralized predictions take the following form:

$$\begin{cases} \hat{\mathbf{X}}(k+1,P|k) = \overline{\mathbf{S}}[\overline{\mathbf{A}}\hat{x}(k|k) + \overline{\mathbf{B}}\mathbf{U}(k,M|k) + \tilde{\mathbf{A}}\hat{\mathbf{X}}(k,P|k-1) \\ \qquad + \tilde{\mathbf{B}}\mathbf{U}(k-1,M|k-1)] \\ \hat{\mathbf{Y}}(k+1,P|k) = \overline{\mathbf{C}}\hat{\mathbf{X}}(k+1,P|k) + \mathbf{T}\tilde{\mathbf{C}}\hat{\mathbf{X}}(k,P|k-1) \end{cases} \tag{5.23}$$

5.2.3.3 Quadratic Program Tansformation

The N-DMPC problem stated in Problem 5.1 will now be formulated as a quadratic program, and then using the minimum principle to obtain the explicit solution of the unconstrained LCO-DMPC.

To simplify, we give the result directly, and the deducing procedure is detailed in Appendix A at the end of this chapter. The readers can also refer to [46].

Problem 5.2 (Quadratic program). Under Assumption 5.1, for each subsystem-based controller C_i, $i = 1, \ldots, m$, the MPC optimization Problem 5.1 at time k can be transferred to the following quadratic program problem:

$$\min_{\Delta \mathbf{u}_i(k,M|k)} [\mathbf{\Delta U}_i^{\mathrm{T}}(k,M|k)\mathbf{H}_i\mathbf{\Delta U}_i(k,M|k) - \mathbf{G}_i^{\mathrm{T}}(k+1,P|k)\Delta\mathbf{U}_i(k,M|k)] \tag{5.24}$$

where the positive definite matrix $\mathbf{H}_i$ has the form

$$\mathbf{H}_i \triangleq \mathbf{N}_i^{\mathrm{T}}\overline{\mathbf{Q}}_i\mathbf{N}_i + \overline{\mathbf{R}}_i \tag{5.25}$$

$$\mathbf{G}_i(k+1,P|k) \triangleq 2\mathbf{N}_i^{\mathrm{T}}\overline{\mathbf{Q}}_i[\mathbf{Y}_i^{\mathrm{d}}(k+1,P|k) - \hat{\mathbf{Z}}_i(k+1,P|k)] \tag{5.26}$$

$$\begin{aligned}\hat{\mathbf{Z}}_i(k+1,P|k) \triangleq{}& \mathbf{S}_i[\overline{\mathbf{B}}_i\overline{\mathbf{\Gamma}}'_i\mathbf{u}_i(k-1) + \overline{\mathbf{A}}_i\hat{\mathbf{x}}_i(k|k) + \hat{\mathbf{W}}_i(k,P|k-1)]\\ &+ \mathbf{T}_i\hat{\mathbf{V}}_i(k,P|k-1)\end{aligned} \tag{5.27}$$

where

$$\begin{aligned}\mathbf{S}_i &\triangleq \overline{\mathbf{C}}_i\overline{\mathbf{S}}_i\\ \mathbf{N}_i &\triangleq \mathbf{S}_i\overline{\mathbf{B}}_i\overline{\mathbf{\Gamma}}_i\\ \overline{\mathbf{Q}}_i &\triangleq \mathrm{diag}_P\{\mathbf{Q}_i\}\\ \overline{\mathbf{R}}_i &\triangleq \mathrm{diag}_P\{\mathbf{R}_i\}\end{aligned} \tag{5.28}$$

$$\begin{aligned}\underset{(M\ \text{block})}{\mathbf{\Gamma}'_i} &\triangleq \begin{bmatrix}\mathbf{I}_{n_{u_i}}\\ \vdots\\ \mathbf{I}_{n_{u_i}}\end{bmatrix}\\ \underset{(M\times M\ \text{blocks})}{\overline{\mathbf{\Gamma}}_i} &\triangleq \begin{bmatrix}\mathbf{I}_{n_{u_i}} & \cdots & 0\\ \vdots & \ddots & \vdots\\ \mathbf{I}_{n_{u_i}} & \cdots & \mathbf{I}_{n_{u_i}}\end{bmatrix}\end{aligned} \tag{5.29}$$

where $\mathbf{Q}_i$ and $\mathbf{R}_i$ are the weight matrices of the cost function (5.5).

In this way, the subsystem-based MPC Problem 5.1 has been transformed into an equivalent unconstrained QP Problem 5.2 which has to be locally solved online at each sampling instant.

5.2.3.4 Solution of Problem 5.1

Then, the explicit solution to the unconstrained LCO-DMPC problem is now stated as a solution to the equivalent quadratic program. By applying minimum principle to Problem 5.2 yields Theorem 5.1.

Theorem 5.1 ***(Closed-loop solution).*** *Under Assumption 5.1, for each controller $\mathcal{C}_i$, $i=1, \ldots, m$, the explicit form of the control law applied at time k by controller $\mathcal{C}_i$ to the subsystem $\mathcal{S}_i$ is given by*

$$\mathbf{u}_i(k) = \mathbf{u}_i(k-1) + \mathbf{K}_i[\mathbf{Y}_i^d(k+1, P|k) - \hat{\mathbf{Z}}_i(k+1, P|k)] \tag{5.30}$$

where

$$\begin{aligned} \mathbf{K}_i &\triangleq \mathbf{\Gamma}_i\overline{\mathbf{K}}_i, \\ \mathbf{\Gamma}_i &\triangleq \begin{bmatrix} \mathbf{I}_{n_{u_i}} & \mathbf{0}_{n_{u_i}\times(m-1)n_{u_i}} \end{bmatrix} \\ \overline{\mathbf{K}}_i &\triangleq \mathbf{H}_i^{-1}\mathbf{N}_i^{\mathrm{T}}\overline{\mathbf{Q}}_i \end{aligned} \tag{5.31}$$

In addition, the expression of the stacked open-loop optimal control sequence at time k is

$$\mathbf{U}_i(k, M|k) = \mathbf{\Gamma}'_1\mathbf{u}_i(k-1) + \overline{\mathbf{\Gamma}}_i\overline{\mathbf{K}}_i[\mathbf{Y}_i^{\mathrm{d}}(k+1, P|k) - \hat{\mathbf{Z}}_i(k+1, P|k)] \tag{5.32}$$

and its complete stacked expression is

$$\begin{aligned} \mathbf{U}(k, m|k) = &\ \mathbf{\Psi}\mathbf{U}(k-1, m|k-1) \\ &+ \mathbf{\Theta}\hat{\mathbf{x}}(k|k) + \mathbf{\Phi}\hat{\mathbf{X}}(k, p|k-1) + \mathbf{\Xi}\mathbf{Y}^d(k+1, p|k) \end{aligned} \tag{5.33}$$

where

$$\begin{aligned} \mathbf{\Gamma}' &\triangleq \mathrm{diag}\{\mathbf{\Gamma}'_1, \ldots, \mathbf{\Gamma}'_m\} \\ \mathbf{\Gamma} &\triangleq \mathrm{diag}\{\mathbf{\Gamma}_1, \ldots, \mathbf{\Gamma}_m\} \end{aligned} \tag{5.34}$$

$$\begin{aligned} \mathbf{S} &\triangleq \mathrm{diag}\{\mathbf{S}_1, \ldots, \mathbf{S}_m\} \\ \mathbf{T} &\triangleq \mathrm{diag}\{\mathbf{T}_1, \ldots, \mathbf{T}_m\} \end{aligned} \tag{5.35}$$

$$\begin{aligned} \mathbf{\Xi} &\triangleq \mathrm{diag}\{\overline{\mathbf{\Gamma}}_1\overline{\mathbf{K}}_1, \ldots, \overline{\mathbf{\Gamma}}_m\overline{\mathbf{K}}_m\} \\ \mathbf{\Theta} &\triangleq -\mathbf{\Xi}\mathbf{S}\overline{\mathbf{A}} \end{aligned} \tag{5.36}$$

$$\begin{aligned} \mathbf{\Phi} &\triangleq -\mathbf{\Xi}(\mathbf{S}\tilde{\mathbf{A}} + \mathbf{T}\tilde{\mathbf{C}}) \\ \mathbf{\Psi} &\triangleq \mathbf{\Gamma}'\mathbf{\Gamma} - \mathbf{\Xi}\mathbf{S}(\overline{\mathbf{B}}\mathbf{\Gamma}'\mathbf{\Gamma} + \tilde{\mathbf{B}}) \end{aligned} \tag{5.37}$$

Proof. The proof can be found in Appendix B.

In Equation (5.32), the complete stacked vectors $\mathbf{U}(k-1, M|k-1)$ built by $\mathcal{C}_i$ are included to calculate the current optimal manipulated variables. Thus, all the $\mathbf{U}_j(k-1, M|k-1)$ of its upstream neighborhood subsystems $\mathcal{S}_j$, $j \in \mathcal{P}_{+i}$ are used for computing the current closed-loop stacked control sequence $\mathbf{U}_i(k, M|k)$. In addition, the stacked control sequence

$\mathbf{U}_i(k, M|k)$ is used only by the controllers $\mathcal{S}_j$, $j \in \mathcal{P}_{-i}$; therefore, in the real implementation of the LCO-DMPC, each subsystem-based controller $\mathcal{C}_i$ can build its $\mathbf{U}(k, M|k)$ with nonzero blocks $\mathbf{U}_j(k-1, M|k-1), j \in \mathcal{P}_{+i}$ and with zero blocks $\mathbf{U}(k, M|k)$ in the other parts.

Consider that only the first element of $\mathbf{U}_i(k, M|k)$, that is, $\mathbf{u}_i(k), i = 1, \ldots, m$, is applied at time k to subsystem $\mathcal{S}_i$, and it is only this element that defines the feedback control law at time k. Therefore, the feedback control law of controller $\mathcal{C}_i$ can also be expressed as

$$\mathbf{u}_i(k) = \mathbf{\Gamma}_i \mathbf{U}_i(k, M|k), i = 1, \ldots, m \tag{5.38}$$

and the complete feedback control law computed by all controllers as

$$\mathbf{u}(k) = \mathbf{\Gamma} \mathbf{U}(k, M|k) \tag{5.39}$$

So far, the stacked control sequence over the control horizon M computed by controller $\mathcal{C}_i$ and the complete stacked control sequence formed by all these stacked control sequences computed by all controllers $\mathcal{C}_i$, $i = 1, \ldots, m$, are provided in Theorem 5.1, the elements up to time $k + M - 1$ of which is broadcasted to its downstream neighborhood $\mathcal{S}_j$, $j \in \mathcal{P}_{-i}$ by the subsystem-based controller $\mathcal{C}_i$ and is used to evaluate the interactions among the subsystems, and then to get the complete stacked control sequence $\mathbf{U}(k, M|k)$ for the next control step.

5.2.3.5 Unconstrained LCO-DMPC Algorithm

Algorithm 5.1 Let the desired output $\mathbf{Y}_i^d(k+1, P|k)$ is provided; then the LCO-DMPC algorithm for the subsystem-based MPC controller $\mathcal{C}_i$ at each time instant k is as follows.

Step 1. Communication and interaction calculation

- Sent $\mathbf{U}_i(k-1, M|k-1)$ and $\hat{\mathbf{X}}_i(k, P|k-1)$ to its downstream neighbors's controller $\mathcal{C}_j$, $j \in \mathcal{P}_{-i}$.
- Get the estimation of the future state trajectories $\hat{\mathbf{X}}_j(k, P|k-1)$ and control inputs $\mathbf{U}_j(k-1, M|k-1)$ from its upstream neighbors' controller $\mathcal{C}_j$, $j \in \mathcal{P}_{+i}$ through network information exchange.
- Set the desired trajectory $\mathbf{Y}^d(k+1, P|k)$ over the horizon P according to the MPC's configuration.
- Get the measurement of $\mathbf{x}_i(k)$ through field instruments or a designed observer.
- Build $\hat{\mathbf{X}}(k, P|k-1)$ and $\mathbf{U}(k, P|k)$ by combining the local state trajectory $\hat{\mathbf{X}}_i(k, P|k-1)$ and control input $\mathbf{U}(k, P|k)$ with the acquired upstream neighbors' information of $\hat{\mathbf{X}}_j(k, P|k-1)$ $\mathbf{U}_j(k-1, M|k-1), j \in \mathcal{P}_{+i}$, and compute the corresponding predictions of the interactions:

$$\hat{\mathbf{W}}_i(k, P|k-1) = \tilde{\mathbf{A}}_i \hat{\mathbf{X}}(k, P|k-1) + \tilde{\mathbf{B}}_i \mathbf{U}(k-1, M|k-1)$$

$$\hat{\mathbf{V}}_i(k, P|k-1) = \hat{\mathbf{C}}_i \hat{\mathbf{X}}(k, P|k-1)$$

Step 2. Compute control law and apply it

- Compute the optimal control sequence according to Equation (5.32), that is,

$$\mathbf{U}_i(k, M|k) = \mathbf{\Gamma}'_1 \mathbf{u}_i(k-1) + \overline{\mathbf{\Gamma}}_i \overline{\mathbf{K}}_i [\mathbf{Y}_i^d(k+1, P|k) - \hat{\mathbf{Z}}_i(k+1, P|k)]$$

- Apply the first element $\mathbf{u}_i(k)$ of the optimal sequence $\mathbf{U}_i(k, M|k)$ as a control input to the physical system $\mathcal{S}_i$.

Step 3. Estimate the future state

- Compute the estimation of the future state trajectory of subsystem S_i over the horizon P by the following equation:

$$\hat{\mathbf{X}}_i(k+1,P|k) = \overline{\mathbf{S}}_i[\overline{\mathbf{A}}_i\hat{x}_i(k|k) + \overline{\mathbf{B}}_i\mathbf{U}_i(k,M|k) + \tilde{\mathbf{A}}_i\hat{\mathbf{X}}(k,P|k-1) + \tilde{\mathbf{B}}_i\mathbf{U}(k-1,M|k-1).$$

Step 4. Go to the next time instant

- At time $k+1$, let $k+1 \rightarrow k$; then go to Step 1 and repeat the algorithm.

Remark 5.2 *According to the fault-tolerant control proposed in [91–93], the LCO-DMPC control solution is able to manage also eventual subsystem faults. For example, for controller $\mathcal{C}_i$, if a fault occurs and leads to a structural or parametric change on the model of subsystem $\mathcal{S}_i$, by model-based techniques [94], then controllers $\mathcal{C}_i$ detects the occurred fault and determines the new configuration of LCO-DMPC and broadcasts the new configuration to its downstreaming neighbors's controller $\mathcal{C}_j$, $j \in \mathcal{P}_{-i}$. The controller $\mathcal{C}_j$ switches to a new MPC policy according to the configuration and goes on controlling its corresponding subsystem. Here, the local fault-detection system has to have the functions of detecting faults, determining the fault types and automatic selecting configuration. Is should also have the function of informing the downstream neighbor controller's new configuration. In addition, the fault-detection algorithm should have the online MPC switching policy which does not affect the normal operating of the corresponding subsystem-based controller. Finally, the different MPC policies should be designed primarily, which have to guarantee the stability of the overall closed-loop system. The different MPC polices for each controller $\mathcal{C}_j$, $j \neq i$, could be obtained by changing the tuning parameters of the subsystem-based MPC in the new configuration.*

5.2.3.6 Computational Complexity

By Theorem 5.1, the resulting computational complexity to obtain the explicit solution for the local subsystem $\mathcal{S}_i$ is mainly given by the inversion of matrix $\mathbf{H}_i$. If using a Gauss–Jordan algorithm for this task and considering that the size of matrix $\mathbf{H}_i$ is equal to $M \cdot n_{u_i}$, the complexity of the inversion algorithm is $\mathcal{O}(M^3 \cdot n_{u_i}^3)$. Thus, if the LCO-DMPC is implemented in a distributed framework, the computational complexity of each subsystem-based MPC in each control cycle is $\mathcal{O}(M^3 \cdot n_{u_i}^3)$. If the centralized control structure is employed, the computational complexity of the centralized MPC in each control cycle is $\mathcal{O}\left(M^3 \cdot \left(\sum_{i=1}^{m} n_{u_i}\right)^3\right)$. Obviously, the computational complexity of the distributed implementation is much less than the centralized one, especially for the large-scale system where there is a large number of subsystems.

5.2.4 Stability Analysis

In this section, the stability condition of the entire closed-loop system is deduced by analyzing the entire closed-loop system's dynamic matrix which could be specified on the basis of the closed-loop solution stated by Theorem 5.1.

For a simple reason, we directly give the stability condition, Theorem 5.2. If the readers want to study the detail of the deducing procedure, they can refer to Appendix C at the end of this chapter or [46].

Theorem 5.2 ***(Distributed stability).*** *The closed-loop system given by the system $\mathcal{S}$ with feedback distributed control laws given by (5.30) is asymptotically stable, if and only if,*

$$\left|\lambda_j\left\{\begin{bmatrix} \mathbf{A} & \mathbf{0} & \mathbf{B\Gamma} & 0 \\ \overline{\mathbf{S}}\,\overline{\mathbf{A}} & \overline{\mathbf{S}}\tilde{\mathbf{A}} & \overline{\mathbf{S}}\,\overline{\mathbf{B}} & \overline{\mathbf{S}}\tilde{\mathbf{B}} \\ \left(\mathbf{\Theta A}+\mathbf{\Phi}\overline{\mathbf{S}}\,\overline{\mathbf{A}}\right) & \mathbf{\Phi}\overline{\mathbf{S}}\tilde{\mathbf{A}} & (\mathbf{\Psi}+\mathbf{\Theta B\Gamma}+\mathbf{\Phi}\overline{\mathbf{S}}\,\overline{\mathbf{B}}) & \mathbf{\Phi}\overline{\mathbf{S}}\tilde{\mathbf{B}} \\ \mathbf{0} & \mathbf{0} & \mathbf{I}_{Mn_u} & 0 \end{bmatrix}\right\}\right| < 1$$

$$\forall j = 1, \dots, m_N \tag{5.40}$$

where $m_N = Pn_x + n_x + 2Mn_u$ of the global closed-loop system.

Remark 5.3 *It can be seen from Equation (5.40) that the first two block rows of the dynamic matrix $\mathbf{A}_N$ depend on the system matrix $\mathbf{A}$ (in the first two block columns) and the system matrix $\mathbf{B}$ (in the last two block columns), the third block row depends on process matrices $\mathbf{A}$, $\mathbf{B}$, and $\mathbf{C}$, weight matrices $\mathbf{Q}_i$ and $\mathbf{R}_i$, and horizons P and M. Therefore, the stability of unconstrained LCO-DMPC introduced in this chapter could be designed and tuned by adjusting the weight matrices $\mathbf{Q}_i$, $\mathbf{R}_i$ and horizons P and M which introduce significant modifications on the third block row of matrix $\mathbf{A}_N$.*

5.2.5 Simulation Results

In this section, the following two input and two output nonminimum phase plant $\mathcal{S}$ is taken as a control objective to investigate the introduced unconstrained LCO-DMPC; the transfer function of system $\mathcal{S}$ is

$$\begin{bmatrix} y_1(z) \\ y_2(z) \end{bmatrix} = \begin{bmatrix} \dfrac{-0.024\,(z-1.492)\,(z+0.810)}{(z-0.819)(z^2-1.922z+0.961)} & \alpha\dfrac{0.018(z+0.935)}{(z^2-1.676z+0.819)} \\ \alpha\dfrac{0.126}{(z-0.368)} & \dfrac{0.147(z-0.668)}{(z^2-1.572z+0.670)} \end{bmatrix} \begin{bmatrix} u_1(z) \\ u_2(z) \end{bmatrix}$$

A state-space realization for $\mathcal{S}$ has the form

$$\begin{cases} \mathbf{x}(k+1) = \mathbf{Ax}(k) + \mathbf{Bu}(k) \\ \mathbf{y}(k) = \mathbf{Cx}(k) \end{cases} \tag{5.41}$$

with matrices

$$\mathbf{A} = \begin{bmatrix} \mathbf{A}_{11} & 0 \\ 0 & \mathbf{A}_{22} \end{bmatrix}$$

$$\mathbf{A}_{11} = \begin{bmatrix} 2.74 & -1.27 & 0.97 & 0 \\ 2 & 0 & 0 & 0 \\ 0 & 0.5 & 0 & 0 \\ 0 & 0 & 0 & 0.37 \end{bmatrix}$$

$$\mathbf{A}_{22} = \begin{bmatrix} 1.68 & -0.82 & 0 & 0 \\ 1 & 0 & 0 & 0 \\ 0 & 0 & 1.57 & -0.67 \\ 0 & 0 & 1 & 0 \end{bmatrix}$$

$$\mathbf{B} = \begin{bmatrix} \mathbf{B}_{11} & 0 \\ 0 & \mathbf{B}_{22} \end{bmatrix}$$

$$\mathbf{B}_{11} = \begin{bmatrix} 0.25 \\ 0 \\ 0 \\ 0.5 \end{bmatrix}, \mathbf{B}_{22} = \begin{bmatrix} 0.25 \\ 0 \\ 0.5 \\ 0 \end{bmatrix}$$

$$\mathbf{C} = \begin{bmatrix} \mathbf{C}_{11} & \mathbf{C}_{12} \\ \mathbf{C}_{21} & \mathbf{C}_{22} \end{bmatrix}$$

$$\begin{aligned} \mathbf{C}_{11} &= \begin{bmatrix} -0.1 & 0.03 & 0.12 & 0 \end{bmatrix} \\ \mathbf{C}_{12} &= \alpha \begin{bmatrix} 0.07 & 0.07 & 0 & 0 \end{bmatrix} \\ \mathbf{C}_{21} &= \alpha \begin{bmatrix} 0 & 0 & 0 & 2.25 \end{bmatrix} \\ \mathbf{C}_{22} &= \begin{bmatrix} 0 & 0 & 0.29 & -0.20 \end{bmatrix} \end{aligned} \tag{5.42}$$

Decompose $\mathcal{S}$ into two SISO subsystems, $\mathcal{S}_1$ and $\mathcal{S}_2$. The corresponding state-space models of $\mathcal{S}_1$ and $\mathcal{S}_2$ have the form (5.1) and are expressed as (5.43) and (5.44), respectively, where the constant parameter α is used to study the interactions between $\mathcal{S}_1$ and $\mathcal{S}_2$:

Subsystem $\mathcal{S}_1$**:**

$$\begin{cases} \mathbf{x}_1(k+1) = \mathbf{A}_{11}\mathbf{x}_1(k) + \mathbf{A}_{12}\mathbf{x}_2(k) + \mathbf{B}_{11}u_1(k) + \mathbf{B}_{12}u_2(k) \\ \mathbf{y}_1(k) = \mathbf{C}_{11}\mathbf{x}_1(k) + \mathbf{C}_{12}\mathbf{x}_2(k) \end{cases} \tag{5.43}$$

Subsystem $\mathcal{S}_2$**:**

$$\begin{cases} \mathbf{x}_2(k+1) = \mathbf{A}_{22}\mathbf{x}_2(k) + \mathbf{A}_{21}\mathbf{x}_1(k) + \mathbf{B}_{22}u_1(k) + \mathbf{B}_{21}u_1(k) \\ \mathbf{y}_2(k) = \mathbf{C}_{22}\mathbf{x}_2(k) + \mathbf{C}_{21}\mathbf{x}_1(k) \end{cases} \tag{5.44}$$

Similar to the centralized MPC, the feasibility depends on the possibility of satisfying Theorem 5.2 for a specific set of parameters of P, M, $\mathbf{Q}$, and $\mathbf{R}_i$, $i=1, \ldots, m$. Here, for simplifying the graphical representations of the results, we choose $P=M$, $\mathbf{R}=\gamma\mathbf{I}_u$, and $\mathbf{Q}=\mathbf{I}_y$.

The three-dimensional graphs of Figures 5.1 and 5.3 show the maximum eigenvalues of the corresponding closed-loop systems calculated for different combinations of γ and P, respectively. In Figures 5.1 and 5.3, the Z axis represents the maximum eigenvalues, the X and Y axes represent the logarithm of γ and P, respectively. The control performance of the closed-loop system is plotted in Figures 5.2, and 5.4, where the black dashed lines correspond to the desired outputs, and the blue solid lines correspond to the system outputs and inputs using LCO-DMPC.

It can be seen from these figures that the stability depends on the choice of the tuning parameters γ and P. For weak interactions, the range of tuning parameters in LCO-DMPC is acceptable. And a good global performance of the closed-loop system is obtained by the control of the introduced LCO-DMPC where the subsystems exhibit interactions (see Figures 5.2, and 5.4). The MSEs of outputs with LCO-DMPC are 0.2568 and 0.2277 when $\alpha=0.1$ and $\alpha=1$, respectively.

In conclusion, for the given example, unconstrained LCO-DMPC provides acceptable regions of tuning parameters. Usually, the stable regions are associated with a big prediction horizon P and small weight γ. The LCO-DMPC can achieve a good performance of the entire closed-loop system when the interactions among subsystems are not very stronger. Furthermore, the cost of computation is very small as compared with the centralized (see the analysis at the end of Section 5.2.3).

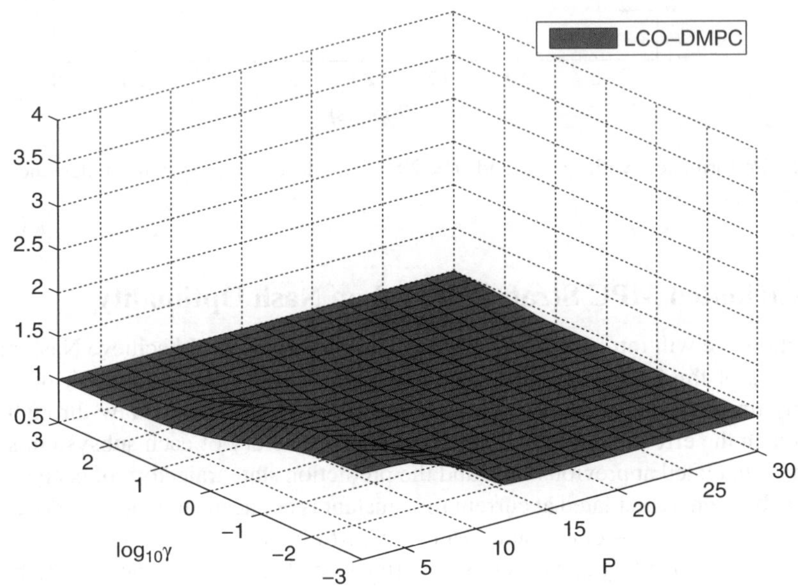

Figure 5.1 Maximum closed-loop eigenvalues with LCO-DMPC when $\alpha=0.1$

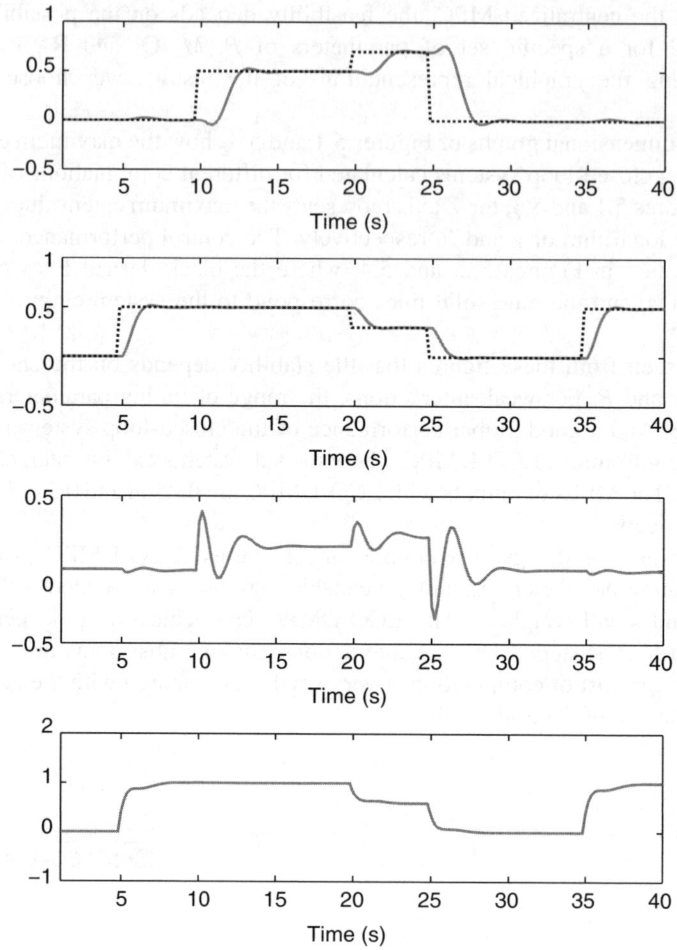

Figure 5.2 Performance with $\gamma = 1$ and $P = 20$ of a closed-loop system under the control of LCO-DMPC with $\alpha = 0.1$

5.3 Distributed MPC Strategy Based on Nash Optimality

In this section, we will introduce the DMPC algorithm which could achieve Nash optimality; the main idea of this method is that: each subsystem-based MPC communicates with each other many times a control period and computes the optimal control law through iteration, through which the error between the presumed state trajectory of each subsystem's upstream subsystems calculated at previous time and the prediction state trajectory of each subsystem's upstream subsystems calculated at current time instant is deduced, and then improve the global performance of the closed-loop system. Here the cost function of each subsystem-based MPC optimizes its corresponding subsystem's cost function. The detail of the DMPC-based Nash optimality is detailed in the following context.

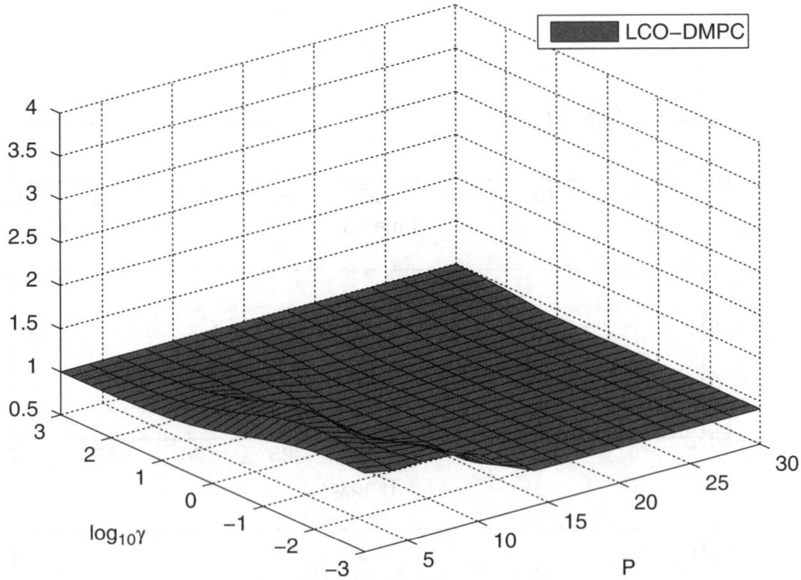

Figure 5.3 Maximum closed-loop eigenvalues with LCO-DMPC when $\alpha = 1$

5.3.1 Formulation

5.3.1.1 Model Predictive Control

As stated in Chapter 3, MPC is formulated as resolving an online open loop optimal control problem in moving horizon style. Using the current state, an input sequence is calculated to minimize a performance index while satisfying some specified constraints. Only the first element of the sequence is taken as a controller output. At the next sampling time, the optimization is resolved with new measurements from the plant. Thus, both the control horizon and the prediction horizon move or recede ahead by one step at next sampling time.

This is the reason why MPC is also sometimes referred to as receding horizon control (RHC) or moving horizon control (MHC). The purpose of taking new measurements at each sampling time is to compensate for unmeasured disturbances and model inaccuracy, both of which cause the system output to be different from its prediction. Suppose that the prediction output model of the whole system is described as

$$\mathbf{Y}(k+j|k) = \mathbf{f}(\mathbf{Y}(k), \Delta\mathbf{u}_M(k|k)) \quad (j = 1, \ldots, P) \tag{5.45}$$

where $\Delta\mathbf{u}_M(k|k) = \left[\Delta\mathbf{u}_{1,M}^{\mathrm{T}}(k|k) \ \cdots \ \Delta\mathbf{u}_{m,M}^{\mathrm{T}}(k|k)\right]^{\mathrm{T}}$ is the increment of the manipulated (the controller output, also the input to plant) variables of the system, P denotes the prediction horizon, M denotes the control horizon, $\mathbf{f}$ is the mapping function vector, where the element $\mathbf{f}_i$ satisfied some smooth condition. The performance index of the whole system is

$$\min_{\Delta\mathbf{u}_M(k|k)} J = \sum_{l=1}^{P} \mathrm{L}[\mathbf{y}(k+l|k), \Delta\mathbf{u}_M(k|k)] \tag{5.46}$$

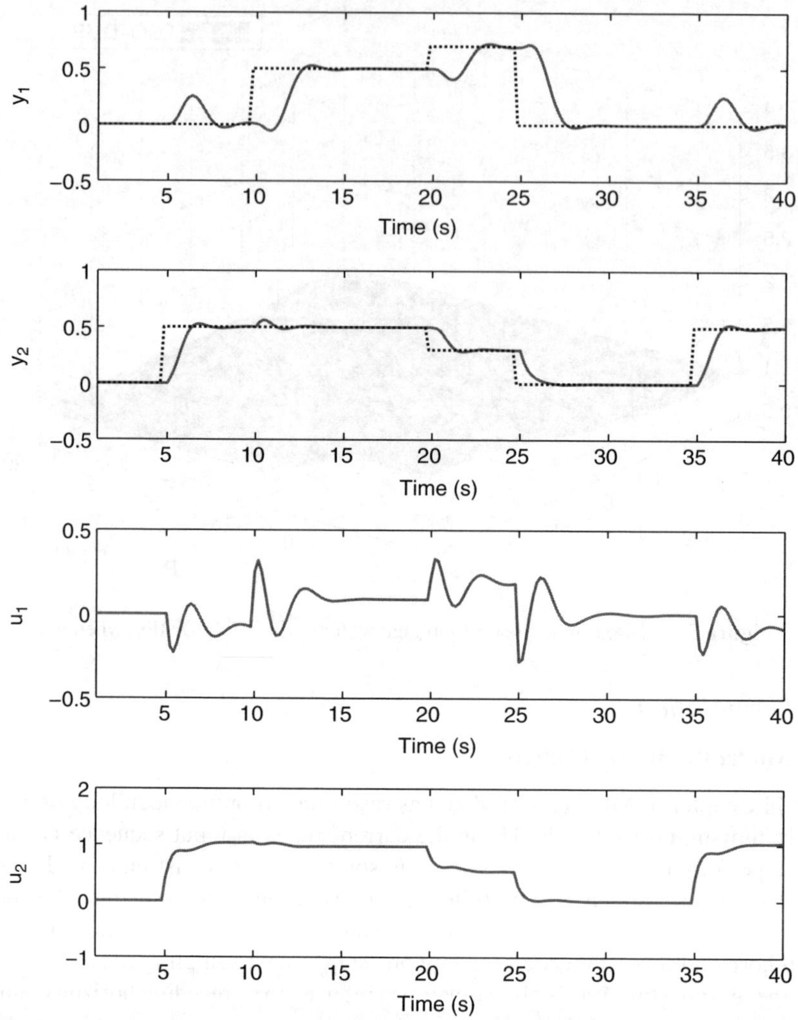

Figure 5.4 Performance with $\gamma = 1$ and $P = 20$ of a closed-loop system under the control of LCO-DMPC with $\alpha = 1$

where L is the nonlinear function of the input and output variables. The objective of the whole system is to regulate the system output to the expected values while keeping the performance index minimal.

5.3.1.2 Distributed Model Predictive Control Based on Nash Optimality

For large-scale systems to avoid the prohibitively high online computational demand and improve the global performance comparing to the noniterative LCO-DMPC, this section will introduce the DMPC control based on Nash optimality.

It is assumed that the behavior of the whole system is described by m subsystems and the nonlinear performance function L is decomposable in the distributed system. The local performance index for the ith controller can be expressed as

$$\min_{\Delta u_{i,M}(k|k)} J_i = \sum_{j=1}^{P} L_i[\mathbf{y}_i(k+j|k)\Delta\mathbf{u}_{i,M}(k|k)] \quad (i = 1, \ldots, m) \tag{5.47}$$

where L_i is the nonlinear function of $\mathbf{y}_i(k+j|k)$ and $\Delta\mathbf{u}_{i,M}(k|k)$. This indicates that the global performance index of the whole system is

$$\min J = \sum_{i=1}^{m} J_i \tag{5.48}$$

At time instant k, the future predictive output of the ith controller can be expressed as

$$\mathbf{y}_i(k+j|k) = \mathrm{f}_i(\mathbf{y}_i(k), \Delta\mathbf{u}_{1,M}(k|k), \ldots, \Delta\mathbf{u}_{m,M}(k|k)) \; (j = 1, \ldots, P) \tag{5.49}$$

It can be seen that the global performance index can be decomposed into a number of local performance indexes, but the output of each subsystem is still related to all the input variables due to the input coupling. Such distributed control problem with different goals can be resolved by means of the Nash optimal concept [95]. Concretely speaking, the group of control decisions $\mathbf{u}^N(t) = \{\mathbf{u}_1^N(t), \ldots, \mathbf{u}_m^N(t)\}$ is called to be the *Nash optimal solution* if for all u_i, $i = 1, \ldots, m$ the following relations are held:

$$J_i^*(\mathbf{u}_1^N, \ldots, \mathbf{u}_i^N, \ldots, \mathbf{u}_m^N) \le J_i(\mathbf{u}_1^N, \ldots, \mathbf{u}_{i-1}^N, \mathbf{u}_i, \mathbf{u}_{i+1}^N, \ldots, \mathbf{u}_m^N) \tag{5.50}$$

If the Nash optimal solution is adopted, each controller does not change its control decision $\mathbf{u}_i$ because it has achieved the locally optimal objective under the above condition; otherwise, the local performance index J_i will degrade. Each controller optimizes its objective (local performance index) only using its own control decision assuming that other controllers' Nash optimal solutions have been known, that is,

$$\min_{u_i} J_i|_{u_j^N (j\neq i)} \tag{5.51}$$

Inspecting Equation (5.51) to obtain the Nash optimal solution $\mathbf{u}_i$ of the ith subsystem, it is necessary to know other subsystems' Nash optimal solutions $\mathbf{u}_j^N(j \neq i)$, so that the whole system could arrive at Nash optimal equilibrium in this coupling decision process. By Nash optimal equilibrium, the global optimization problem can be decomposed into a number of local optimization problems.

An iterative algorithm is developed on the basis of the work of Du *et al.* [89] to seek the Nash optimal solution of the whole system at each sampling time. Since the mutual communication and the information exchange are adequately taken into account, each controller resolves local optimal problem provided that the other subsystem-based MPCs' optimal solutions have been known. Then each subsystem-based MPC compares the newly computed optimal solution with that obtained in last iteration and checks if the terminal condition is satisfied. If the algorithm is convergent, all the terminal conditions of the m agents will be satisfied, and the whole system will arrive at Nash equilibrium at this time. This Nash-optimization process will be repeated at the next sampling time.

5.3.2 *Algorithm*

Algorithm 5.2

Step 1: At sampling time instant k, each controller makes initial estimation of the input variables and announces it to the other controllers; let the iterative index $l=0$,

$$\Delta\bar{\mathbf{u}}^{l}_{i,M}(k) = [\Delta\bar{\mathbf{u}}^{l}_{i}(k), \Delta\bar{\mathbf{u}}^{l}_{i}(k+1), \ldots, \Delta\bar{\mathbf{u}}^{l}_{i}(k+M-1)]^{\mathrm{T}} \quad (i = 1, \ldots, m)$$

Step 2: Each controller resolves its optimal problem simultaneously to obtain its solution $\Delta\mathbf{u}^{*}_{i,M}(k) \quad (i = 1, \ldots, m)$

Step 3: Each controller checks if its terminal iteration condition is satisfied, that is, for the given error accuracy ε_i $(i=1, \ldots, m)$, if there exist

$$\|\Delta\mathbf{u}^{(l+1)}_{i,M}(k) - \Delta\mathbf{u}^{(l)}_{i,M}(k)\| \le \varepsilon_i \qquad (i = 1, \ldots, m)$$

If all the terminal conditions satisfied, then end the iteration and go to Step 4. Otherwise, let $l=l+1, \Delta\mathbf{u}^{l}_{i,M}(k) = \Delta\mathbf{u}^{*}_{i,M}(k)$ $(i = 1, \ldots, m)$, all controllers communicate to exchange this information, and take the latest solution to Step 2;

Step 4: Computes the instant control law

$$\Delta\mathbf{u}_i(k) = [\mathbf{I}\ 0\ 0\ \cdots\ 0]\ \Delta\mathbf{u}^{*}_{i,M}(k) \quad (i = 1, \ldots, m)$$

and takes them as the controller output of each agent;

Step 5: Move horizon to the next sampling time, that is, $k+1 \to k$, and go to Step 1, repeat the above process.

5.3.3 *Computational Convergence for Linear Systems*

Consider this DMPC of linear dynamic plants. At sampling time instant k, the output prediction model of the ith agent can be described as

$$\tilde{\mathbf{y}}_{i,PM}(k) = \mathbf{y}_{i,P}(k) + \mathbf{A}_{ii}\Delta\mathbf{u}_{i,M}(k) + \sum_{j=1,j\neq i}^{m} \mathbf{A}_{ij}\Delta\mathbf{u}_{j,M}(k) \quad (i = 1, \ldots, m) \tag{5.52}$$

where $\mathbf{A}_{ii}$ and $\mathbf{A}_{ij}$ are the dynamic matrix of the ith subsystem and the step response matrix of the ith subsystem stimulated by the jth subsystem, respectively. They are expressed in terms of the matrix

$$\mathbf{A}_{ij} = \begin{bmatrix} a_{ij}(1) & \cdots & 0 \\ \vdots & \ddots & \vdots \\ a_{ij}(M) & \cdots & a_{ij}(1) \\ \vdots & \vdots & \vdots \\ a_{ij}(P) & \cdots & a_{ij}(P-M+1) \end{bmatrix}$$

$$\mathbf{A} = \begin{bmatrix} \mathbf{A}_{11} & \cdots & \mathbf{A}_{1m} \\ \vdots & \ddots & \vdots \\ \mathbf{A}_{m1} & \cdots & \mathbf{A}_{mm} \end{bmatrix}$$

where $a_{ij}(k)(k = 1, \ldots, P; j = 1, \ldots, m)$ is the sampled output value of the ith subsystem for the jth subsystem unit step input at the sampling time instant k. The local performance index for the ith subsystem can be expressed as

$$\min_{\Delta \mathbf{u}_{i,M}(k)} J_i(k) = \|\boldsymbol{\omega}_{i,P}(k) - \tilde{\mathbf{y}}_{i,PM}(k)\|^2_{\mathbf{Q}_i} + \|\Delta \mathbf{u}_{i,M}(k)\|^2_{\mathbf{R}_i} \ (i = 1, \ldots, m) \tag{5.53}$$

where

$$\boldsymbol{\omega}_i(k) = \left[\boldsymbol{\omega}_i^{\mathrm{T}}(k+1) \cdots \boldsymbol{\omega}_i^{\mathrm{T}}(k+P)\right]^{\mathrm{T}}$$

are the expected output values of the ith subsystem.

$$\tilde{\mathbf{y}}^{\mathrm{T}}_{i,PM}(k) = [\tilde{\mathbf{y}}^{\mathrm{T}}_{i,M}(k+1|k)\ \mathbf{L}\ \tilde{\mathbf{y}}^{\mathrm{T}}_{i,M}(k+P|k)]^{\mathrm{T}}$$

$$\tilde{\mathbf{y}}^{\mathrm{T}}_{i,P0}(k) = [\tilde{\mathbf{y}}^{\mathrm{T}}_{i,0}(k+1|k)\ \mathbf{L}\ \tilde{\mathbf{y}}^{\mathrm{T}}_{i,0}(k+P|k)]^{\mathrm{T}}$$

and

$$\Delta \mathbf{u}^{\mathrm{T}}_{i,PM}(k) = [\Delta \mathbf{u}_{i,M}(k+1|k)\ \mathbf{L}\ \Delta \mathbf{u}_{i,M}(k+M-1|k)]^{\mathrm{T}}$$

According to Nash optimality and extremum necessary condition

$$\frac{\partial J_i}{\partial \Delta \boldsymbol{u}_{i,M}(k)} = 0$$

at the sampling time instant k, the Nash optimal solution of the ith agent can be derived as

$$\Delta \mathbf{u}^{(l+1)}_{i,M}(k) = \mathbf{D}_{ii}\left[\boldsymbol{\omega}_{i,P}(k) - \mathbf{y}_{i,P}(k) - \sum_{\substack{j=1\\ j\neq i}}^{m} \mathbf{A}_{ij}\Delta \mathbf{u}^{(l)}_{j,M}(k)\right] (i = 1, \ldots, m) \tag{5.54}$$

with $\mathbf{D}_{ii} = (\mathbf{A}^{\mathrm{T}}_{ii}\overline{\mathbf{Q}}_i\mathbf{A}_{ii} + \overline{\mathbf{R}}_i)^{-1}\mathbf{A}^{\mathrm{T}}_{ii}\overline{\mathbf{Q}}_i$. If the algorithm is convergent, the Nash optimal solution of the whole system can be written as

$$\Delta \mathbf{u}_M(k) = \mathbf{D}_0\Delta \mathbf{u}_M(k) + \mathbf{D}_1[\boldsymbol{\omega}(k) - \tilde{\mathbf{y}}_{P0}(k)] \tag{5.55}$$

with

$$\mathbf{D}_0 = \begin{bmatrix} \mathbf{0} & -\mathbf{D}_{11}\mathbf{A}_{12} & \cdots & -\mathbf{D}_{11}\mathbf{A}_{1m} \\ -\mathbf{D}_{22}\mathbf{A}_{21} & \mathbf{0} & \cdots & -\mathbf{D}_{22}\mathbf{A}_{2m} \\ \vdots & \vdots & \ddots & \vdots \\ -\mathbf{D}_{mm}\mathbf{A}_{m1} & \cdots & \cdots & \mathbf{0} \end{bmatrix}$$

$$\mathbf{D}_1 = \begin{bmatrix} \mathbf{D}_{11} & & & \mathbf{0} \\ & \mathbf{D}_{22} & & \\ & & \ddots & \\ \mathbf{0} & & & \mathbf{D}_{mm} \end{bmatrix}$$

In the iteration procedure, Equation (5.55) can be expressed as

$$\Delta \mathbf{u}^{(l+1)}_M(k) = \mathbf{D}_0\Delta \mathbf{u}^{l}_M(k) + \mathbf{D}_1[\boldsymbol{\omega}(k) - \tilde{\mathbf{y}}_{P0}(k)] \tag{5.56}$$

At time instant k, $\omega(k)$, and $\tilde{\mathbf{y}}_{P0}(k)$ are known in advance; hence, $\mathbf{D}_1[\boldsymbol{\omega}(k) - \tilde{\mathbf{y}}_{P0}(k)]$ is the constant term irrelevant to the iteration. The convergence of expression (5.56) is then equivalent to that of the following equation:

$$\Delta\mathbf{u}_M^{(l+1)}(k) = \mathbf{D}_0\Delta\mathbf{u}_M^l(k) \tag{5.57}$$

From the above analysis, the convergent condition for the algorithm in application to distributed linear model predictive control is

$$|\rho(\mathbf{D}_0)| < 1 \tag{5.58}$$

That is, the spectrum radius must be less than 1 to guarantee a convergent computation.

5.3.4 Nominal Stability of Distributed Model Predictive Control System

In order to analyze the nominal stability, rewrite the prediction output model of (5.52) in terms of state-space equation [96]. The predictive state-space model of the ith agent at the time instant k can be written as

$$\begin{cases} \mathbf{x}_i(k+1) = \mathbf{S}\mathbf{x}_i(k) + \boldsymbol{a}_{ii}\Delta u_i(k) + \sum\limits_{j=1,\, j\neq i}^{m} \boldsymbol{a}_{ij}\Delta u_j(k) \\ \mathbf{Y}_i(k) = \mathbf{C}\mathbf{S}\mathbf{x}_i(k) + \mathbf{A}_{ii}\Delta\mathbf{u}_{i,M}(k) + \sum\limits_{j=1,\, j\neq i}^{m} \mathbf{A}_{ij}\Delta\mathbf{u}_{j,M}(k) \end{cases} \quad (i = 1, \ldots, m) \tag{5.59}$$

where

$$\Delta u_i(k) = [1\ 0\ \cdots\ 0]\Delta\mathbf{u}_{i,M}(k)$$

$$\mathbf{S} = \begin{bmatrix} 0 & 1 & \cdots & \mathbf{0} \\ \vdots & \ddots & \ddots & \vdots \\ 0 & \cdots & 0 & 1 \\ 0 & \cdots & 0 & 1 \end{bmatrix}_{(N\times N)}$$

where N is the modeling horizon, and

$$\begin{aligned} \mathbf{a}_{ij} &= [a_{ij}(1) \cdots a_{ij}(N)]^{\mathrm{T}}, \\ \mathbf{x}_i(k) &= [x_{i1}(k) \cdots x_{iN}(k)], \\ \mathbf{Y}_i(k) &= [y_i(k+1) \cdots y_i(k+P)]^{\mathrm{T}} \\ \mathbf{C} &= \begin{bmatrix} \mathbf{I}_{P\times P} & \mathbf{0}_{P\times(N-P)} \end{bmatrix} \end{aligned}$$

denote the operation of taking out the first P vectors from the N-dimensional vectors. The Nash optimal solution in the state-space expression of the ith agent at the time instant k is

$$\Delta\mathbf{v}_{i,M}^{(l+1)}(k) = \mathbf{D}_{ii}\left[\boldsymbol{\omega}_{i,P}(k) - \mathbf{y}_{i,P}(k) - \sum_{\substack{j=1 \\ j\neq i}}^{m} \mathbf{A}_{ij}\Delta\boldsymbol{u}_{j,M}^{(l)}(k)\right] \tag{5.60}$$

The integral Nash optimal solution of the whole system, provided that the algorithm is convergent at each sampling time, can be written as

$$\Delta \mathbf{U}^N(k) = (\mathbf{I} - \mathbf{D}_0)^{-1}\mathbf{D}_1[\mathbf{R}(k) - \mathbf{F}_2\mathbf{X}(k)] \tag{5.61}$$

This is the state feedback control law. The instant control law of the whole system is $\Delta \mathbf{u}^N(k) = \mathbf{L}\Delta \mathbf{U}^N(k)$, with

$$\mathbf{L} = \text{block-diag}\left(\underbrace{\mathbf{L}_0 \ \cdots \ \mathbf{L}_0}_{m}\right), \mathbf{L}_0 = \begin{bmatrix} 1 & 0 & \cdots & 0 \end{bmatrix}_{1\times M}$$

$$\mathbf{F}_2 = \text{block-diag}\,\underbrace{(\mathbf{CS}, \dots, \mathbf{CS})}_{m}$$

$$\Delta \mathbf{U}^N(k) = \left[\left(\Delta \mathbf{u}^N_{1,M}(k)\right)^{\mathrm{T}} \ \cdots \ (\Delta \mathbf{u}^N_{m,M}(k))^{\mathrm{T}}\right]^{\mathrm{T}}$$

$$\boldsymbol{\omega}(k) = \begin{bmatrix} \boldsymbol{\omega}_1^{\mathrm{T}}(k) & \cdots & \boldsymbol{\omega}_m^{\mathrm{T}}(k) \end{bmatrix}^T$$

$$\mathbf{X}(k) = \begin{bmatrix} \mathbf{x}_1^{\mathrm{T}}(k) & \cdots & \mathbf{x}_m^{\mathrm{T}}(k) \end{bmatrix}^T$$

Without loss of generality, let the expected outputs be $\boldsymbol{\omega}_i(k) = \mathbf{0}$, $(i = 1, \dots, m)$, then the state-space model of the whole system at the time instant k can be expressed as

$$\mathbf{X}(k+1) = \mathbf{F}_1\mathbf{X}(k) + \mathbf{BL}\Delta \mathbf{U}^N(k) = [\mathbf{F}_1 - \mathbf{BL}(\mathbf{I} - \mathbf{D}_0)^{-1}\mathbf{D}_1\mathbf{F}_2]\mathbf{X}(k) \tag{5.62}$$

with

$$\mathbf{F}_1 = \text{block-diag}\,\underbrace{(\mathbf{S}, \dots, \mathbf{S})}_{m}, \mathbf{B} = \begin{bmatrix} \boldsymbol{a}_{11} & \cdots & \boldsymbol{a}_{1m} \\ \vdots & \ddots & \vdots \\ \boldsymbol{a}_{m1} & \cdots & \boldsymbol{a}_{mm} \end{bmatrix}$$

Expression (5.62) shows the state mapping relationship of the distributed system between the time instants k and k+1. According to the contraction mapping principle [96], the nominal stability of the whole distributed system can be guaranteed, if and only if

$$\left\| \lambda[\mathbf{F}_1 - \mathbf{BL}(\mathbf{I} - \mathbf{D}_0)^{-1}\mathbf{D}_1\mathbf{F}_2] \right\| < 1 \tag{5.63}$$

That is, the norms of eigenvalues of state mapping are less than 1.

5.3.5 *Performance Analysis with Single-step Horizon Control Under Communication Failure*

In distributed control, each controller can work independently to achieve its local objective, but cannot accomplish the whole task on its own. These autonomous agents can communicate and coordinate with each other, and can exchange information through a network in order to accomplish the whole task or objective. If a distributed system is subjected to communication failure, does this strategy work well and what does the performance of the whole system change? In this section, the performance deviation on *single-step horizon* under the communication failure is discussed. Because MPC takes a receding-horizon control policy in which the optimization is resolved online at each sampling time with updated measurements, it is reasonable to focus on the single-step horizon.

To indicate the communication connection among agents, define a connection matrix $\mathbf{E} = (e_{ij})$.

All elements in the main diagonal of E are zeros and other elements in the nonmain diagonal of E are 1. 0.1 denotes the communication connection between two agents existed and 0 shows no communication connection. Under the ideal situation of a communication connection without structural disturbance $e_{ij} = 1(i, j = 1, \ldots, m, i \neq j)$, the output prediction model and the Nash optimal solution of the ith agent at the time instant k can, respectively, be rewritten as

$$\tilde{\mathbf{y}}_{i,PM}(k) = \tilde{\mathbf{y}}_{i,P0}(k) + \mathbf{A}_{ii}\Delta\mathbf{u}_{i,M}(k) + \sum_{j=1,\, j\neq i}^{m} e_{ij}\mathbf{A}_{ij}\Delta\mathbf{u}_{j,M}(k), (i = 1, \ldots, m) \tag{5.64}$$

and

$$\Delta\mathbf{u}_{i,M}^{*}(k) = \mathbf{D}_{ii}\left[\boldsymbol{\omega}_i - \tilde{\mathbf{y}}_{i,P0}(k) - \sum_{j=1,\, j\neq i}^{m} \mathbf{G}_{ij}\Delta\mathbf{u}_{j,M}^{*}(k)\right], (i = 1, \ldots, m) \tag{5.65}$$

where $\mathbf{G} = \mathbf{EA} = [\mathbf{G}_{ij}]$ denotes the dot multiplication with

$$\mathbf{G} = \begin{bmatrix} \mathbf{0} & \mathbf{e}_{12} & \cdots & \mathbf{e}_{1m} \\ \mathbf{e}_{21} & \mathbf{0} & \cdots & \mathbf{e}_{2m} \\ \vdots & \vdots & \mathbf{0} & \vdots \\ \mathbf{e}_{m1} & \mathbf{e}_{m2} & \cdots & 0 \end{bmatrix} \begin{bmatrix} \mathbf{A}_{11} & \mathbf{A}_{12} & \cdots & \mathbf{A}_{1m} \\ \mathbf{A}_{21} & \mathbf{A}_{22} & \cdots & \mathbf{A}_{2m} \\ \vdots & \vdots & 0 & \vdots \\ \mathbf{A}_{m1} & \mathbf{A}_{m2} & \cdots & \mathbf{A}_{mm} \end{bmatrix}$$

$$= \begin{bmatrix} \mathbf{0} & \mathbf{e}_{12}\mathbf{A}_{12} & \cdots & \mathbf{e}_{1m}\mathbf{A}_{1m} \\ \mathbf{e}_{21}\mathbf{A}_{21} & \mathbf{0} & \cdots & \mathbf{e}_{2m}\mathbf{A}_{2m} \\ \vdots & \vdots & 0 & \vdots \\ \mathbf{e}_{m1}\mathbf{A}_{m1} & \mathbf{e}_{m2}\mathbf{A}_{m2} & \cdots & 0 \end{bmatrix}$$

Then the Nash optimal solution of the whole system under convergent computation is

$$\Delta\mathbf{u}_{M}^{*}(k) = (\mathbf{I}\text{-}\mathbf{D}_E)^{-1}[\boldsymbol{\omega}(k) - \tilde{\mathbf{y}}_{P0}(k)] \tag{5.66}$$

with

$$\mathbf{D}_E = -\mathbf{D}_1\mathbf{G} = \begin{bmatrix} \mathbf{0} & -\mathbf{D}_{11}\mathbf{e}_{12}\mathbf{A}_{12} & \cdots & -\mathbf{D}_{11}\mathbf{e}_{1m}\mathbf{A}_{1m} \\ -\mathbf{D}_{22}\mathbf{e}_{21}\mathbf{A}_{21} & \mathbf{0} & \cdots & -\mathbf{D}_{22}\mathbf{e}_{2m}\mathbf{A}_{2m} \\ \vdots & \vdots & \mathbf{0} & \vdots \\ -\mathbf{D}_{mm}\mathbf{e}_{m1}\mathbf{A}_{m1} & -\mathbf{D}_{mm}\mathbf{e}_{m2}\mathbf{A}_{m2} & \cdots & \mathbf{0} \end{bmatrix}$$

In the following analysis, assume that the prediction horizon and the control horizon are equal, and the communication failure is confined within a stable region. To analyze system performance deviation, define a communication failure matrix $\mathbf{T}$. The matrix $\mathbf{T}$ is a diagonal matrix or a block diagonal matrix. For a diagonal matrix, define the elements of its main diagonal as

0 or 1. For a block diagonal matrix, the elements of its main diagonal block are all 0s or all 1s. The value 0 corresponds to the communication failure existed, and 1 for no failure.

Remark 5.4 *Here the communication failure is classified into three cases as follows:*

- Row failure. *In this case, the communication failure happens on the receiving channels. In this case, the agent cannot receive the information coming from other agents, equivalently the corresponding row of matrix* $\mathbf{G}$ *becomes 0 and* $\mathbf{G}$ *becomes* $\mathbf{G}^{\text{dis}}, \mathbf{G}^{\text{dis}} = \mathbf{T_r G}$, *and the corresponding element of the communication failure matrix* $\mathbf{T_r}$ *has changed from 1 to 0.*
- Column failure. *In this case, the communication failure happens on the transmitting channels. In this case, the agent cannot send its information to other agents, equivalently the corresponding column of matrix* $\mathbf{G}$ *becomes 0 and* $\mathbf{G}$ *becomes* $\mathbf{G}^{\text{dis}}, \mathbf{G}^{\text{dis}} = \mathbf{G T_c}$ *and the corresponding element of the communication failure matrix* $\mathbf{T_c}$ *has changed from 1 to 0.*
- Mixed failure. *In this case, both row and column failures exist, the corresponding row and column of matrix* $\mathbf{G}$ *become 0 and* $\mathbf{G}$ *becomes* $\mathbf{G}^{\text{dis}}$.
- $\mathbf{G}^{\text{dis}} = \mathbf{T}_r\mathbf{G}\mathbf{T}_c$, *and the corresponding element of communication failure matrices* $\mathbf{T_r}$ *and* $\mathbf{T_c}$ *has changed from 1 to 0.*

With these preliminaries a theorem is presented.

Theorem 5.3 *For a distributed system, assume that the prediction horizon and the control horizon are equal and the communication failure cannot affect the stability. Its performance at the time instant k under the local communication failure is degrading. The degrading* δ *magnitude of the performance index satisfies* $0 \le \delta \le \delta_{\max}$, *and the upper bound of this magnitude* $\delta_{\max}$ *is*

$$\delta_{\max} = \frac{\|\mathbf{W}_{\max}\|}{\lambda_m(\mathbf{F})}$$

where $\lambda_m(F)$ *denotes the minimal eigenvalue of* $\boldsymbol{F}$, *with*

$$\begin{aligned}
\mathbf{W}_{\max} &= [\mathbf{D}_1^{-1}(\mathbf{I}-\mathbf{D}_E)-\mathbf{A}]^{\mathrm{T}}\mathbf{Q}[\mathbf{A}-\mathbf{A}_0(\mathbf{I}-\mathbf{D}_E)] \\
&\quad + [\mathbf{A}-\mathbf{A}_0(\mathbf{I}-\mathbf{D}_E)]^{\mathrm{T}} \times \mathbf{Q}[\mathbf{D}_1^{-1}(\mathbf{I}-\mathbf{D}_E)-\mathbf{A}] \\
&\quad + [\mathbf{A}-\mathbf{A}_0(\mathbf{I}-\mathbf{D}_E)]^{\mathrm{T}}\mathbf{Q}[\mathbf{A}-\mathbf{A}_0(\mathbf{I}-\mathbf{D}_E)] \\
&\quad - \mathbf{D}_E^{\mathrm{T}}\mathbf{R}\mathbf{D}_E - \mathbf{R}\mathbf{D}_E - \mathbf{D}_E^{\mathrm{T}}\mathbf{R} \\
\mathbf{F} &= [\mathbf{D}_1^{-1}(\mathbf{I}-\mathbf{D}_E)-\mathbf{A}]^{\mathrm{T}}\mathbf{Q}[\mathbf{D}_1^{-1}(\mathbf{I}-\mathbf{D}_E)-\mathbf{A}] + \mathbf{R} \\
\mathbf{A}_0 &= \begin{bmatrix} \mathbf{A}_{11} & & \mathbf{0} \\ & \ddots & \\ \mathbf{0} & & \mathbf{A}_{mm} \end{bmatrix} \\
\mathbf{Q} &= \text{block-diag}(\mathbf{Q}_1, \dots, \mathbf{Q}_m) \\
\mathbf{R} &= \text{block-diag}(\mathbf{R}_1, \dots, \mathbf{R}_m)
\end{aligned}$$

Proof. Without loss of generality, take the mixed failure as an example

$$\mathbf{D}_E^{\text{dis}} = -\mathbf{D}_1\mathbf{G}^{\text{dis}} = -\mathbf{D}_1\mathbf{T}_r\mathbf{G}\mathbf{T}_c = -\mathbf{T}_r\mathbf{D}_1\mathbf{G}\mathbf{T}_c = \mathbf{T}_r\mathbf{D}_E\mathbf{T}_c$$

The Nash optimal solution of the whole system in this case is

$$\Delta\mathbf{u}_M^{\text{dis}}(k) = (\mathbf{I} - \mathbf{T}_r\mathbf{D}_E\mathbf{T}_c)^{-1}\mathbf{D}_1[\boldsymbol{\omega}(k) - \tilde{\mathbf{y}}_{P0}(k)] \tag{5.67}$$

Using the matrix decomposition technique, it gives

$$\begin{aligned}(\mathbf{I} - \mathbf{T}_r\mathbf{D}_E\mathbf{T}_c)^{-1} &= 2[(\mathbf{I} - \mathbf{D}_E) + (\mathbf{I} + \mathbf{D}_E - 2\mathbf{T}_r\mathbf{D}_E\mathbf{T}_c)]^{-1} \\ &= 2(\mathbf{I} - \mathbf{D}_E)^{-1} - 2(\mathbf{I} - \mathbf{D}_E)^{-1}[(\mathbf{I} - \mathbf{D}_E)^{-1} + (\mathbf{I} + \mathbf{D}_E - 2\mathbf{T}_r\mathbf{D}_E\mathbf{T}_c)^{-1}]^{-1}(\mathbf{I} - \mathbf{D}_E)^{-1}\end{aligned} \tag{5.68}$$

In general, $(\mathbf{I} - \mathbf{D}_E)^{-1}$ and $(\mathbf{I} + \mathbf{D}_E - 2\mathbf{T}_r\mathbf{D}_E\mathbf{T}_c)^{-1}$ exist; therefore, the above equation holds. Substitute (5.68) into (5.67) to obtain

$$\begin{aligned}\Delta\mathbf{u}_M^{\text{dis}}(k) &= 2\Delta\mathbf{u}_M^*(k) - 2(\mathbf{I} - \mathbf{D}_E)^{-1}[(\mathbf{I} - \mathbf{D}_E)^{-1} + (\mathbf{I} + \mathbf{D}_E - 2\mathbf{T}_r\mathbf{D}_E\mathbf{T}_c)^{-1}]^{-1}\Delta\mathbf{u}_M^*(k) \\ &= \overline{\mathbf{S}}\Delta\mathbf{u}_M^*(k)\end{aligned} \tag{5.69}$$

with

$$\overline{\mathbf{S}} = 2\mathbf{I} - 2(\mathbf{I} - \mathbf{D}_E)^{-1}[(\mathbf{I} - \mathbf{D}_E)^{-1} + (\mathbf{I} + \mathbf{D}_E - 2\mathbf{T}_r\mathbf{D}_E\mathbf{T}_c)^{-1}]^{-1}$$

From $\Delta\mathbf{u}_M^*(k) = (\mathbf{I}\text{-}\mathbf{D}_E)^{-1}\mathbf{D}\left[\boldsymbol{\omega}(k) - \tilde{\mathbf{y}}_{P0}(k)\right]$, it has

$$\boldsymbol{\omega}(k) - \tilde{\mathbf{y}}_{P0}(k) = \mathbf{D}^{-1}(\mathbf{I} - \mathbf{D}_E)\Delta\mathbf{u}_M^*(k)$$

Then it gives

$$\begin{aligned}J^* &= \|\boldsymbol{\omega}(k) - \tilde{\mathbf{y}}_{P0}(k) - \mathbf{A}\Delta\mathbf{u}_M^*(k)\|_Q^2 + \|\Delta\mathbf{u}_M^*(k)\|_R^2 \\ &= \|\mathbf{D}_1^{-1}(\mathbf{I} - \mathbf{D}_E)\Delta\mathbf{u}_M^*(k) - \mathbf{A}\Delta\mathbf{u}_M^*(k)\|_Q^2 + \|\Delta\mathbf{u}_M^*(k)\|_R^2 \\ &= \|\Delta\mathbf{u}_M^*(k)\|_F^2\end{aligned} \tag{5.70}$$

with $\mathbf{F} = [\mathbf{D}_1^{-1}(\mathbf{I} - \mathbf{D}_E) - \mathbf{A}]^{\mathrm{T}}\mathbf{Q}[\mathbf{D}_1^{-1}(\mathbf{I} - \mathbf{D}_E) - \mathbf{A}] + \mathbf{R}$

Let

$$\mathbf{A}_0 = \begin{bmatrix}\mathbf{A}_{11} & & \mathbf{0} \\ & \ddots & \\ 0 & & \mathbf{A}_{mm}\end{bmatrix}$$

Then the prediction model of the whole distributed system under the mixed failure can be written as

$$\overline{\mathbf{y}}_{PM}^{\text{dis}} = \tilde{\mathbf{y}}_{P0}(k) + (\mathbf{A}_0 + \mathbf{T}_r\mathbf{G}\mathbf{T}_c)\Delta\mathbf{u}_M^{\text{dis}}(k) = \tilde{\mathbf{y}}_{P0}(k) + \overline{\mathbf{A}}\Delta\mathbf{u}_M^{\text{dis}}(k) \tag{5.71}$$

with

$$\overline{\mathbf{A}} = \mathbf{A}_0 + \mathbf{T}_r\mathbf{G}\mathbf{T}$$

Substituting (5.69) and (5.71) into (5.53), we derive

$$\begin{aligned}J^{\text{dis}} &= \|\boldsymbol{\omega}(k) - \tilde{\mathbf{y}}_{P0}(k) - \overline{\mathbf{A}}\,\overline{\mathbf{S}}\Delta\mathbf{u}_M^*(k)\|_{\mathbf{Q}}^2 + \|\overline{\mathbf{S}}\Delta\mathbf{u}_M^*(k)\|_{\mathbf{R}}^2 \\ &= \|\boldsymbol{\omega}(k) - \tilde{\mathbf{y}}_{P0}(k) - \mathbf{A}\Delta\mathbf{u}_M^*(k) + (\mathbf{A} - \overline{\mathbf{A}}\,\overline{\mathbf{S}})\Delta\mathbf{u}_M^*(k)\|_{\mathbf{Q}}^2 \\ &\quad + \|\Delta\mathbf{u}_M^*(k) + (\overline{\mathbf{S}} - \mathbf{I})\Delta\mathbf{u}_M^*(k)\|_{\mathbf{R}}^2 \\ &= J^* + \|\Delta\mathbf{u}_M^*(k)\|_{\mathbf{W}}^2\end{aligned} \tag{5.72}$$

with

$$\mathbf{W} = [\mathbf{D}_1^{-1}(\mathbf{I}-\mathbf{D_E})-\mathbf{A}]^{\mathrm{T}}\mathbf{Q}(\mathbf{A}-\overline{\mathbf{A}}\,\overline{\mathbf{S}})+(\mathbf{A}-\overline{\mathbf{A}}\,\overline{\mathbf{S}})^{\mathrm{T}}\mathbf{Q}[\mathbf{D}_1^{-1}(\mathbf{I}-\mathbf{D}_E)-\mathbf{A}]$$
$$+(\mathbf{A}-\overline{\mathbf{A}}\,\overline{\mathbf{S}})^{\mathrm{T}}\mathbf{Q}(\mathbf{A}-\overline{\mathbf{A}}\,\overline{\mathbf{S}})+(\overline{\mathbf{S}}-\mathbf{I})^{\mathrm{T}}\mathbf{R}(\overline{\mathbf{S}}-\mathbf{I})+\mathbf{R}(\overline{\mathbf{S}}-\mathbf{I})+(\overline{\mathbf{S}}-\mathbf{I})^{\mathrm{T}}\mathbf{R}$$

Then

$$\|\Delta\mathbf{u}_M^*(k)\|_W^2 \le \Delta\mathbf{u}_M^{*\mathrm{T}}(k)\|\mathbf{W}\|\Delta\mathbf{u}_M^*(k) = \|\mathbf{W}\|\|\Delta\mathbf{u}_M^*(k)\|^2$$
$$\le \frac{\|\mathbf{W}\|}{\lambda_m(F)}\|\Delta\mathbf{u}_M^*(k)\|_F^2 = \frac{\|\mathbf{W}\|}{\lambda_m(F)}J^*$$

Here, $\lambda_m(F)$ is the minimal eigenvalue of $\mathbf{F}$. From the above derivations, the performance relationship between the communication failure free and communication failure can be expressed as

$$J^{\mathrm{dis}} \le J^* + \frac{\|\mathbf{W}\|}{\lambda_m(F)}J^* = \left[1+\frac{\|\mathbf{W}\|}{\lambda_m(F)}\right]J^* = (1+\delta)J^* \tag{5.73}$$

where $\delta = \|\mathbf{W}\|/\lambda_m(F)$ denotes the degrading magnitude of the performance index under the local communication failure.

Inspection of (5.72) shows that $\|\mathbf{W}\|$ depends on $\mathbf{G}^{\mathrm{dis}}$ and $\mathbf{D}_E^{\mathrm{dis}}$, while $\mathbf{G}^{\mathrm{dis}}$ and $\mathbf{D}_E^{\mathrm{dis}}$ are affected by the communication failure matrix $\mathbf{T}_r$ and $\mathbf{T}_c$. So in the case of all existed communication failures, $\|\mathbf{W}\|$ can arrive at the maximal value, at this time, $\mathbf{T}_r\mathbf{D}_E\mathbf{T}_c = 0, \mathbf{G}^{\mathrm{dis}} = 0, \mathbf{D}_E^{\mathrm{dis}} = 0, \overline{\mathbf{A}} = \mathbf{A}_0, \overline{\mathbf{S}} = \mathbf{I}-\mathbf{D}_E$, and

$$\mathbf{W}_{\max} = [\mathbf{D}_1^{-1}(\mathbf{I}-\mathbf{D}_E)-\mathbf{A}]^{\mathrm{T}}\mathbf{Q}[\mathbf{A}-\mathbf{A}_0(\mathbf{I}-\mathbf{D}_E)]+[\mathbf{A}-\mathbf{A}_0(\mathbf{I}-\mathbf{D}_E)]^{\mathrm{T}}$$
$$\times\ \mathbf{Q}[D_1^{-1}(\mathbf{I}-\mathbf{D}_E)-\mathbf{A}]+[\mathbf{A}-\mathbf{A}_0(\mathbf{I}-\mathbf{D}_E)]^{\mathrm{T}}\mathbf{Q}[\mathbf{A}-\mathbf{A}_0(\mathbf{I}-\mathbf{D}_E)]$$
$$-\ \mathbf{D}_E^{\mathrm{T}}\mathbf{R}\mathbf{D}_E-\mathbf{R}\mathbf{D}_E-\mathbf{D}_E^{\mathrm{T}}\mathbf{R}$$

Therefore, the upper bound of the performance deviation under the local communication failure is

$$\delta_{\max} = \frac{\|W_{\max}\|}{\lambda_m(F)}$$

Theorem 5.4 *The convergent condition of the distributed linear model predictive control system under the communication failure is* $|\rho(\mathbf{T}_r\mathbf{D}_E\mathbf{T})| < 1$, *where* $\mathbf{D}_E, \mathbf{T}_r$, *and* $\mathbf{T}_c$ *are the same as defined before.*

Proof. The output prediction model of the ith agent under the communication failure at the time instant k can be described as

$$\tilde{\mathbf{y}}_{i,PM}^{\mathrm{dis}} = \tilde{\mathbf{y}}_{i,P0}(k)+\mathbf{A}_{ii}\Delta\mathbf{u}_{i,M}^{\mathrm{dis}}(k)+\sum_{j=1,\,j\ne i}^{m}\mathbf{G}_{ij}^{\mathrm{dis}}\Delta\mathbf{u}_{j,M}^{\mathrm{dis}}(k)\ \ (i=1,\ldots,m) \tag{5.74}$$

The local performance index for the ith agent can be expressed as

$$\min J_i^{\mathrm{dis}} = \|\boldsymbol{\omega}_i(k)-\tilde{\mathbf{y}}_{i,PM}^{\mathrm{dis}}(k)\|_{Q_i}^2+\|\Delta\mathbf{u}_{i,M}^{\mathrm{dis}}(k)\|_{R_i}^2\ \ (i=1,\ldots,m) \tag{5.75}$$

According to Nash optimality, at the sampling time instant k, the Nash optimal solution of the ith agent can be derived as

$$\Delta \mathbf{u}_{i,M}^{\mathrm{dis}}(k) = \mathbf{D}_{ii}\left[\boldsymbol{\omega}_i - \tilde{\mathbf{y}}_{i,P0}(k) - \sum_{j=1,\, j\neq i}^{m} \mathbf{G}_{ij}^{\mathrm{dis}} \Delta \mathbf{u}_{j,M}^{\mathrm{dis}}(k)\right] \quad (i = 1, \ldots, m) \tag{5.76}$$

If the algorithm is convergent, the Nash optimal solution of the whole system can be written as

$$\Delta \mathbf{u}_M^{\mathrm{dis}}(k) = \mathbf{D}_1[\omega - \tilde{\mathbf{y}}_{P0}(k)] + \mathbf{D}_E^{\mathrm{dis}} \Delta \mathbf{u}_M^{\mathrm{dis}}(k) \tag{5.77}$$

In the iteration procedure, Equation (5.77) can be expressed as

$$\Delta \mathbf{u}_M^{\mathrm{dis}}(k)|_{l+1} = \mathbf{D}_1[\boldsymbol{\omega}(k) - \tilde{\mathbf{y}}_{P0}(k)] + (\mathbf{T}_r \mathbf{D}_E \mathbf{T}_c)\Delta \mathbf{u}_M^{\mathrm{dis}}(k)|_l \ \ (l = 0, 1, \ldots) \tag{5.78}$$

At the time instant k, $\omega(k)$ and $\tilde{\mathbf{y}}_{P0}(k)$ are known in advance; hence, $\mathbf{D}_1[\boldsymbol{\omega}(k) - \tilde{\mathbf{y}}_{P0}(k)]$ is the constant term irrelevant to the iteration. The convergence of expression (5.78) is then equivalent to that of the following:

$$\Delta \mathbf{u}_M^{\mathrm{dis}}(k)|_{l+1} = (\mathbf{T}_r \mathbf{D}_E \mathbf{T}_c)\Delta \mathbf{u}_M^{\mathrm{dis}}(k)|_l \ \ (l = 0, 1, \ldots) \tag{5.79}$$

Therefore, the convergent condition of the distributed linear model predictive control system under the communication failure is

$$|\rho(\mathbf{T}_r \mathbf{D}_E \mathbf{T})| < 1.$$

Remark 5.5 *Under the communication failure, each agent cannot exchange information properly. In the extreme case,* $\mathbf{T}_r \mathbf{D}_E \mathbf{T}_c = 0$, $|\rho(\mathbf{T}_r \mathbf{D}_E \mathbf{T})| < 1$ *is always satisfied, which corresponds to the full decentralized architecture.*

5.3.6 Simulation Results

Consider the Shell heavy oil fractionator benchmark control problem as shown in Figure 5.5. The heavy oil fractionator is characterized by three product draws and three side circulating loops. Product specifications for the top and side draw streams are determined by economics and operating requirements. There is no product specification for the bottom draw, but there is an operating constraint on the temperature in the lower part of the column. The three circulating loops remove heat to achieve the desired product separation. The heat exchangers in these loops re-boil columns in other parts of the plant. Therefore, they have varying heat duty requirements. The bottom loop has an enthalpy controller which regulates heat removal in the loop by adjusting steam make. Its heat duty can be used as a manipulated variable to control the column. The heat duties of the other two loops act as disturbances to the column.

Prett and Morari [97] presented a model for a heavy oil fractionator as the benchmark process for the Shell standard control problem

$$\mathbf{y} = \mathbf{G(s)u} + \mathbf{G}_d(\mathbf{s})\mathbf{d}$$

where $\mathbf{u} = [u_1\ u_2\ u_3]^{\mathrm{T}}$ are manipulated variables to control the process, u_1 represents the product draw rate from the top of the column, u_2 represents the product draw rate from the side of

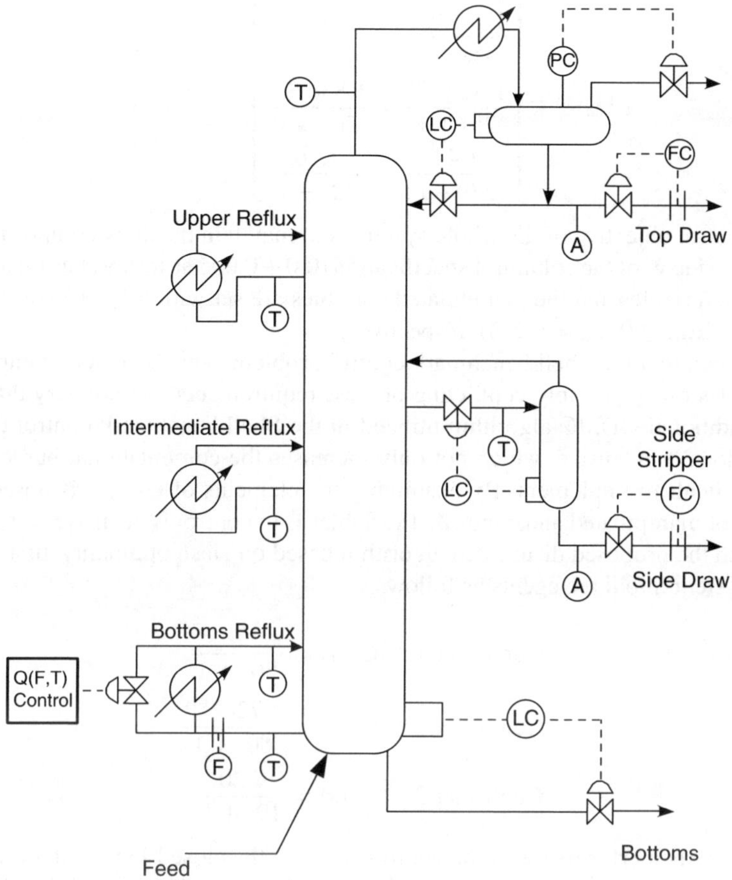

Figure 5.5 Shell heavy oil fractionator benchmark control problem

the column, and u_3 represents the reflux heat duty for the bottom of the column; $\mathbf{d} = [d_1\ d_2]^T$ are unmeasured but bounded disturbances entering the column, d_1 represents the reflux heat duty for the intermediate section of the column, $d_2$represents the reflux heat duty for the top of the column, with $|d_1| \le 0.5$ and $|d_2| \le 0.5$; $\mathbf{y} = [y_1\ y_2\ y_3]^T$ are output variables, y_1 represents the draw composition from the top of the column, y_2 represents the draw composition from the side of the column, and y_3 represents the reflux temperature at the bottom of the column; transfer function matrices $\mathbf{G}(s)$ and $\mathbf{G}_d(s)$ are

$$\mathbf{G}(s) = \begin{bmatrix} \dfrac{4.05e^{-27s}}{50s+1} & \dfrac{1.77e^{-28s}}{60s+1} & \dfrac{5.88e^{-27s}}{50s+1} \\ \dfrac{5.39e^{-18s}}{50s+1} & \dfrac{5.72e^{-14s}}{60s+1} & \dfrac{6.90e^{-15s}}{40s+1} \\ \dfrac{4.38e^{-20s}}{33s+1} & \dfrac{4.42e^{-22s}}{44s+1} & \dfrac{7.20}{19s+1} \end{bmatrix}$$

$$\mathbf{G}_d(s) = \begin{bmatrix} \frac{1.20e^{-27s}}{45s+1} & \frac{1.44e^{-27s}}{40s+1} \\ \frac{1.52e^{-15s}}{25s+1} & \frac{1.83e^{-15s}}{20s+1} \\ \frac{1.44}{27s+1} & \frac{1.26}{32s+1} \end{bmatrix}$$

The main control objective of the whole system is to maintain the draw composition from the top y_1 and the side y_2 of the column at specification (0.0 ± 0.005 at steady state). The constraints of the output variables and the manipulated variables are set with $|y_i| \leq 0.5$ $(i=1,2)$, $y_3 \geq 0.5$ and $|u_i| \leq 0.5$, $|\Delta u_i| \leq 0.2$ $(i=1,2,3)$, respectively.

It can be seen that the Shell benchmark control problem is an extremely complex problem which includes many possibly conflicting process requirements that are very difficult to satisfy. The traditional QDMC algorithm utilized in the Shell benchmark control problem [97] is computationally intensive, which not only increases the computational burden but is also relatively difficult to implement. By examining the elements of $\mathbf{G}(s)$, it is observed that the best pairing of manipulated and controlled variables is to control y_1 with u_1, y_2 with u_2, and y_3 with u_3. With the proposed distributed algorithm based on Nash optimality, first of all, divide the whole system into three agents as follows:

$$\text{Controller 1}: \; G_1(s) = \frac{4.05e^{-27}}{50s+1}$$

$$\text{Controller 2}: \; G_2(s) = \frac{5.72e^{-14s}}{60s+1}$$

$$\text{Controller 3}: \; G_3(s) = \frac{7.20}{19s+1}$$

In order to test the performance of the control scheme, the closed-loop system was subjected to disturbance patterns $\mathbf{d}^1 = [0.5\ 0.5]^{\mathrm{T}}$ and $\mathbf{d}^2 = [-0.5 - 0.5]^{\mathrm{T}}$, This means that $\mathbf{d}^1$ and $\mathbf{d}^2$ represent the worst-case scenarios, since d_1 and d_2 are at the extremes of ± 0.5 and have the same sign.

The MATLAB-based simulation results are shown in Figures 5.6–5.9. The tuning parameters for each agent are set with $P=8$, $M=3$, $\mathbf{Q}_1 = \mathbf{I}_{P\times P}$, $\mathbf{Q}_2 = \mathbf{I}_{P\times P}$, $\mathbf{Q}_3 = 0.1\mathbf{I}_{P\times P}$, $\mathbf{R}_i = 0.5\mathbf{I}_{M\times M}$, $(i=1,2,3)$, using a sampling time of 4 min, and $\varepsilon_i = 0.01, (i=1,2,3)$. Figures 5.6 and 5.7 show closed-loop system output responses and manipulated/control signals with no communication failure under the disturbance pattern $\mathbf{d}^1 = [0.5\ 0.5]^{\mathrm{T}}$and the disturbance pattern $\mathbf{d}^2 = [-0.5 - 0.5]^{\mathrm{T}}$, respectively. Figures 5.8 and 5.9 show closed-loop system output responses and manipulated/control signals with the mixed communication failure (the second row shows the communication failure and the third column shows the communication failure) under the disturbance pattern $\mathbf{d}^1 = [0.5\ 0.5]^{\mathrm{T}}$ and the disturbance pattern $\mathbf{d}^2 = [-0.5 - 0.5]^{\mathrm{T}}$, respectively. It can be observed that under two disturbance test patterns, each agent in this distributed structure can properly meet the steady-state specifications, outputs y_1 and y_2 are rapidly stabilized to zero, and all manipulated variables are within the saturation and rate limit constraints. By contrasting Figure 5.8 with Figure 5.6 and Figure 5.9 with Figure 5.7, we have that, although performance of the whole system under the local communication failure is degrading, each agent can meet the steady-state specifications and acquire satisfactory control results. In addition, the design parameters for

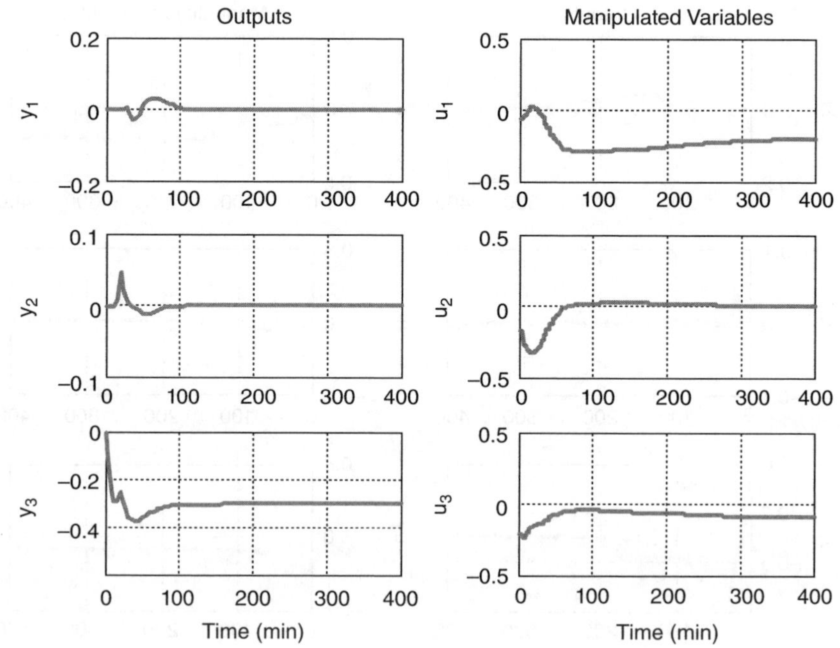

Figure 5.6 Closed-loop system output responses and manipulated/control signals with no communication failure under the disturbance pattern $\mathbf{d}^1 = [0.5\ 0.5]^T$

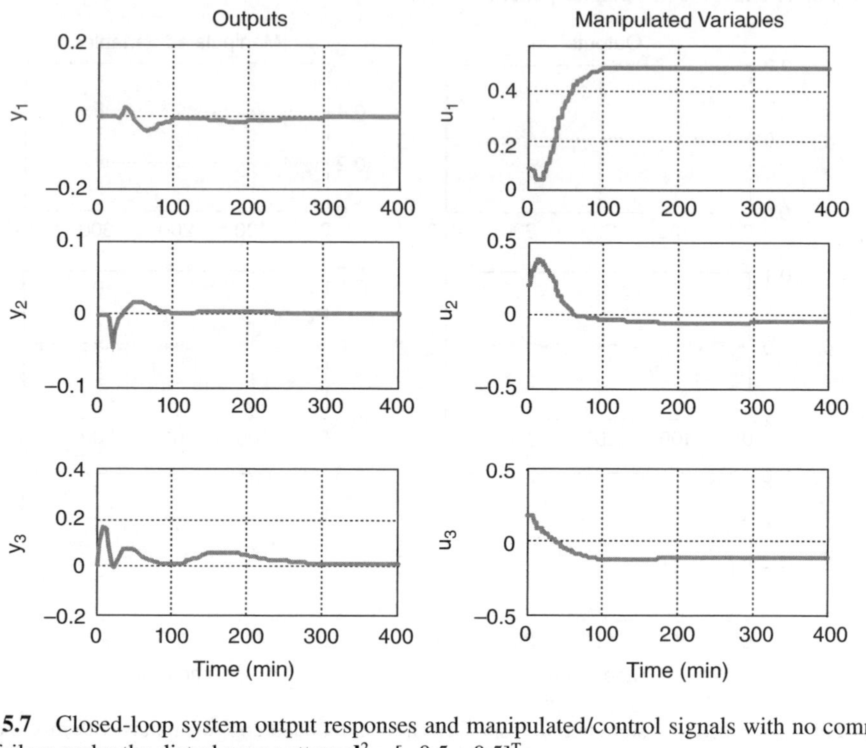

Figure 5.7 Closed-loop system output responses and manipulated/control signals with no communication failure under the disturbance pattern $\mathbf{d}^2 = [-0.5 - 0.5]^T$

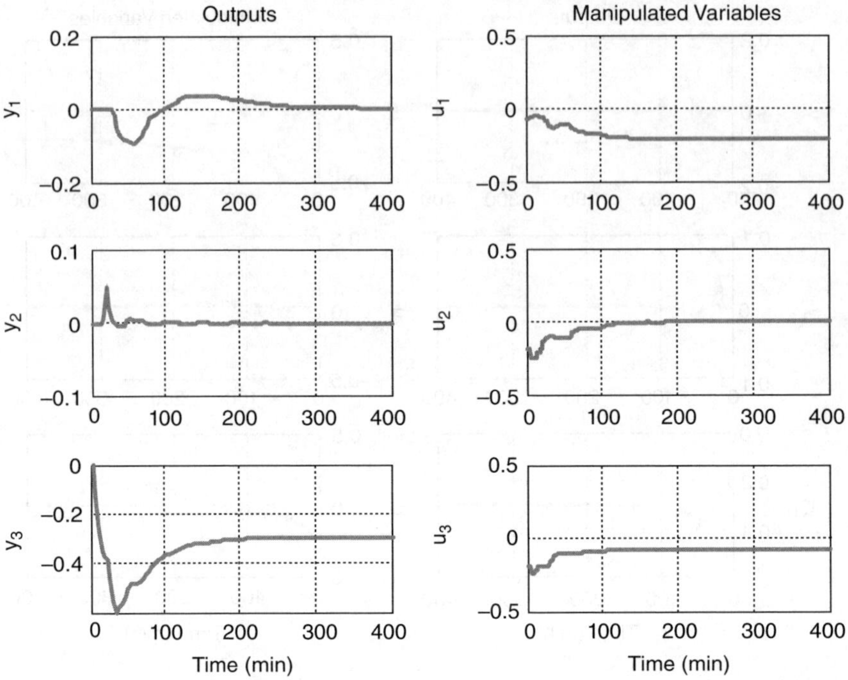

Figure 5.8 Closed-loop system output responses and manipulated/control signals with mixed communication failure under the disturbance pattern $\mathbf{d}^1 = [0.5\ 0.5]^T$

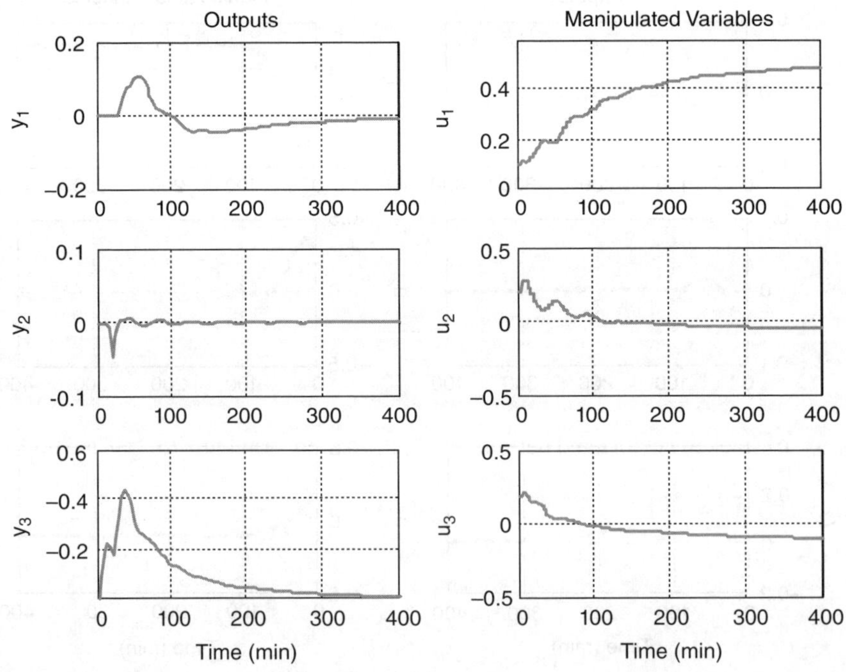

Figure 5.9 Closed-loop system output responses and manipulated/control signals with mixed communication failure under the disturbance pattern $\mathbf{d}^2 = [-0.5 - 0.5]^T$

each agent such as prediction horizon, control horizon, weighting matrices, and sample time, etc. can all be designed and tuned separately, which is superior to the centralized control and can significantly reduce the online computational burden and be simple to implement. Notice that the proposed scheme is not limited to the Shell benchmark control problem and can be used in a wide range of real-world complex control problem.

5.4 Conclusion

In this chapter, the DMPC methods based on local cost optimization and Nash optimality are developed for large-scale linear systems. To avoid the prohibitively high online computational demand, the MPC is implemented in a distributed scheme with the inexpensive controllers within the network environment. These controllers can cooperate and communicate with each other to achieve the objective of the whole system. Coupling effects among the agents are fully taken into account in this scheme, which is superior to other traditional decentralized control methods.

The main advantage of this scheme is that the online optimization of a large-scale system can be converted to that of several small-scale systems, thus can significantly reduce the computational complexity while keeping satisfactory performance. Furthermore, the design parameters for each subsystem-based controller such as the prediction horizon, control horizon, weighting matrix, sample time, etc. can all be designed and tuned separately, which provides more flexibility for the analysis and applications. And these methods maintain the control system integrity under component or system failure and the reduction on the computational load.

The first part of the chapter presents the LCO-DMPC, including the closed-loop solution and the stability conditions. This method provides acceptable regions of tuning parameters for which the stability is guaranteed and the performances are satisfactory. Usually, the stable regions are associated with big prediction horizon P and small weight $\mathbf{R}$. The second part of this chapter provides the Nash-optimization-based DMPC and investigates the performance of the distributed control scheme. The nominal stability and the performance deviation on the single-step horizon under the communication failure are analyzed. These will provide users a better understanding to the developed algorithm and sensible guidance in applications. In addition, some simulation examples are presented to verify the efficiency and practicality of the distributed MPC algorithms.

Appendix

Appendix A. QP problem transformation

Making use of stacked vectors and definitions (5.28), the cost function (5.7) in the optimization problem of the MPC controller C_i can be expressed in the equivalent form

$$J_i = \|\hat{\mathbf{Y}}_i(k+1,P|k) - \mathbf{Y}_i^{\mathrm{d}}(k+1,P|k)\|^2_{\overline{\mathbf{Q}}_i} + \|\Delta\mathbf{U}_i(k,M|k)\|^2_{\overline{\mathbf{R}}_i}$$

The concatenated local output prediction $\hat{\mathbf{Y}}_i(k+1,P|k)$ is a function of the manipulated variables; therefore, in order to express J_i as a function of $\Delta\,\mathbf{U}_i(k,M|k)$, an explicit expression for such a prediction is needed.

Considering the relationship between the manipulated variable and its increment

$$\mathbf{u}_i(k+h|k) = \mathbf{u}_i(k-1) + \sum_{r=0}^{h} \Delta\mathbf{u}_i(k+r|k)$$

using the local stacked control sequence $\mathbf{u}_i(k, M|k)$ together with (5.15) and (5.25)–(5.27), we can obtain the output prediction in the form

$$\hat{\mathbf{Y}}_i(k+1,p|k) = \mathbf{N}_i\Delta\mathbf{U}_i(k,m|k) + \hat{\mathbf{Z}}_i(k+1,p|k)$$

By substituting this expression into the local cost function J_i, the optimization Problem 5.1 takes the form (5.24). The positive definiteness of matrices $\mathbf{Q}_i$ and $\mathbf{R}_i$ implies the same property for matrix $\mathbf{H}_i$.

Appendix B. Proof of Theorem 5.1

Problem 5.2 states that a solution to the LCO-DMPC problem minimizes the cost function (5.24) with respect to the control sequence $\Delta\mathbf{u}_i(k, M|k)$. By the minimum principle, this solution has the form

$$\Delta\mathbf{u}_i(k,M|k) = \frac{1}{2}\mathbf{H}_i^{-1}\mathbf{G}_i(k+1,P|k)$$

which is, by (5.26), equivalent to

$$\Delta\mathbf{U}_i(k,m|k) = \overline{\mathbf{K}}_i[\mathbf{Y}_i^{\mathrm{d}}(k+1,p|k) - \hat{\mathbf{Z}}_i(k+1,p|k)]$$

In MPC algorithms, only the first element of the optimal sequence is actually applied to the process; thus, the control law at the time instant k can be expressed as

$$\mathbf{u}_i(k) = \mathbf{u}_i(k-1) + \mathbf{\Gamma}_i\Delta\mathbf{U}_i(k,m|k)$$

which gives the final explicit form (5.30).

In addition, considering that the local control action applied at time $k-1$ is given by

$$\mathbf{u}_i(k-1) = \mathbf{\Gamma}_i\mathbf{U}_i(k-1,M|k-1)$$

the optimal control sequence $\mathbf{U}_i(k, M|k)$ for subsystem $\mathcal{S}_i$ at time k is

$$\mathbf{U}_i(k,m|k) = \mathbf{\Gamma}_1'\mathbf{\Gamma}_i\mathbf{U}_i(k-1,m|k-1) + \overline{\mathbf{\Gamma}}_i\Delta\mathbf{U}_i(k,m|k)$$

By (5.30), it directly follows (5.32). Then, by (5.27), (5.32), (5.16), and (5.34)–(5.37), the complete stacked optimal manipulated variable sequence has the form (5.33).

Appendix C. Proof of Theorem 5.2

Combining the process equation (5.3), the feedback control law (5.39), the global prediction equation (5.23), and the controller equation (5.33), the closed-loop state-space representation of the entire closed-loop system under the control of unconstrained LCO-DMPC is derived as

$$\begin{aligned}
\mathbf{x}(k) &= \mathbf{A}\mathbf{x}(k-1) + \mathbf{B}\boldsymbol{\Gamma}\mathbf{U}(k-1, P|k-1) \\
\hat{\mathbf{X}}(k, P|k-1) &= \overline{\mathbf{S}}[\overline{\mathbf{A}}\mathbf{x}(k-1) + \tilde{\mathbf{A}}\hat{\mathbf{X}}(k-1, P|k-2) \\
&\quad + \overline{\mathbf{B}}\mathbf{U}(k-1, M|k-1) + \tilde{\mathbf{B}}\mathbf{U}(k-2, M|k-2)] \\
\mathbf{U}(k, M|k) &= \boldsymbol{\Theta}\mathbf{x}(k) + \boldsymbol{\Phi}\hat{\mathbf{X}}(k, P|k-1) \\
&\quad + \boldsymbol{\Psi}\mathbf{U}(k-1, M|k-1) + \boldsymbol{\Xi}\mathbf{Y}^d(k+1, P|k) \\
\mathbf{y}(k) &= \mathbf{C}\mathbf{x}(k)
\end{aligned}$$

where $\tilde{\mathbf{x}}(k|k)$ has been substituted with $\mathbf{x}(k)$ due to the assumption of fully accessible state. Defining the extended state

$$\mathbf{X}_N(k) \triangleq [\mathbf{x}^{\mathrm{T}}(k) \;\; \hat{\mathbf{X}}^T(k, P|k-1) \;\; \mathbf{U}^{\mathrm{T}}(k, M|k) \;\; \mathbf{U}^{\mathrm{T}}(k-1, M|k-1)]$$

the closed-loop state-space representation has the form

$$\begin{aligned}
\mathbf{X}_N(k) &= \mathbf{A}_N\mathbf{X}_N(k-1) + \mathbf{B}_N\mathbf{Y}^d(k+1, P|k), \\
\mathbf{y}(k) &= \mathbf{C}_N\mathbf{X}_N(k)
\end{aligned}$$

where the closed-loop dynamic matrix

$$\mathbf{A}_N = \begin{bmatrix}
\mathbf{A} & \mathbf{0} & \mathbf{B}\boldsymbol{\Gamma} & 0 \\
\overline{\mathbf{S}}\,\overline{\mathbf{A}} & \overline{\mathbf{S}}\tilde{\mathbf{A}} & \overline{\mathbf{S}}\,\overline{\mathbf{B}} & \overline{\mathbf{S}}\tilde{\mathbf{B}} \\
\left(\boldsymbol{\Theta}\mathbf{A} + \boldsymbol{\Phi}\overline{\mathbf{S}}\,\overline{\mathbf{A}}\right) & \boldsymbol{\Phi}\overline{\mathbf{S}}\tilde{\mathbf{A}} & (\boldsymbol{\Psi} + \boldsymbol{\Theta}\mathbf{B}\boldsymbol{\Gamma} + \boldsymbol{\Phi}\overline{\mathbf{S}}\,\overline{\mathbf{B}}) & \boldsymbol{\Phi}\overline{\mathbf{S}}\tilde{\mathbf{B}} \\
\mathbf{0} & \mathbf{0} & \mathbf{I}_{Mn_u} & 0
\end{bmatrix}$$

is equal to the matrix in Equation (5.40). This proves the theorem.

6

Cooperative Distributed Predictive Control

6.1 Introduction

As has been introduced in Chapter 5, the optimization performance of the closed-loop system under the control of distributed predictive control is usually not as good as that under the control of centralized predictive control, especially when the strong coupling exists among subsystems. As mentioned previously, the iterative algorithm is employed in solving each subsystem-based predictive control, in which, each subsystem-based predictive control communicates several times with its neighbors and solves the *Quadratic Programming* problem several times in each control period. Essentially, it improves the global performance through minimizing the computational error, which refers to the difference between the input sequence calculated at previous iterative and the input sequence calculated in current computation. However, the research direction of the whole optimization problem is not the gradient of the entire cost function, and the optimal solution calculated by this method is *Nash optimality* but not the global optimality.

Is there any other strategy to improve the global performance of the closed-loop system under the control of distributed predictive control? The authors of [89, 98] proposed a strategy where each subsystem-based predictive control optimizes not only the cost function of the subsystem it corresponded, but also that of the whole system to improve the performance of the entire closed-loop system. The advantage of improving the optimization performance of the entire closed-loop system has been proved by the authors of [37, 48, 53], and some applications are also presented to validate this strategy [43, 48, 99]. To introduce the concept more clearly, the unconstrained DPC [44, 53], both iterative and noniterative algorithms, based on this coordination strategy are presented. In this strategy, each subsystem-based MPC requires to be able to access to the required information of all subsystems for calculating its optimal solution.

In this chapter, the first part provides the noniterative cooperative DPC formulation, the closed-loop solution, the stability condition, and the analysis of why this coordination strategy could improve the global performance. The second part presents the iterative cooperative DPC

Distributed Model Predictive Control for Plant-Wide Systems, First Edition. Shaoyuan Li and Yi Zheng.

by which the Pareto optimality is accomplished. In addition, some simulations are provided to illustrate the efficiency of the proposed methods.

6.2 Noniterative Cooperative DMPC

6.2.1 System Description

Without loss of generality, suppose that the whole system $\mathcal{S}$ is composed of m linear, discrete-time subsystems $\mathcal{S}_i$, $i = 1, \ldots, m$, and each subsystem interacts with others by both inputs and states. Then, the state-space representation of $\mathcal{S}_i$ can be expressed as

$$\begin{cases} \mathbf{x}_i(k+1) = \mathbf{A}_{ii}\mathbf{x}_i(k) + \mathbf{B}_{ii}\mathbf{u}_i(k) + \sum\limits_{\substack{j=1,\ldots,m;\\ j\neq i}} \mathbf{A}_{ij}\mathbf{x}_j(k) + \sum\limits_{\substack{j=1,\ldots,m;\\ j\neq i}} \mathbf{B}_{ij}\mathbf{u}_j(k) \\ \mathbf{y}_i(k) = \mathbf{C}_{ii}\mathbf{x}_i(k) + \sum\limits_{\substack{j=1,\ldots,m,\\ j\neq i}} \mathbf{C}_{ij}\mathbf{x}_j(k) \end{cases} \tag{6.1}$$

where $\mathbf{x}_i \in \mathbb{R}^{n_{x_i}}$, $\mathbf{u}_i \in \mathbb{R}^{n_{u_i}}$, and $\mathbf{y}_i \in \mathbb{R}^{n_{y_i}}$ are the local state, input, and output vectors, respectively. The model of $\mathcal{S}$ can be expressed as

$$\mathbf{u}_i(k+h|k) = \mathbf{u}_i(k-1) + \sum_{r=0}^{h} \Delta\mathbf{u}_i(k+r|k) \tag{6.2}$$

where $\mathbf{x} \in \mathbb{R}^{n_x}$, $\mathbf{u} \in \mathbb{R}^{n_u}$, and $\mathbf{y} \in \mathbb{R}^{n_y}$ are state, input, and output vectors of S, respectively. $\mathbf{A}$,$\mathbf{B}$, and $\mathbf{C}$ are system matrices.

The control objective minimizes the following global performance index:

$$J(k) = \sum_{i=1}^{m}\left(\sum_{l=1}^{P}\left\|\mathbf{y}_i(k+l) - \mathbf{y}_i^d(k+l)\right\|_{\mathbf{Q}_i}^2 + \sum_{l=1}^{M}\|\Delta\mathbf{u}_i(k+l-1)\|_{\mathbf{R}_i}^2\right) \tag{6.3}$$

where $\mathbf{y}_i^d$ and $\Delta\mathbf{u}_i(k)$ are output set-point and input increment of $\mathcal{S}_j$, and $\Delta\mathbf{u}_i(k) = \mathbf{u}_i(k) - \mathbf{u}_i(k-1)$. $\mathbf{Q}_i$ and $\mathbf{R}_i$ are weight matrices, and P and N, $P, M \in \mathbb{N}$, $P \geq M$, are predictive horizon and control horizon, respectively.

The problem is to design a coordination strategy for improving the global performance of the closed-loop system in the distributed framework.

6.2.2 Formulation

The proposed control architecture is based on a set of independent MPC controllers C_i for each $\mathcal{S}_i$, $i = 1, 2, \ldots, m$. These MPCs could exchange information with its neighbors through a network. To clearly discuss the proposed control methodology, the simplifying Assumption 6.1 and the notations defined in Table 6.1 are adopted in this chapter.

Assumption 6.1

- controllers are synchronous, since the sampling interval is usually rather long compared with the computational time in process control;

Table 6.1 Notations definition

Notations	Explanations
$\text{diag}_a\{\mathbf{A}\}$	Diagonal block matrix made by a blocks equal to A
$\lambda_j\{\mathbf{A}\}$	The jth eigenvalue of a square matrix A
$\mathcal{O}(a)$	To be proportional to a
$\mathbf{0}_{a\times b}$	An $a\times b$ null matrix
$\mathbf{0}_a$	An $a\times a$ null matrix
$\mathbf{I}_a$	An $a\times a$ identity matrix
$\hat{\mathbf{x}}_i(l\|h)$	Predictions of $\mathbf{x}_i(l)$ computed by $\mathcal{C}_i$ at time h
$\hat{\mathbf{y}}_i(l\|h)$	Predictions of $\mathbf{y}_i(l)$ computed by $\mathcal{C}_i$ at time h
$\mathbf{u}_i(l\|h)$	Input $\mathbf{u}_i(l)$ computed by $\mathcal{C}_i$ at time h
$\Delta\mathbf{u}_i(l\|h)$	Input increment $\Delta\mathbf{u}_i(l)$ computed by $\mathcal{C}_i$ at time h
$\mathbf{y}_i^d(l\|h)$	Set-point of $\mathbf{y}_i(l\|h)$
$\mathbf{y}^d(l\|h)$	Set-point of $\mathbf{y}(l\|h)$
$\hat{\mathbf{x}}^i(l\|h)$	Predictions of $\mathbf{x}(l)$ computed by $\mathcal{C}_i$ at time h
$\hat{\mathbf{y}}^i(l\|h)$	Predictions of $\mathbf{y}(l)$ computed by $\mathcal{C}_i$ at time h
$\mathbf{U}_i(l,p\|h)$	A complete input vector $\mathbf{U}_i(l,p\|h)=[\mathbf{u}_i^{\mathrm{T}}(l\|h)\quad \mathbf{u}_i^{\mathrm{T}}(l+1\|h)\quad \cdots\quad \mathbf{u}_i^{\mathrm{T}}(l+p\|h)]^{\mathrm{T}}$
$\Delta\mathbf{U}_i(l,p\|h)$	Input increment sequence vector, $\Delta\mathbf{U}_i(l,p\|h)=\left[\Delta\mathbf{u}_i^{\mathrm{T}}(l\|h)\quad \Delta\mathbf{u}_i^{\mathrm{T}}(l+1\|h)\quad \cdots\quad \Delta\mathbf{u}_i^{\mathrm{T}}(l+p\|h)\right]^{\mathrm{T}}$
$\mathbf{U}(l,p\|h)$	Complete stacked input vector, $\mathbf{U}(l,p\|h)=\left[\mathbf{u}^{\mathrm{T}}(l\|h)\quad \mathbf{u}^{\mathrm{T}}(l+p\|h)\quad \cdots\quad \mathbf{u}^{\mathrm{T}}(l+p\|h)\right]^{\mathrm{T}}$
$\hat{\mathbf{X}}^i(l,p\|h)$	Sacked distributed state vector, $\hat{\mathbf{X}}^i(l,p\|h)=\left[\hat{\mathbf{x}}^{i\mathrm{T}}(l\|h)\quad \hat{\mathbf{x}}^{i\mathrm{T}}(l+1\|h)\quad \cdots\quad \hat{\mathbf{x}}^{i\mathrm{T}}(l+p\|h)\right]^{\mathrm{T}}$
$\hat{\mathbf{X}}_i(l,p\|h)$	Stacked distributed state vector, $\hat{\mathbf{X}}_i(l,p\|h)=\left[\hat{\mathbf{x}}_i^{\mathrm{T}}(l\|h)\quad \hat{\mathbf{x}}_i^{\mathrm{T}}(l+1\|h)\quad \cdots\quad \hat{\mathbf{x}}_i^{\mathrm{T}}(l+p\|h)\right]^{\mathrm{T}}$
$\hat{\mathbf{X}}(l,p\|h)$	Complete stacked state vector, $\hat{\mathbf{X}}(l,p\|h)=\left[\hat{\mathbf{x}}^{\mathrm{T}}(l\|h)\quad \hat{\mathbf{x}}^{\mathrm{T}}(l+1\|h)\quad \cdots\quad \hat{\mathbf{x}}^{\mathrm{T}}(l+p\|h)\right]^{\mathrm{T}}$
$\hat{\mathbf{Y}}^i(l,p\|h)$	Stacked distributed state vector, $\hat{\mathbf{Y}}^i(l,p\|h)=\left[\hat{\mathbf{y}}^{i\mathrm{T}}(l\|h)\quad \hat{\mathbf{y}}^{i\mathrm{T}}(l+1\|h)\quad \cdots\quad \hat{\mathbf{y}}^{i\mathrm{T}}(l+p\|h)\right]^{\mathrm{T}}$
$\mathbf{Y}(l,p\|h)$	Complete stacked state vector, $\mathbf{Y}(l,p\|h)=\left[\mathbf{y}^{\mathrm{T}}(l\|h)\quad \mathbf{y}^{\mathrm{T}}(l+1\|h)\quad \cdots\quad \mathbf{y}^{\mathrm{T}}(l+p\|h)\right]^{\mathrm{T}}$
$\mathbf{Y}^d(l,p\|h)$	Set-point of $\mathbf{Y}(l,p\|h)$;
$\hat{\mathbb{X}}(l,p\|h)$	Complete stacked state vector, $\hat{\mathbb{X}}(l,p\|h)=\left[\hat{\mathbf{X}}_1^{\mathrm{T}}(l,p\|h)\quad \cdots\quad \hat{\mathbf{X}}_m^{\mathrm{T}}(l,p\|h)\right]^{\mathrm{T}}$
$\mathbb{Y}^d(l,p\|h)$	$\mathbb{Y}^d(l,p\|h)=diag_m(\mathbf{Y}^d)$;
$\mathbb{U}(l,p\|h)$	Complete stacked state vector, $\mathbb{U}(l,p\|h)=\left[\mathbf{U}_1^{\mathrm{T}}(l,p\|h)\quad \cdots\quad \mathbf{U}_m^{\mathrm{T}}(l,p\|h)\right]^{\mathrm{T}}$

1. a and b are constants; p, l, h are positive integers, and $h < l$; A is a matrix.

- communication channel introduces a delay of a single sampling time interval, since an instantaneous data transfer is not possible in real situations;
- controllers communicate only once within a sampling time interval;
- local states $\mathbf{x}_i(k), i = 1, 2, \ldots, m$, are accessible.

6.2.2.1 Performance Index

Since the optimal control decision of $\mathcal{S}_i$ affects, or even destroys, the performance of other subsystems, the performance of other subsystems should be considered in finding the optimal of the solution of $\mathcal{S}_i$. To improve the global performance of the whole closed-loop system, the following so-called global performance index is adopted in each $\mathcal{C}_i$, $i = 1, \ldots, m$:

$$\overline{J}_i(k) = \sum_{l=1}^{P} \|\hat{\mathbf{y}}^i(k+l|k) - \mathbf{y}^d(k+l|k)\|_{\mathbf{Q}}^2 + \sum_{l=1}^{M} \|\Delta\mathbf{u}_i(k+l-1|k)\|_{\mathbf{R}_i}^2 \tag{6.4}$$

where $\mathbf{Q} = \mathrm{diag}\{\mathbf{Q}_1, \mathbf{Q}_2, \ldots, \mathbf{Q}_m\}$. It should be noticed that $\Delta\mathbf{u}_j(k+l-1|k)$ is excluded in the performance index, since it is independent of the future inputs sequence of $\mathcal{S}_j$.

6.2.2.2 Prediction Model

Since the state evolution of other subsystems is affected by $\mathbf{u}_i(k)$ after one or several control periods, to improve the prediction precision, this influence is considered in $\mathcal{C}_i$ when predicting the future states of all subsystems. In addition, due to the unit delay introduced by the network, the information of other subsystems is available only after one sampling time interval. Therefore, in $\mathcal{C}_i$, the states and outputs of all subsystems in l-step ahead are predicted by

$$\begin{cases} \hat{\mathbf{x}}^i(k+l+1|k) = \mathbf{A}^l\mathbf{L}_i\mathbf{x}(k) + \mathbf{A}^l\mathbf{L}_i'\hat{\mathbf{x}}(k|k-1) \\ \qquad + \sum_{s=1}^{l} \mathbf{A}^{s-1}\mathbf{B}_i\mathbf{u}_i(k+l|k) + \sum_{\substack{j\in\{1,\ldots,m\}\\ j\neq i}} \sum_{s=1}^{l} \mathbf{A}^{s-1}\mathbf{B}_j\mathbf{u}_j(k+l|k-1) \\ \hat{\mathbf{y}}^i(k+l+1|k) = \mathbf{C}\hat{\mathbf{x}}^i(k+l+1|k) \end{cases} \tag{6.5}$$

where

$$\mathbf{L}_i = \begin{bmatrix} \mathbf{0}_{n_{xi}\times\sum_{j=1}^{i-1} n_{xj}} & \mathbf{I}_{n_{xi}} & \mathbf{0}_{n_{xi}\times\sum_{j=i+1}^{m} n_{xj}} \end{bmatrix}$$

$$\mathbf{L}_i' = \mathrm{diag}\left\{ \mathbf{I}_{\sum_{j=1}^{i-1} n_{xj}}, \mathbf{0}_{n_{xi}}, \mathbf{I}_{\sum_{j=i+1}^{m} n_{xj}} \right\}$$

$$\mathbf{B}_i = \begin{bmatrix} \mathbf{B}_{1i}^{\mathrm{T}} & \mathbf{B}_{2i}^{\mathrm{T}} & \cdots & \mathbf{B}_{mi}^{\mathrm{T}} \end{bmatrix}^{\mathrm{T}}$$

Remark 6.1 *It should be noticed that the input of this neighborhood model is still the input of $\mathcal{S}_i$, and the inputs and states of other subsystem are regarded as disturbances. The estimations of future states and outputs of all subsystems (except $\mathcal{S}_i$) are only used in the controller $\mathcal{C}_i$, and these estimations are different from that estimated by the controller $\mathcal{C}_j$ itself.*

6.2.2.3 Optimization Problem

Problem 6.1 For each independent controller $\mathcal{C}_i$, $i=1, \ldots, m$, the unconstrained C-DMPC problem with the prediction horizon P and control horizon M, $M<P$, at time k is to minimize the performance index (6.4) with the system equation constraint (6.5), that is,

$$\min_{\Delta\mathbf{U}_i(k,M|k)} \sum_{l=1}^{P} \|\hat{\mathbf{y}}^i(k+l|k) - \mathbf{y}^d(k+l|k)\|_{\mathbf{Q}}^2 + \sum_{l=1}^{M} \|\Delta\mathbf{u}_i(k+l-1|k)\|_{\mathbf{R}_i}^2$$

$$st. \quad Eq.(7) \tag{6.6}$$

At time k, based on the exchanged information $\hat{\mathbf{x}}_j(k|k-1)$, $\mathbf{U}_j(k+l|k-1)$, together with $\mathbf{x}(k)$, the optimization problem (6.6) is solved in each independent $\mathcal{C}_i$. The first element of the optimal solution is selected and $\mathbf{u}_i(k)=\mathbf{u}_i(k-1)+\Delta\mathbf{u}_i(k|k)$ is applied to $\mathcal{S}_j$. Then, by Equation (6.5), each local controller estimates the future state at time $k+1$ and broadcasts it in the network together with the optimal control sequence over the control horizon. At time $k+1$, each local controller uses this information to repeat the whole procedure.

6.2.3 *Closed-Form Solution*

The main result of this subsection is the computation of a closed-form solution to the C-DMPC problem. For this purpose, the C-DMPC Problem 6.1 is first transformed into a quadratic program (QP) problem which has to be locally solved online at each sampling instant.

Define

$$\tilde{\mathbf{T}}_i = \text{diag}\left\{\mathbf{I}_{\sum_{j=1}^{i-1} n_{uj}}, \mathbf{0}_{n_{ui}}, \mathbf{I}_{\sum_{j=i+1}^{M} n_{uj}}\right\} \tag{6.7}$$

$$\tilde{\mathbf{B}}_i = \left[\begin{array}{c:c} \mathbf{0}_{(M-1)n_x \times n_u} & \text{diag}_{M-1}\{\mathbf{B}\tilde{\mathbf{T}}_i\} \\ \hdashline \mathbf{0}_{n_x \times (M-1)n_u} & \mathbf{B}\tilde{\mathbf{T}}_i \\ \vdots & \vdots \\ \mathbf{0}_{n_x \times (M-1)n_u} & \mathbf{B}\tilde{\mathbf{T}}_i \end{array}\right] \tag{6.8}$$

$$\overline{\mathbf{S}} = \begin{bmatrix} \mathbf{A}^0 & \mathbf{0} & \cdots & \mathbf{0} \\ \mathbf{A}^1 & \mathbf{A}^0 & \ddots & \vdots \\ \vdots & \ddots & \ddots & \mathbf{0} \\ \mathbf{A}^{P-1} & \cdots & \mathbf{A}^1 & \mathbf{A}^0 \end{bmatrix}, \overline{\mathbf{A}}_a = \begin{bmatrix} \mathbf{A} \\ \mathbf{0} \\ \vdots \\ \mathbf{0} \end{bmatrix} \tag{6.9}$$

$$\overline{\mathbf{C}}_a = \text{diag}_P\{\mathbf{C}\} \tag{6.10}$$

$$\overline{\mathbf{B}}_i = \left[\begin{array}{c:c} \mathbf{0}_{(M-1)n_x \times n_{ui}} & \text{diag}_{M-1}\{\mathbf{B}_i\} \\ \hdashline \mathbf{0}_{n_x \times (M-1)n_{ui}} & \mathbf{B}_i \\ \vdots & \vdots \\ \mathbf{0}_{n_x \times (M-1)n_{ui}} & \mathbf{B}_i \end{array}\right]$$

$$\overline{\boldsymbol{\Gamma}}_i = \begin{bmatrix} \mathbf{I}_{n_{ui}} & \mathbf{0}_{n_{ui}} & \cdots & \mathbf{0}_{n_{ui}} \\ \mathbf{I}_{n_{ui}} & \mathbf{I}_{n_{ui}} & \ddots & \vdots \\ \vdots & \ddots & \ddots & \mathbf{0}_{n_{ui}} \\ \mathbf{I}_{n_{ui}} & \cdots & \mathbf{I}_{n_{ui}} & \mathbf{I}_{n_{ui}} \end{bmatrix}, \boldsymbol{\Gamma}'_i = \begin{bmatrix} \overbrace{\mathbf{I}_{n_{ui}} \quad \cdots \quad \mathbf{I}_{n_{ui}}}^{M} \end{bmatrix}^{\mathrm{T}} \tag{6.11}$$

$$\mathbf{N}_i = \overline{\mathbf{C}}_a \overline{\mathbf{S}}\, \overline{\mathbf{B}}_i \overline{\boldsymbol{\Gamma}}_i, \overline{\mathbf{Q}} = \mathrm{diag}_P\{\mathbf{Q}\}, \overline{\mathbf{R}}_i = \mathrm{diag}_M\{\mathbf{R}_i\} \tag{6.12}$$

The following Lemma can be obtained based on Equations (6.5) and (6.7)–(6.12).

Lemma 6.1 (Quadratic program) *Under Assumption 6.1, each independent controller $\mathcal{C}_i$, $i = 1, \ldots, m$, has to solve at time k the following optimization problem:*

$$\min_{\Delta\mathbf{U}_i(k,M|k)} [\Delta\mathbf{U}_i^{\mathrm{T}}(k, M|k)\mathbf{H}_i\Delta\mathbf{U}_i(k, M|k) - \mathbf{G}_i(k+1, P|k)\Delta\mathbf{U}_i(k, M|k)] \tag{6.13}$$

where the positive definite matrix $\boldsymbol{H}_i$ has the form

$$\mathbf{H}_i = \mathbf{N}_i^{\mathrm{T}}\overline{\mathbf{Q}}\mathbf{N}_i + \overline{\mathbf{R}}_i \tag{6.14}$$

and

$$\mathbf{G}_i(k+1, P|k) = 2\mathbf{N}_i^{\mathrm{T}}\overline{\mathbf{Q}}[\mathbf{Y}^d(k+1, P|k) - \hat{\mathbf{Z}}_i(k+1, P|k)] \tag{6.15}$$

with

$$\begin{aligned} \hat{\mathbf{Z}}_i(k+1, P|k) = \overline{\mathbf{C}}_a\overline{\mathbf{S}}[&\overline{\mathbf{B}}_i\boldsymbol{\Gamma}'_i\mathbf{u}_i(k-1) + \overline{\mathbf{A}}_a\mathbf{L}_i x_i(k|k) \\ &+ \overline{\mathbf{A}}_a\mathbf{L}'_i\hat{\mathbf{x}}(k|k-1) + \tilde{\mathbf{B}}_i\mathbf{U}(k-1, M|k-1)] \end{aligned} \tag{6.16}$$

Proof. According to Equations (6.5) and (6.7)–(6.12), the stacked predictions of states and outputs of S calculated by subsystem $\mathcal{S}_i$ at time k is

$$\begin{cases} \hat{\mathbf{X}}^i(k+1, P|k) = \overline{\mathbf{S}}[\overline{\mathbf{A}}_a\mathbf{L}_i\mathbf{x}_i(k) + \overline{\mathbf{B}}_i\mathbf{U}_i(k, M|k) \\ \qquad\qquad + \overline{\mathbf{A}}_a\mathbf{L}'_i\hat{\mathbf{x}}(k|k-1) + \tilde{\mathbf{B}}_i\mathbf{U}(k-1, M|k-1)] \\ \hat{\mathbf{Y}}^i(k+1, P|k) = \overline{\mathbf{C}}_a\hat{\mathbf{X}}^i(k+1, P|k) \end{cases} \tag{6.17}$$

where the last $P - M + 1$ samples of $\hat{\mathbf{U}}(k-1, P|k-1)$ and $\mathbf{U}_i(k, P|k)$ are assumed to be equal to the last element of $\mathbf{U}(k-1, M|k-1)$ and $\mathbf{U}_i(k, M|k)$, respectively.

By

$$\mathbf{u}_i(k+h|k) = \mathbf{u}_i(k-1) + \sum_{r=0}^{h} \Delta\mathbf{u}_i(k+r|k)$$

and (6.11), it has

$$\mathbf{U}_i(k, M|k) = \boldsymbol{\Gamma}'_i\mathbf{u}_i(k-1) + \overline{\boldsymbol{\Gamma}}_i\Delta\mathbf{U}_i(k, M|k) \tag{6.18}$$

Then the QP problem (6.13) can be deduced by substituting (6.7)–(6.12), (6.17) into (6.6). This concludes the proof.

According to Equation (6.13), the solution to problem (6.6) can be deduced as

$$\Delta\mathbf{U}_i(k,M|k) = 1/2\cdot\mathbf{H}_i^{-1}\mathbf{G}_i(k+1,P|k)$$

By noting that only the first element of the optimal sequence is actually applied to the process, Theorem 6.1 is obtained.

Theorem 6.1 (Closed-form solution) *Under Assumption 6.1, for each controller C_i, $i=1,\ \ldots,m$, the closed-form control law applied at time k to subsystem S_i is given by*

$$\mathbf{u}_i(k) = \mathbf{u}_i(k-1) + \mathbf{K}_i[\mathbf{Y}^d(k+1,P|k) - \hat{\mathbf{Z}}_i(k+1,P|k)] \tag{6.19}$$

where

$$\begin{aligned}
\mathbf{K}_i &= \boldsymbol{\Gamma}_i\overline{\mathbf{K}}_i\\
\overline{\mathbf{K}}_i &= \mathbf{H}_i^{-1}\mathbf{N}_i^{\mathrm{T}}\overline{\mathbf{Q}}\\
\boldsymbol{\Gamma}_i &= \begin{bmatrix}\mathbf{I}_{n_{u_i}} & \mathbf{0}_{n_{u_i}\times(M-1)n_{u_i}}\end{bmatrix}
\end{aligned} \tag{6.20}$$

Remark 6.2 *In C_i, the complexity to obtain the closed-form solution is mainly incurred by the inversion of $\boldsymbol{H}_{\mathrm{i}}$. By using the Gauss–Jordan algorithm for this task and considering that the size of $\boldsymbol{H}_{\mathrm{i}}$ equals $M\cdot n_{u_i}$, the complexity of the inversion algorithm is $O(M^3\cdot n_{u_i}^3)$. Therefore, the total computational complexity of solving* ***C-DMPC*** *is only $O\left(M^3\cdot\sum_{i=1}^{n} n_{u_i}^3\right)$, while the computational complexity of the centralized control strategy equals $O\left(M^3\cdot\left(\sum_{i=1}^{n} n_{u_i}\right)^3\right)$.*

6.2.4 Stability and Performance Analysis

6.2.4.1 Stability Analysis

On the basis of the closed-form solution stated by Theorem 6.4, the closed-loop dynamics can be specified and the stability condition can be verified by analyzing the closed-loop dynamic matrix. Define

$$\begin{aligned}
\boldsymbol{\Omega} &= \begin{bmatrix}\boldsymbol{\Omega}_1^{\mathrm{T}} & \cdots & \boldsymbol{\Omega}_P^{\mathrm{T}}\end{bmatrix}^{\mathrm{T}}\\
\boldsymbol{\Omega}_l &= \mathrm{diag}\{\boldsymbol{\Omega}_{1l},\ldots,\boldsymbol{\Omega}_{ml}\}\\
\boldsymbol{\Omega}_{il} &= \begin{bmatrix}\mathbf{0}_{n_{x_i}\times(l-1)n_{x_i}} & \mathbf{I}_{n_{x_i}} & \mathbf{0}_{n_{x_i}\times(P-l)n_{x_i}}\end{bmatrix}\\
&(i=1,\ldots,m,\ l=1,\ldots,P);
\end{aligned} \tag{6.21}$$

$$\begin{aligned}
\boldsymbol{\Pi} &= \begin{bmatrix}\boldsymbol{\Pi}_1^{\mathrm{T}} & \cdots & \boldsymbol{\Pi}_M^{\mathrm{T}}\end{bmatrix}^{\mathrm{T}}\\
\boldsymbol{\Pi}_l &= \mathrm{diag}\{\boldsymbol{\Pi}_{1l},\ldots,\boldsymbol{\Pi}_{ml}\}\\
\boldsymbol{\Pi}_{il} &= \begin{bmatrix}\mathbf{0}_{n_{u_i}\times(l-1)n_{u_i}} & \mathbf{I}_{n_{u_i}} & \mathbf{0}_{n_{u_i}\times(M-l)n_{u_i}}\end{bmatrix}\\
&(i=1,\ldots,m,\ l=1,\ldots,M);
\end{aligned} \tag{6.22}$$

$$\begin{aligned}
\overline{\mathbf{A}} &= \mathrm{diag}_m\{\overline{\mathbf{A}}_a\}\\
\overline{\mathbf{B}} &= \mathrm{diag}\{\overline{\mathbf{B}}_1, \dots, \overline{\mathbf{B}}_m\}\\
\overline{\mathbf{C}} &= \mathrm{diag}_m\{\overline{\mathbf{C}}_a\}\\
\tilde{\mathbf{B}} &= \left[\tilde{\mathbf{B}}_1^{\mathrm{T}} \cdots \tilde{\mathbf{B}}_m^{\mathrm{T}}\right]^{\mathrm{T}},
\end{aligned} \tag{6.23}$$

$$\begin{aligned}
\overline{\mathbf{L}}_i &= \mathrm{diag}_P\{\mathbf{L}_i^{\mathrm{T}}\}\\
\mathbf{L} &= \mathrm{diag}\{\mathbf{L}_1, \dots, \mathbf{L}_m\}\\
\overline{\mathbf{L}} &= \mathrm{diag}\{\overline{\mathbf{L}}_1, \dots, \overline{\mathbf{L}}_m\}\\
\mathbf{L}' &= \left[\mathbf{L}_1^{\prime\mathrm{T}} \cdots \mathbf{L}_m^{\prime\mathrm{T}}\right]^{\mathrm{T}}\\
\tilde{\mathbf{L}} &= \mathbf{L}'\left[\mathbf{I}_{n_x}\ \mathbf{0}_{n_x\times(P-1)n_x}\right]
\end{aligned} \tag{6.24}$$

$$\begin{aligned}
\boldsymbol{\Gamma}' &= \mathrm{diag}\{\boldsymbol{\Gamma}_1', \dots, \boldsymbol{\Gamma}_m'\}\\
\boldsymbol{\Gamma} &= \mathrm{diag}\{\boldsymbol{\Gamma}_1, \dots, \boldsymbol{\Gamma}_m\}\\
\mathbf{S} &= \mathrm{diag}_m\{\overline{\mathbf{S}}\}\\
\boldsymbol{\Xi} &= \mathrm{diag}\{\overline{\boldsymbol{\Gamma}}_1\overline{\mathbf{K}}_1, \dots, \overline{\boldsymbol{\Gamma}}_m\overline{\mathbf{K}}_m\}
\end{aligned} \tag{6.25}$$

$$\begin{aligned}
\boldsymbol{\Theta} &= -\boldsymbol{\Xi}\overline{\mathbf{C}}\mathbf{S}\overline{\mathbf{A}}\mathbf{L}\\
\boldsymbol{\Phi} &= -\boldsymbol{\Xi}\overline{\mathbf{C}}\mathbf{S}\overline{\mathbf{A}}\tilde{\mathbf{L}}\boldsymbol{\Omega}\\
\boldsymbol{\Psi} &= \boldsymbol{\Gamma}'\boldsymbol{\Gamma} - \boldsymbol{\Xi}\overline{\mathbf{C}}\mathbf{S}(\overline{\mathbf{B}}\boldsymbol{\Gamma}'\boldsymbol{\Gamma} + \tilde{\mathbf{B}}\boldsymbol{\Pi})
\end{aligned} \tag{6.26}$$

Then, Theorem 6.2 can be deduced.

Theorem 6.2 ***(C-DMPC stability)*** *The closed-loop system given by the feedback connection of plant $\mathcal{S}$ with the set of independent controller C_i, $i = 1, \dots, m$, whose closed-form control laws are given by Equation (6.19), is asymptotically stable if and only if*

$$|\lambda_j\{\mathbf{A}_N\}| < 1, \forall j = 1, \dots, n_N \tag{6.27}$$

where

$$\mathbf{A}_N = \begin{bmatrix}
\mathbf{A} & \mathbf{0} & \mathbf{B}\boldsymbol{\Gamma} & \mathbf{0}\\
\overline{\mathbf{L}}\mathbf{S}\overline{\mathbf{A}}\mathbf{L} & \overline{\mathbf{L}}\mathbf{S}\overline{\mathbf{A}}\tilde{\mathbf{L}}\boldsymbol{\Omega} & \overline{\mathbf{L}}\mathbf{S}\overline{\mathbf{B}} & \overline{\mathbf{L}}\mathbf{S}\tilde{\mathbf{B}}\boldsymbol{\Pi}\\
\boldsymbol{\Theta}\mathbf{A} + \boldsymbol{\Phi}\overline{\mathbf{L}}\mathbf{S}\overline{\mathbf{A}}\mathbf{L} & \boldsymbol{\Phi}\overline{\mathbf{L}}\mathbf{S}\overline{\mathbf{A}}\tilde{\mathbf{L}}\boldsymbol{\Omega} & \boldsymbol{\Theta}\mathbf{B}\boldsymbol{\Gamma} + \boldsymbol{\Phi}\overline{\mathbf{L}}\mathbf{S}\overline{\mathbf{B}} + \boldsymbol{\Psi} & \boldsymbol{\Phi}\overline{\mathbf{L}}\mathbf{S}\tilde{\mathbf{B}}\boldsymbol{\Pi}\\
\mathbf{0} & \mathbf{0} & \mathbf{I}_{Mn_u} & \mathbf{0}
\end{bmatrix}$$

$n_N = Pn_x + n_x + 2Mn_u$ is the order of the whole closed-loop system.

Proof. By Equations (6.7) and (6.13), the stacked prediction of the states $\mathcal{S}_i$ in C_i at time k is expressed as

$$\begin{aligned}
\hat{\mathbf{X}}_i(k+1, P|k) = \overline{\mathbf{L}}_i^{\mathrm{T}}\overline{\mathbf{S}}[&\overline{\mathbf{A}}_a\mathbf{L}_i\mathbf{x}_i(k) + \overline{\mathbf{B}}_i\mathbf{U}_i(k, M|k)\\
&+ \overline{\mathbf{A}}_i\mathbf{L}_i'\hat{\mathbf{x}}(k|k-1) + \tilde{\mathbf{B}}_i\mathbf{U}(k-1, M|k-1)]
\end{aligned} \tag{6.28}$$

Through Equation (14), it has

$$\hat{\mathbf{X}}(k, P|k-1) = \boldsymbol{\Omega}\hat{\mathbb{X}}(k, P|k-1) \tag{6.29}$$

$$\mathbf{U}(k, M|k-1) = \boldsymbol{\Pi}\mathbb{U}(k, M|k-1) \tag{6.30}$$

By (6.29), (6.30), (6.23)–(6.25), and (6.28), the stacked prediction of the states of all subsystems at time k is expressed as

$$\begin{aligned}\hat{\mathbb{X}}(k+1, P|k) = \overline{\mathbf{L}}\mathbf{S}[\overline{\mathbf{A}}\mathbf{L}\mathbf{x}(k) + \overline{\mathbf{B}}\mathbb{U}(k, M|k) \\ + \overline{\mathbf{A}}\tilde{\mathbf{L}}\boldsymbol{\Omega}\hat{\mathbb{X}}(k, P|k-1) + \tilde{\mathbf{B}}\boldsymbol{\Pi}\mathbb{U}(k-1, M|k-1)]\end{aligned} \tag{6.31}$$

Noting that $\mathbf{u}_i(k-1) = \Gamma_i\mathbf{U}_i(k-1, M|k-1)$, by Equations (6.10) and (6.13e), it has

$$\begin{aligned}\mathbf{U}_i(k, M|k) = \boldsymbol{\Gamma}'_i\boldsymbol{\Gamma}_i\mathbf{U}_i(k-1, M|k-1) \\ + \overline{\boldsymbol{\Gamma}}_i\overline{\mathbf{K}}_i[\mathbf{Y}^d(k+1, P|k) - \hat{\mathbf{Z}}_i(k+1, P|k)]\end{aligned} \tag{6.32}$$

Substituting (6.16) into (6.32), by (6.12), (6.23)–(6.25), (6.29), and (6.30), the complete stacked open-loop optimal sequence becomes

$$\begin{aligned}\mathbb{U}(k, M|k) = \boldsymbol{\Psi}\mathbb{U}(k-1, M|k-1) \\ + \boldsymbol{\Theta}\mathbf{x}(k) + \boldsymbol{\Phi}\hat{\mathbb{X}}(k, P|k-1) + \boldsymbol{\Xi}\mathbf{Y}^d(k+1, P|k)\end{aligned} \tag{6.33}$$

Noting that the complete feedback control law computed by all controllers is

$$\mathbf{u}(k) = \boldsymbol{\Gamma}\mathbb{U}(k, M|k) \tag{6.34}$$

Combining (6.2), (6.31), (6.33), and (6.34), the closed-loop state-space representation for the distributed case is derived as

$$\mathbf{x}(k) = \mathbf{A}\mathbf{x}(k-1) + \mathbf{B}\boldsymbol{\Gamma}\mathbb{U}(k-1, M|k-1) \tag{6.35}$$

$$\begin{aligned}\hat{\mathbb{X}}(k, P|k-1) = \overline{\mathbf{L}}\mathbf{S}[\overline{\mathbf{A}}\mathbf{L}\mathbf{x}(k-1) + \overline{\mathbf{B}}\mathbb{U}(k-1, M|k-1) \\ + \overline{\mathbf{A}}\tilde{\mathbf{L}}\boldsymbol{\Omega}\hat{\mathbb{X}}(k-1, P|k-2) + \tilde{\mathbf{B}}\boldsymbol{\Pi}\mathbb{U}(k-2, M|k-2)]\end{aligned} \tag{6.36}$$

$$\begin{aligned}\mathbb{U}(k, M|k) = \boldsymbol{\Theta}[\mathbf{A}\mathbf{x}(k-1) + \mathbf{B}\boldsymbol{\Gamma}\mathbb{U}(k-1, M|k-1)] \\ + \boldsymbol{\Phi}\overline{\mathbf{L}}\mathbf{S}[\overline{\mathbf{A}}\mathbf{L}\mathbf{x}(k-1) + \overline{\mathbf{B}}\mathbb{U}(k-1, M|k-1) \\ + \overline{\mathbf{A}}\tilde{\mathbf{L}}\boldsymbol{\Omega}\hat{\mathbb{X}}(k-1, P|k-2) + \tilde{\mathbf{B}}\boldsymbol{\Pi}\mathbb{U}(k-2, M|k-2)] \\ + \boldsymbol{\Psi}\mathbb{U}(k-1, M|k-1) + \boldsymbol{\Xi}\mathbb{Y}^d(k+1, P|k)\end{aligned} \tag{6.37}$$

$$\mathbf{y}(k) = \mathbf{C}\mathbf{x}(k) \tag{6.38}$$

where $\hat{\mathbf{x}}(k|k)$ is substituted with $\mathbf{x}(k)$ due to the assumption of the fully accessible state.

Define the extended state as

$\mathbf{X}_N(k) = [\mathbf{x}^{\mathrm{T}}(k), \hat{\mathbf{X}}^{\mathrm{T}}(k, P|k-1), \mathbf{U}^{\mathrm{T}}(k, M|k), \mathbf{U}^{\mathrm{T}}(k-1, M|k-1)]^{\mathrm{T}}$ Then by Equations (6.35)–(6.38), Theorem 6.2 is deduced.

Remark 6.3 *It should be noticed that the first two block rows of the dynamic matrix $\mathbf{A}_{\mathrm{N}}$ included in Equation (6.27) depend on the elements of matrix $\mathbf{A}$ and the elements of matrix $\mathbf{B}$, while the third block row depends on $\mathbf{A}$, $\mathbf{B}$, $\mathbf{C}$, $\mathbf{Q}$, $\mathbf{R}_{\mathrm{i}}$, P, and M. The degree of freedom available to the designer is based on the choices of $\mathbf{Q}$, $\mathbf{R}_{\mathrm{i}}$, P, and M in the C-DMPC design phase, which introduces significant modifications only on the third block row of $\mathbf{A}_{\mathrm{N}}$ used for the stability test.*

6.2.4.2 Performance Analysis

To explain the essential difference between the C-MPC and the centralized MPC, for each C_i, $i = 1, \ldots, m$, the optimization problem of C-DMPC (6.6) is rewritten as

$$
\begin{aligned}
&\min_{\Delta\mathbf{U}_i(k,M|k)} \sum_{l=1}^{P} \|\hat{\mathbf{y}}^i(k+l|k) - \mathbf{y}^d(k+l|k)\|_{\mathbf{Q}}^2 + \sum_{l=1}^{M} \|\Delta\mathbf{u}_i(k+l-1|k)\|_{\mathbf{R}_i}^2 \\
&\text{s.t.} \begin{bmatrix} \hat{\mathbf{x}}_i^i(k+l+1|k) \\ \vdots \\ \hat{\mathbf{x}}_{i-1}^i(k+l+1|k) \\ \hat{\mathbf{x}}_i(k+l+1|k) \\ \hat{\mathbf{x}}_{i+1}^i(k+l+1|k) \\ \vdots \\ \hat{\mathbf{x}}_m^i(k+l+1|k) \end{bmatrix} = \mathbf{A}^l \begin{bmatrix} \hat{\mathbf{x}}_1(k|k-1) \\ \vdots \\ \hat{\mathbf{x}}_{i-1}(k|k-1) \\ \mathbf{x}_i(k|k) \\ \hat{\mathbf{x}}_{i+1}(k|k-1) \\ \vdots \\ \hat{\mathbf{x}}_m(k|k-1) \end{bmatrix} + \sum_{s=1}^{l} \mathbf{A}^{s-1}\mathbf{B}\tilde{\mathbf{U}}(k,l|k) \\
&\tilde{\mathbf{U}}(k,M|k) = \big[\mathbf{u}_1^{\mathrm{T}}(k|k-1) \ \ldots \ \mathbf{u}_{i-1}^{\mathrm{T}}(k|k-1) \ \ \mathbf{u}_i(k|k) \\
&\qquad \mathbf{u}_{i+1}^{\mathrm{T}}(k|k-1) \ \cdots \ \mathbf{u}_m^{\mathrm{T}}(k|k-1) \ \cdots \\
&\qquad \mathbf{u}_1^{\mathrm{T}}(k+l|k-1) \ \ldots \ \mathbf{u}_{i-1}^{\mathrm{T}}(k+l|k-1) \ \ \mathbf{u}_i(k+l|k) \\
&\qquad \mathbf{u}_{i+1}^{\mathrm{T}}(k+l|k-1) \ \cdots \ \mathbf{u}_m^{\mathrm{T}}(k+l|k-1)\big]^{\mathrm{T}} \\
&\hat{\mathbf{y}}^i(k+l|k) = \mathbf{C}\hat{\mathbf{x}}^i(k+l|k)
\end{aligned}
\tag{6.39}
$$

The optimization problem of the centralized MPC can be written as

$$
\begin{aligned}
&\min_{\Delta\mathbf{U}_i(k,M|k)} \sum_{l=1}^{P} \|\hat{\mathbf{y}}(k+l|k) - \mathbf{y}^d(k+l|k)\|_{\mathbf{Q}}^2 + \sum_{l=1}^{M} \|\Delta\mathbf{u}_i(k+l-1|k)\|_{\mathbf{R}_i}^2 \\
&\text{s.t.} \begin{bmatrix} \hat{\mathbf{x}}_1(k+l+1|k) \\ \vdots \\ \hat{\mathbf{x}}_{i-1}(k+l+1|k) \\ \hat{\mathbf{x}}_i(k+l+1|k) \\ \hat{\mathbf{x}}_{i+1}(k+l+1|k) \\ \vdots \\ \hat{\mathbf{x}}_m(k+l+1|k) \end{bmatrix} = \mathbf{A}^l \begin{bmatrix} \mathbf{x}_1(k|k) \\ \vdots \\ \mathbf{x}_{i-1}(k|k) \\ \mathbf{x}_i(k|k) \\ \mathbf{x}_{i+1}(k|k) \\ \vdots \\ \mathbf{x}_m(k|k) \end{bmatrix} + \sum_{s=1}^{l} \mathbf{A}^{s-1}\mathbf{B}\mathbf{U}(k,l|k) \\
&\tilde{\mathbf{U}}(k,M|k) = \big[\mathbf{u}_1^{\mathrm{T}}(k|k) \ \ldots \ \mathbf{u}_{i-1}^{\mathrm{T}}(k|k) \ \ \mathbf{u}_i(k|k) \\
&\qquad \mathbf{u}_{i+1}^{\mathrm{T}}(k|k) \ \cdots \ \mathbf{u}_m^{\mathrm{T}}(k|k) \ \cdots \\
&\qquad \mathbf{u}_1^{\mathrm{T}}(k+l|k) \ \ldots \ \mathbf{u}_{i-1}^{\mathrm{T}}(k+l|k) \ \ \mathbf{u}_i(k+l|k) \\
&\qquad \mathbf{u}_{i+1}^{\mathrm{T}}(k+l|k) \ \cdots \ \mathbf{u}_m^{\mathrm{T}}(k+l|k)\big]^{\mathrm{T}} \\
&\hat{\mathbf{y}}(k+l|k) = \mathbf{C}\hat{\mathbf{x}}(k+l|k)
\end{aligned}
\tag{6.40}
$$

It can be seen that the performance indices in (6.39) and (6.40) are identical. The state evolution models are also similar. The only difference between these two problems is that, in C-DMPC, the initial states and future control sequences of other subsystems at time k are substituted by the estimations calculated at time $k-1$. If there is disturbance, model mismatch or set-point change, the future input sequences of subsystems calculated at time k are not equal to that calculated at time $k-1$, which induces estimation errors of future states between two optimization strategies. This affects the final performance of the closed-loop system. Although this difference exists, the optimization problem of C-DMPC is still very close to that of the centralized MPC.

6.2.5 *Example*

In this section, the performance of the proposed C-DMPC is investigated and compared with that of the LCO-MPC presented in Chapter 5. Consider the nonminimum phase plant $\mathcal{S}$ appeared in Chapter 5. We discretize this plant with a sampling time of 0.2 s, which yields

$$\begin{bmatrix} y_1(z) \\ y_2(z) \end{bmatrix} = \begin{bmatrix} \dfrac{-0.024\,(z-1.492)\,(z+0.810)}{(z-0.819)(z^2-1.922z+0.961)} & \alpha\dfrac{0.018(z+0.935)}{(z^2-1.676z+0.819)} \\ \alpha\dfrac{0.126}{(z-0.368)} & \dfrac{0.147(z-0.668)}{(z^2-1.572z+0.670)} \end{bmatrix} \begin{bmatrix} u_1(z) \\ u_2(z) \end{bmatrix}$$

A state-space realization for $\mathcal{S}$ has the form (6.2), with matrices

$$\mathbf{A} = \begin{bmatrix} \mathbf{A}_{11} & 0 \\ 0 & \mathbf{A}_{22} \end{bmatrix}$$

$$\mathbf{A}_{11} = \begin{bmatrix} 2.74 & -1.27 & 0.97 & 0 \\ 2 & 0 & 0 & 0 \\ 0 & 0.5 & 0 & 0 \\ 0 & 0 & 0 & 0.37 \end{bmatrix}$$

$$\mathbf{A}_{22} = \begin{bmatrix} 1.68 & -0.82 & 0 & 0 \\ 1 & 0 & 0 & 0 \\ 0 & 0 & 1.57 & -0.67 \\ 0 & 0 & 1 & 0 \end{bmatrix}$$

$$\mathbf{B} = \begin{bmatrix} \mathbf{B}_{11} & 0 \\ 0 & \mathbf{B}_{22} \end{bmatrix}$$

$$\mathbf{B}_{11} = \begin{bmatrix} 0.25 \\ 0 \\ 0 \\ 0.5 \end{bmatrix} \mathbf{B}_{22} = \begin{bmatrix} 0.25 \\ 0 \\ 0.5 \\ 0 \end{bmatrix}$$

$$\mathbf{C} = \begin{bmatrix} \mathbf{C}_{11} & \mathbf{C}_{12} \\ \mathbf{C}_{21} & \mathbf{C}_{22} \end{bmatrix}$$

$$\mathbf{C}_{11} = \begin{bmatrix} -0.1 & 0.03 & 0.12 & 0 \end{bmatrix}$$

$$\mathbf{C}_{12} = \alpha \begin{bmatrix} 0.07 & 0.07 & 0 & 0 \end{bmatrix}$$

$$\mathbf{C}_{21} = \alpha \begin{bmatrix} 0 & 0 & 0 & 2.25 \end{bmatrix}$$

$$\mathbf{C}_{22} = \begin{bmatrix} 0 & 0 & 0.29 & -0.20 \end{bmatrix}$$

Decompose $\mathcal{S}$ into two SISO subsystems, $\mathcal{S}_1$ and $\mathcal{S}_2$. The corresponding state-space realizations of $\mathcal{S}_1$ and $\mathcal{S}_2$ have the form (5.1), with matrices $\{\mathbf{A}_{11}, \mathbf{B}_{11}, \mathbf{C}_{11}\}$ and $\{\mathbf{A}_{22}, \mathbf{B}_{22}, \mathbf{C}_{22}\}$, respectively. The constant parameter α is used to study the interactions between $\mathcal{S}_1$ and $\mathcal{S}_2$.

Similar to N-DMPC, the computational complexity of C-DMPC depends on the plant size, in particular on the number of inputs of independent controllers, while the feasibility depends on the possibility of satisfying Theorem 6.2 for a specific set of parameters P, M, $\mathbf{Q}$, and $\mathbf{R}_i$, $i=1, \ldots, m$. Different weighting matrices and horizons can be selected. Here, for simplifying the graphical representations of the results, we choose $P=M$, $\mathbf{R}=\gamma\mathbf{I}_u$, and $\mathbf{Q}=\mathbf{I}_y$.

The maximum eigenvalues of the corresponding closed-loop systems are computed and plotted in three-dimensional graphs of Figures 6.1a, 6.2a, and 6.3a for different combinations of γ and P. (More degrees of freedom can be used if the achieved performances and stability are not satisfactory.) The Z axis represents the maximum eigenvalues, and the X and Y axes represent the logarithms of γ and P, respectively. The control performance of the closed-loop system is plotted in Figures 6.1b, 6.2b, and 6.3b, where the black dashed lines correspond to the desired outputs, the blue solid lines correspond to the system outputs and inputs using LCO-DMPC, and the red dashed lines represent the system outputs and inputs using C-DMPC.

The stability depends on the choice of the tuning parameters γ and P. For weak interactions (see Figures 6.1a and 6.2a), the ranges of tuning parameters in C-DMPC are similar to that of LCO-DMPC. At the same time, a wider range of tuning parameters in C-DMPC, than that in LCO-DMPC, is available for strong interactions (see Figure 6.3a).

Most importantly, a better global performance of the closed-loop system is observed for the proposed C-DMPC where the subsystems exhibit interactions (see Figures 6.1b and 6.2b). The MSEs of outputs with C-DMPC are less than that using LCO-DMPC when $\alpha=0.1$ and $\alpha=1$, which are (0.2086, 0.2034) for C-DMPC and (0.2568, 0.2277) for N-DMPC, respectively. The closed-loop system is unstable using N-DMPC, while it is stable using C-DMPC when $\alpha=10$, $\gamma=1$, and $P=20$ (see Figure 6.3).

In conclusion, for the given example, C-DMPC provides larger regions of tuning parameters than LCO-DMPC. Usually, the stable regions are associated with the big prediction horizon P and small weight γ. The C-DMPC can achieve a satisfactory global performance whether the interactions among subsystems are stronger or not. Furthermore, the cost of computation is very small as compared with the centralized.

6.3 Distributed Predictive Control based on Pareto Optimality

In this section, we will introduce the DMPC algorithm which could achieve Pareto optimality, the main idea of this method is that: each subsystem-based MPC communicates with each other many times a control period and computes the optimal control law through iteration, and each subsystem-based MPC optimizes the cost of the entire closed-loop system, then improves the global performance of the closed-loop system. The detail of the DMPC-based Nash optimality is detailed in the following context.

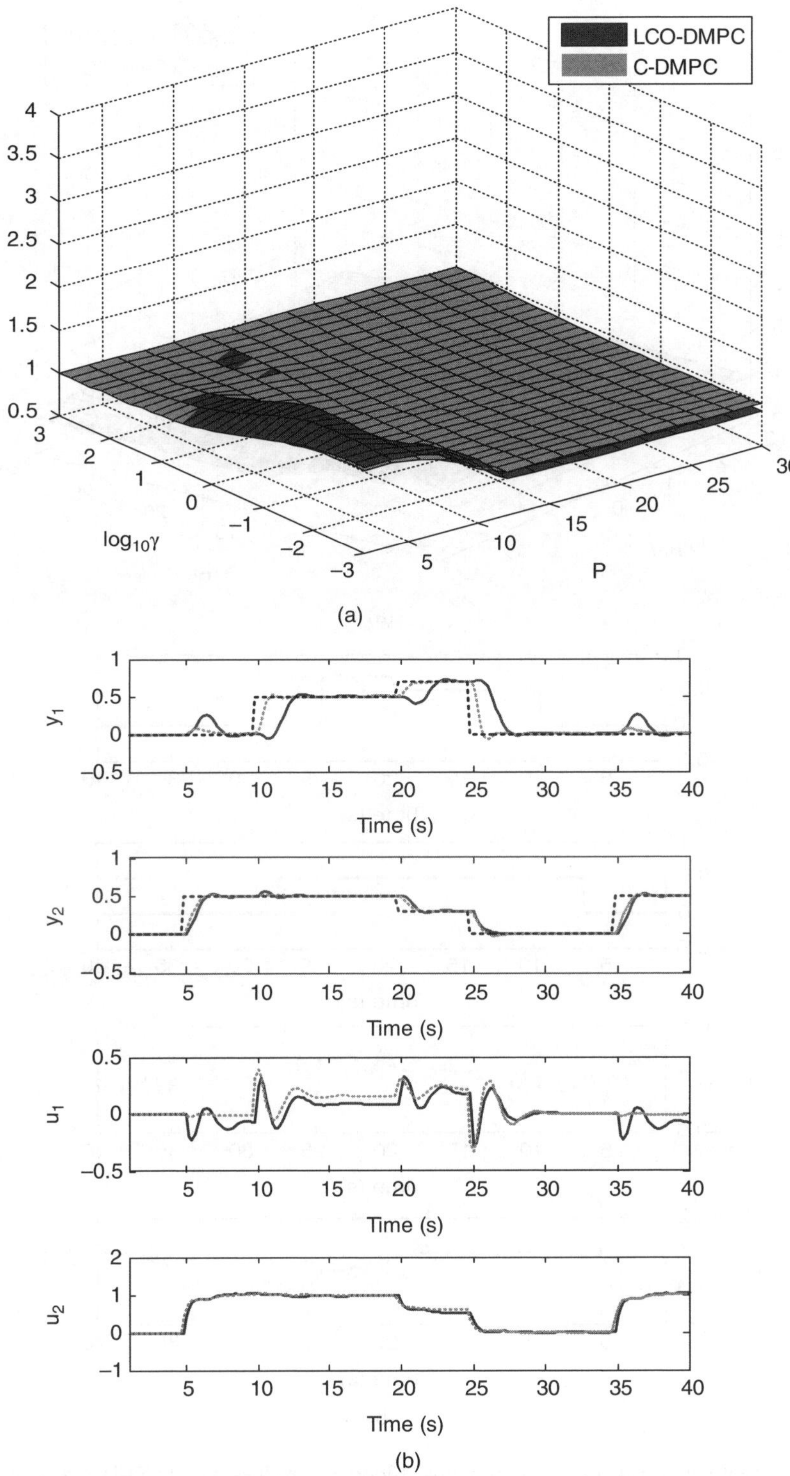

Figure 6.1 Plant with $\alpha = 0.1$: (a) maximum closed-loop eigenvalues with LCO-DMPC and C-DMPC; (b) control performance with $\gamma = 1$ for LCO-DMPC (blue line, MSE = 0.2568) and C-DMPC (red line, MSE = 0.2086).

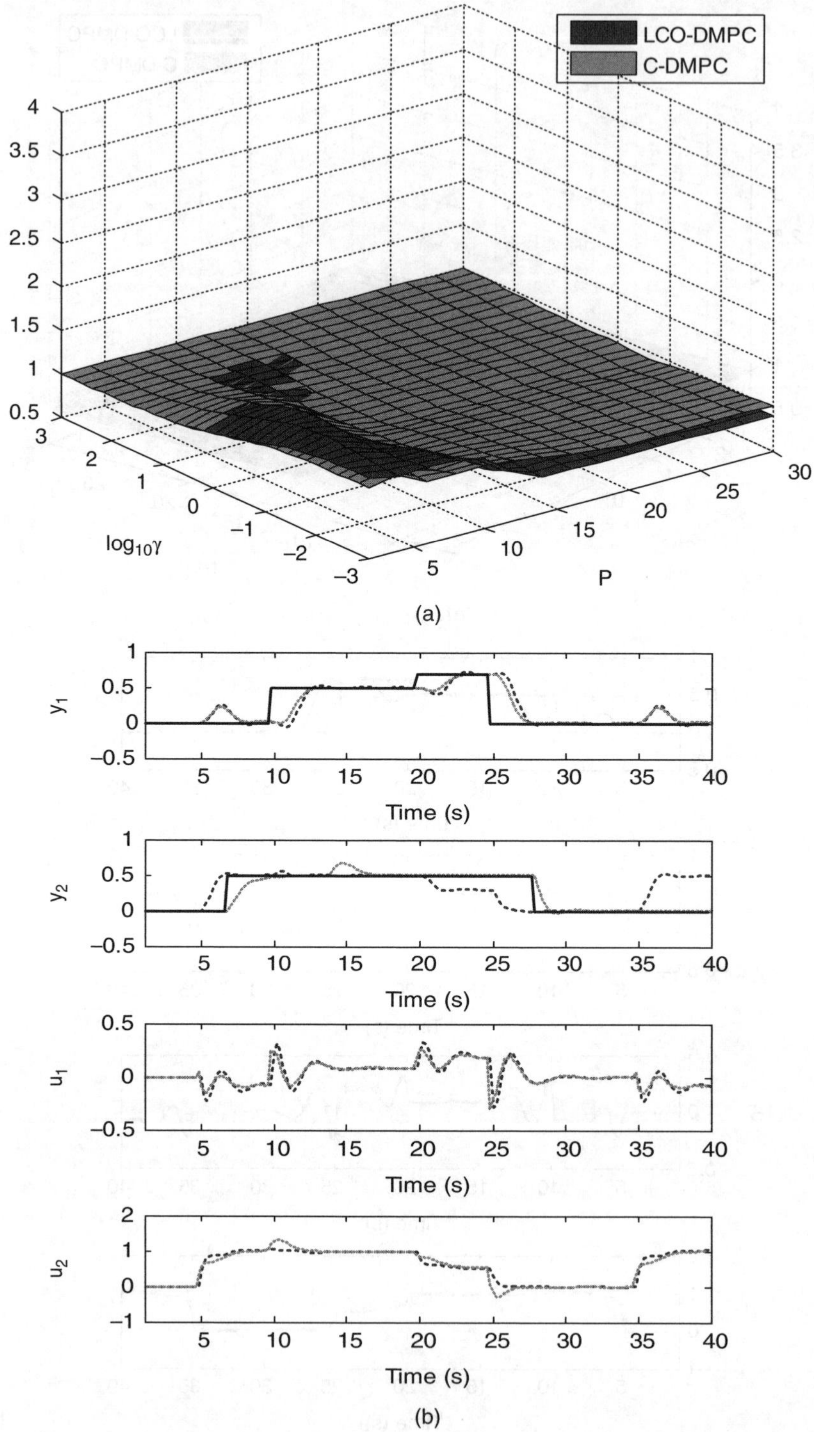

Figure 6.2 Plant with $\alpha = 1$: (a) maximum closed-loop eigenvalues with LCO-DMPC and C-DMPC; (b) control performance with $\gamma = 1$ for LCO-DMPC (blue line, MSE = 0.2277) and C-DMPC (red line, MSE = 0.2034).

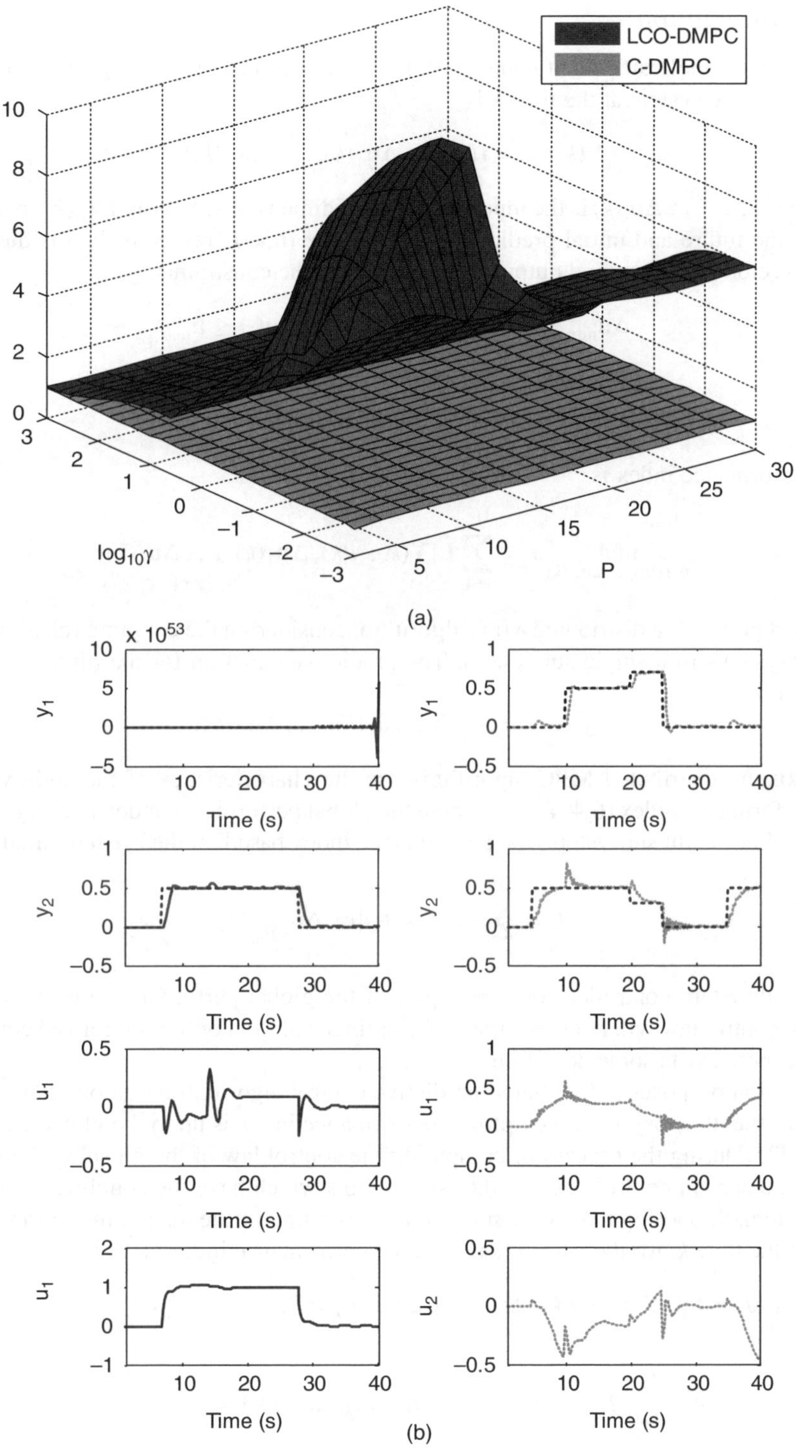

Figure 6.3 Plant with $\alpha = 10$: (a) maximum closed-loop eigenvalues with LCO-DMPC and C-DMPC; (b) control performance with $\gamma = 1$ for LCO-DMPC (blue line, unstable) and C-DMPC (red line, MSE = 0. 1544).

6.3.1 Formulation

Considering a complicated plant wide control of n subsystems, the j step ahead output prediction of the whole system at the time k is

$$\mathbf{Y}(k+j) = \mathbf{f}(\mathbf{y}_0(k), \Delta\mathbf{u}_1(k), \ldots, \Delta\mathbf{u}_m(k)) \tag{6.41}$$

where $j = 1, 2, \ldots, P$, $\Delta\mathbf{u}_i(k)$ is the input vector of M_i dimensions, $\mathbf{Y}(k)$ and $\mathbf{Y}_0(k)$ are predictive output in the future and initial predictive output at the time k, respectively, f is the mapping function vector. The input and output all have to meet the constraints:

$$\Delta\mathbf{u}_{\min} \leq \Delta\mathbf{u}(\cdot) \leq \Delta\mathbf{u}_{\max}, \mathbf{u}_{\min} \leq \mathbf{u}(\cdot) \leq \mathbf{u}_{\max}$$
$$\mathbf{Y}_{\min} \leq \mathbf{Y}(\cdot) \leq \mathbf{Y}_{\max}$$
$$\mathbf{u}(k+1) = \mathbf{u}(k) + \Delta\mathbf{u}(k)$$

The performance index is

$$\min_{\Delta\mathbf{u}_1(k),\ldots,\Delta\mathbf{u}_m(k)} J = \sum_{j=1}^{P} \mathbf{L}[\mathbf{Y}(k+j|k), \Delta\mathbf{u}_1(k), \ldots, \Delta\mathbf{u}_n(k)] \tag{6.42}$$

This part proposes a distributed MPC algorithm, considering the coupling relationship from other subsystems to a single subsystem. The predictive equation for the ith ($i = 1, 2, \ldots, m$) subsystem is

$$\mathbf{y}_i(k) = f_i[\mathbf{y}_{i,0}(k), \Delta\mathbf{u}_1(k), \ldots, \Delta\mathbf{u}_m(k)] \tag{6.43}$$

The existing distributed MPC algorithms use the characteristics of the additivity of the global performance index (6.42) to distribute the global performance index to every subsystem controller. For the ith subsystem, the performance index based on the Nash optimality is

$$J_i = \sum_{j=1}^{P} \mathbf{L}_i[\mathbf{y}_i(k+j|k), \Delta\mathbf{u}_{i,M_i}(k)] \tag{6.44}$$

Every subsystem controller contains a part of the global performance index (6.42). As a result, the control law will not converge to the optimal solution of the centralized control algorithm and there exists some deviation.

This section proposes a distributed predictive control algorithm based on plant-wide optimality. For the ith subsystem, the optimal performance index is up to the global performance index (6.42). During the process of computing the control law of the ith subsystem, we need to know the current control law of other subsystems which have the coupling input with the ith subsystem. Supposing that every subsystem can communicate many times during a control period, at the time k, for the $l+1$th iteration, the performance index is

$$\min_{\Delta\mathbf{u}_i^{l+1}(k)} J_i = \mathbf{L_i}[\mathbf{y}_i^{l+1}(k+j|k), \Delta\mathbf{u}_i^{l+1}(k), \Delta\mathbf{u}_j^{l}(k), j = 1, 2, \ldots, m, j \neq i]$$
$$+ \sum_{j=1, j\neq i}^{P} \mathbf{L}_j[\mathbf{y}_j^{l+1}(k+j|k), \Delta\mathbf{u}_i^{l+1}(k), \Delta\mathbf{u}_n^{l}(k), n = 1, 2, \ldots, m, n \neq i] \tag{6.45}$$

where the first item is the local performance index in some existing research; the second item considers the impact to the global performance by computing the local optimal problem. The $\mathbf{y}_i^{l+1}(k)$ and $\mathbf{y}_j^{l+1}(k)$ can be concluded by the following predictive equation:

$$\mathbf{y}_i^{l+1}(k) = \mathbf{f}_i[\mathbf{y}_{i,0}(k), \Delta\mathbf{u}_i^{l+1}(k), \Delta\mathbf{u}_i^{l+1}(k), j = 1, 2, \ldots, m, j \neq i] \tag{6.46}$$

$$\mathbf{y}_j^{l+1}(k) = \mathbf{f}_j[\mathbf{y}_{j,0}(k), \Delta\mathbf{u}_i^{l+1}(k), \Delta\mathbf{u}_n^{l}(k), n = 1, 2, \ldots, m, n \neq i] \tag{6.47}$$

There is a trick in Equation (6.47) that we use other subsystems' control law and the ith subsystem in the last iteration to represent the unknown other subsystems' predictive output $\mathbf{y}_j^{l+1}(k)$.

6.3.2 Algorithm

The algorithm can be concluded as follows.

Algorithm 6.1

Step 1. At time k, each subsystem transmits the initial predictive value $\mathbf{y}_{i,0}(k)$ to other subsystems and receives the initial predictive values $\mathbf{y}_{j,0}(k)(j = 1, 2, \ldots, m, j \neq i)$ from other subsystems. Then it sends the estimator of the control law to other subsystems. Let the iteration number $l = 0$.

Step 2. Each subsystem computes the optimal problem in parallel with last optimal control laws from other subsystems, performance index, and the predictive equations are (6.45)–(6.47). The constraints are the same as mentioned before. We can get the optimal solution $\Delta\mathbf{u}_i^{l+1}(k)(i = 1, 2, \ldots, m)$ at the iteration.

Step 3. Check the convergence condition of all the subsystems. If the precision $\varepsilon_i(i = 1, 2, \ldots, m)$ of all the subsystems meets the condition

$$\|\Delta\mathbf{u}_i^{l+1}(k) + \Delta\mathbf{u}_i^{l}(k)\| \leq \varepsilon_i$$

we can get $\Delta\mathbf{u}_i^{*}(k) = \Delta\mathbf{u}_i^{l+1}(k)(i = 1, 2, \ldots, m)$ and this iteration ends, and it goes to Step 4; otherwise, let $\Delta\mathbf{u}_i^{l}(k) = \Delta\mathbf{u}_i^{l+1}(k)(i = 1, 2, \ldots, m), l = l + 1$ and go to Step 2

Step 4. Compute the control law at time k

$$\Delta\mathbf{u}_i(k) = [I 0 \cdots 0]\Delta\mathbf{u}_{i,M}^{*}(k) i = 1, 2, \ldots, m$$

Step 5. Let $k + 1 \rightarrow k$ and go to Step 1.

During the optimal control time, each subsystem can get a local MPC control law without considering the coupling relationship between subsystems. The local MPC control law will be taken as the iteration initial value.

6.3.3 The DMPC Algorithm Based on Plant-Wide Optimality

Without generality, the control horizons of all subsystems are the same as M. The whole system predictive output model is

$$\mathbf{Y}_{PM}(k) = \mathbf{Y}_{P0}(k) + \mathbf{A}\Delta\mathbf{u}_M(k) \tag{6.48}$$

If the system structure is centralized control, the performance index is

$$\min J(k) = \|\mathbf{w}(k) - \mathbf{Y_{PM}}(k)\|^2_{\mathbf{Q}} + \|\mathbf{\Delta u}_M(k)\|^2_{\mathbf{R}} \tag{6.49}$$

The optimal control law is

$$\mathbf{\Delta u}_M = (\mathbf{A}^{\mathrm{T}}\mathbf{QA} + \mathbf{R})^{-1}\mathbf{A}^{\mathrm{T}}\mathbf{Q}[\mathbf{w}(k) - \mathbf{y}_{p0}(k)] \tag{6.50}$$

If the system structure is distributed, the input of the ith subsystem is independent from the inputs of other subsystems. So, we can obtain the performance index based on the plant-wide optimality:

$$\min_{\mathbf{\Delta u}^{l+1}_{i,M}(k)} J_i = \|\mathbf{w}_i(k) - \mathbf{Y}_{i,PM}(k)\|^2_{\mathbf{Q}_i} + \|\mathbf{\Delta u}_{i,M}(k)\|^2_{\mathbf{R}_i} + \sum_{j=1,j\neq i}^{n} \|\mathbf{w}_j(k) - \mathbf{Y}_{j,PM}(k)\|^2_{\mathbf{Q}_j} \tag{6.51}$$

At time k, the predictive equation for the $l+1$iteration is

$$\mathbf{y}^{l+1}_{i,PM}(k) = \mathbf{y}_{i,P0}(k) + \mathbf{A}_{ii}\mathbf{\Delta u}^{l+1}_{i,M}(k) + \sum_{j=1,j\neq i}^{m} \mathbf{A}_{ij}\mathbf{\Delta u}^{l}_{j,M}(k) \tag{6.52}$$

$$\begin{aligned}\mathbf{y}^{l+1}_{j,PM}(k) = {} & \mathbf{y}_{j,P0}(\mathrm{k}) + \mathbf{A}_{jj}\mathbf{\Delta u}^{l+1}_{i,M}(k) \\ & + \sum_{\substack{n=1 \\ n\neq i, n\neq j}}^{m} \mathbf{A}_{jn}\mathbf{\Delta u}^{l}_{n,M}(k) + \mathbf{A}_{ji}\mathbf{\Delta u}^{l+1}_{i,M}(k)\end{aligned} \tag{6.53}$$

Without the constraints, we can obtain the explicit solution

$$\begin{aligned}&\left(\sum_{j=1}^{n}\left(\mathbf{A}^{\mathrm{T}}_{ji}\mathbf{Q}_j\mathbf{A}_{ji} + \mathbf{R}_i\right)\right)\mathbf{\Delta u}^{l+1}_{i,M}(k) \\ &\quad = \sum_{j=1}^{m}\left\{\mathbf{A}^{\mathrm{T}}_{ji}\mathbf{Q}_j\left[\mathbf{w}_j(k) - \mathbf{y}_{j,p0}(k) - \sum_{n=1,n\neq i}^{m}\mathbf{A}_{jn}\mathbf{\Delta u}^{l}_{n,M}(k)\right]\right\}\end{aligned} \tag{6.54}$$

Let

$$\mathbf{D} \triangleq \mathbf{A}^{\mathrm{T}}\mathbf{QA} = \mathbf{D}_d + \mathbf{D}_{nd}$$

where $\mathbf{D}_d$ is a diagonal matrix made up of diagonal elements in the matrix $\mathbf{D}$, and $\mathbf{D}_{nd}$ is a matrix made up of nondiagonal elements in the matrix $\mathbf{D}$. From Equation (14), we can obtain the optimal control law

$$\begin{aligned}\mathbf{\Delta u}^{l+1}_{M}(k) = {} & (\mathbf{D}_d + \mathbf{R})^{-1}\mathbf{A}^{\mathrm{T}}\mathbf{Q}[\mathbf{w}(k) - \mathbf{y}_{p0}(k)] \\ & - (\mathbf{D}_d + \mathbf{R})^{-1}\mathbf{D}_{nd}\mathbf{\Delta u}^{l}_{M}(k)\end{aligned} \tag{6.55}$$

6.3.4 The Convergence Analysis of the Algorithm

The convergence is the key for an algorithm. Based on Equation (6.55), which is the whole system control law, we can obtain the convergence condition for the DMPC based on plant-wide optimality:

$$\rho[(\mathbf{D}_d + \mathbf{R})^{-1}\mathbf{D}_{nd}] < 1 \tag{6.56}$$

where $\rho[\cdot]$ is the spectrum radius of a matrix. If the algorithm is convergence, then we can obtain the optimal control law without constraints:

$$\Delta\mathbf{u}_M^{l+1}(k) = \mathbf{D}_1[\mathbf{w}(k) - \mathbf{y}_{p0}(k)] - \mathbf{D}_0\Delta\mathbf{u}_M^l(k) \tag{6.57}$$

where

$$\mathbf{D}_1 = \begin{bmatrix} \mathbf{D}_{11} & \mathbf{0} & \cdots & \mathbf{0} \\ \mathbf{0} & \mathbf{D}_{11} & \ddots & \vdots \\ \vdots & \ddots & \ddots & \mathbf{0} \\ \mathbf{0} & \cdots & \mathbf{0} & \mathbf{D}_{mm} \end{bmatrix}$$

$$\mathrm{D}_1 = \begin{bmatrix} 0 & -\mathrm{D}_{11}\mathrm{A}_{12} & \cdots & -\mathrm{D}_{11}\mathrm{A}_{1m} \\ -\mathrm{D}_{22}\mathrm{A}_{21} & 0 & \cdots & -\mathrm{D}_{22}\mathrm{A}_{2m} \\ \vdots & \vdots & \ddots & \vdots \\ -\mathrm{D}_{mm}\mathrm{A}_{m1} & -\mathrm{D}_{mm}\mathrm{A}_{m2} & \cdots & 0 \end{bmatrix}$$

The control law mentioned in Equation (6.57) is not constricted to the optimal control law for the centralized control. If the algorithm is convergence, from Equation (6.53), we can obtain

$$\sum_{j=1}^{m} \mathbf{A}_{ji}^{\mathrm{T}}\mathbf{Q}_j\mathbf{A}_{ji}\Delta\mathbf{u}_{j,M}^*(k) + \mathbf{R}_i\Delta\mathbf{u}_{i,M}^*(k) = \sum_{j=1}^{m} \mathbf{A}_{ji}^{\mathrm{T}}\mathbf{Q}_j[\mathbf{w}_j(k) - \mathbf{y}_{j,p0}(k)] \tag{6.58}$$

where $\Delta\mathbf{u}_{i,M}^*(k)$ and $\Delta\mathbf{u}_{j,M}^*(k)$ are the convergence values of the subsystems' optimal control law at time k.

Let $\Delta\mathbf{u}_M^*(k) = [\Delta\mathbf{u}_{1,M}^*(k), \ldots, \Delta\mathbf{u}_{m,M}^*(k)]$, we can obtain a new form of the optimal control law

$$\Delta\mathbf{u}_M^*(k) = (\mathbf{A}^{\mathrm{T}}\mathbf{Q}\mathbf{A} + \mathbf{R})^{-1}\mathbf{A}^{\mathrm{T}}\mathbf{Q}[\mathbf{w}(k) - \mathbf{y}_{p0}(k)] \tag{6.59}$$

which is the same as Equation (6.42). That is to say that the algorithm mentioned in this section is constricted to the optimal control law.

6.4 Simulation

Consider a three input and three output system whose transfer function is

$$G(s) = \begin{bmatrix} \dfrac{4.05e^{-27s}}{50s+1} & \dfrac{1.77e^{-28s}}{60s+1} & \dfrac{5.88e^{-27s}}{50s+1} \\ \dfrac{5.39e^{-18s}}{50s+1} & \dfrac{5.72e^{-14s}}{60s+1} & \dfrac{6.90e^{-15s}}{40s+1} \\ \dfrac{4.38e^{-20s}}{33s+1} & \dfrac{4.42e^{-22s}}{44s+1} & \dfrac{7.20}{19s+1} \end{bmatrix}$$

The transfer functions of these three subsystems are, respectively,

$$G_1(s) = \frac{4.05e^{-27s}}{50s+1}$$

$$G_2(s) = \frac{5.72e^{-14s}}{60s+1}$$

$$G_3(s) = \frac{7.20}{19s+1}$$

Both input and output constraints are $[-0.5 \quad 0.5]$, and the sample period is 4 min. Let the iterative condition be $\varepsilon = 0.01$, prediction condition $P = 8$, and control horizon $M = 3$.

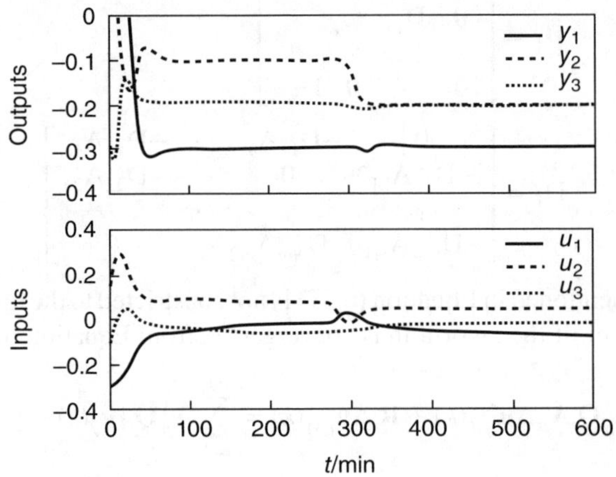

Figure 6.4 The outputs and inputs of each subprocess

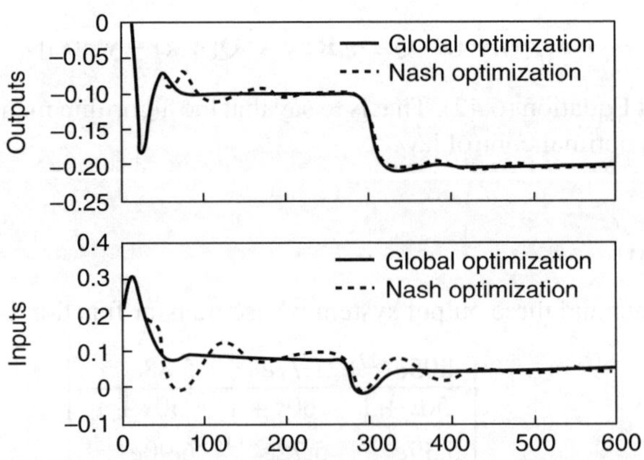

Figure 6.5 The outputs and inputs of the second subprocess

The simulation duration is 600 min. And in the first 300 min, the set-point of system outputs is $(-0.3, -0.1, -0.2)$, and in the last 300 minutes, the set-point of system outputs is $(-0.3, -0.2, -0.2)$. The simulation results are shown in Figure 6.4. Each subsystem can track the set-point. Figure 6.5 is the comparison between the global cost optimization-based DMPC and the Nash optimization-based DMPC.

6.5 Conclusions

In this chapter, the DMPC methods based on global cost optimization and Pareto optimality are developed for large-scale linear systems when the global information is accessible by each subsystem. The main advantage of this scheme is that the online optimization of a large-scale system can be converted to that of several small-scale systems, and thus can significantly reduce the computational complexity while keeping satisfactory performance.

The first part of the chapter gives the idea of global cost optimization-based DMPC, the closed-loop solution, and the stability conditions. This method provides acceptable regions of tuning parameters for which the stability is guaranteed and the performances are satisfactory. Usually, the stable regions are associated with big prediction horizon P and small weight $\mathbf{R}$. The second part of this chapter provides the Pareto optimization-based DMPC and investigates the performance of the distributed control scheme. The nominal stability and the convergence of the DMPC algorithm is analyzed. These will provide users a better understanding to the developed algorithm and sensible guidance in applications. In addition, some simulation examples are presented to verify the efficiency and practicality of the distributed MPC algorithms.

7

Networked Distributed Predictive Control with Information Structure Constraints

7.1 Introduction

The majority advantage of distributed model predictive control (DMPC) is that it has the characteristics of good flexibility and error tolerance. This characteristic is based on the fact that the controllers are relevantly independent from each other. It means that the number of systems that each subsystem-based MPC communicates with will decrease, and then the flexibility and the ability of the error-tolerance of the whole closed-loop control system will improve. In addition, in some fields or processes, the global information are unavailable to controllers (e.g., in a multi-intelligent vehicle system) for the management or the system scale reasons. Thus, designing a DMPC which could significantly improve the global performance of the closed-loop system with limited information structure constraints is valuable.

In the previous chapters, we introduced the basic DMPC and the Nash optimality-based DMPC where each subsystem-based controller pursues the performance of a local subsystem. Chapter 6 presents a method to improve global optimality based on global information. This chapter will propose a coordination strategy which could improve the global performance using appropriate network resources, where the optimization objective of each subsystem-based MPC considers not only the performance of the corresponding local subsystem but also those it has a direct impact on. In the optimization, each local controller takes into account not only the impacts coming from its neighbors but also the impacts applied to its neighbors for improving global performance. Both the algorithms, where each subsystem-based controller communicates with each other once a control period and the iterative algorithm are designed in this chapter.

For the noniterative algorithm, the closed-loop stability analysis is also provided for guiding local MPC's tuning. Moreover, the performance of the closed-loop system using the proposed distributed MPC is analyzed and the application to the accelerated cooling and controlled (ACC) process is presented to validate the efficiency of this method. For the iterative algorithm where each subsystem-based MPC exchanges information several times during it solves its

Distributed Model Predictive Control for Plant-Wide Systems, First Edition. Shaoyuan Li and Yi Zheng.

local optimization problem, the optimality of the iteration-based networked MPC algorithm is analyzed and the nominal stability is derived for distributed control systems without the control and output constraints. An illustrative example is provided to verify the optimality of the networked MPC algorithm.

The contents are organized as follows. Section 7.1 describes the noniterative networked DPC and gives its closed-form solution, as well as the stability condition of the closed-loop system. Section 7.2 details the networked DMPC with an iterative algorithm based on neighborhood optimization.

7.2 Noniterative Networked DMPC

7.2.1 Problem Description

Without losing generality, suppose that the whole system is composed of n linear, discrete-time subsystems $\mathcal{S}_i$, $i = 1, 2, \ldots, m$, and each subsystem interacts with each other by both inputs and states; then the state-space model of subsystem $\mathcal{S}_i$ can be expressed as

$$\begin{cases} \mathbf{x}_i(k+1) = \mathbf{A}_{ii}\mathbf{x}_i(k) + \mathbf{B}_{ii}\mathbf{u}_i(k) + \sum\limits_{j=1(j\neq i)}^{m} \mathbf{A}_{ij}\mathbf{x}_j(k) + \sum\limits_{j=1(j\neq i)}^{m} \mathbf{B}_{ij}\mathbf{u}_j(k) \\ \mathbf{y}_i(k) = \mathbf{C}_{ii}\mathbf{x}_i(k) + \sum\limits_{j=1(j\neq i)}^{n} \mathbf{C}_{ij}\mathbf{x}_j(k) \end{cases} \tag{7.1}$$

where vectors $\mathbf{x} \in \mathbb{R}^{n_x}$, $\mathbf{u} \in \mathbb{R}^{n_u}$, and $\mathbf{y} \in \mathbb{R}^{n_y}$ are the local state, control input, and output vectors, respectively. When at least one of the matrices $\mathbf{A}_{ij}$, $\mathbf{B}_{ij}$, $\mathbf{C}_{ij}$ is not null, it is said that $\mathcal{S}_j$ interacts with $\mathcal{S}_i$. The whole system can be expressed as

$$\begin{cases} \mathbf{x}(k+1) = \mathbf{A}\mathbf{x}(k) + \mathbf{B}\mathbf{u}(k) \\ \mathbf{y}(k) = \mathbf{C}\mathbf{x}(k) \end{cases} \tag{7.2}$$

where $\mathbf{x} \in \mathbb{R}^{n_x}$, $\mathbf{u} \in \mathbb{R}^{n_u}$, and $\mathbf{y} \in \mathbb{R}^{n_y}$ are the state, control input, and output vectors, respectively. The control objective of this system is to minimize a global performance index $J(k)$ at time k, and

$$J(k) = \sum_{i=1}^{n} \left[\sum_{l=1}^{P} \left\| \mathbf{y}_i(k+l) - \mathbf{y}_i^d(k+l) \right\|_{\mathbf{Q}_i}^2 + \sum_{l=1}^{M} \left\| \Delta\mathbf{u}_i(k+l-1) \right\|_{\mathbf{R}_i}^2 \right] \tag{7.3}$$

where $\mathbf{Q}_i$ and $\mathbf{R}_i$ are weight matrices, $P, M \in \mathbb{N}$ are the predictive horizon and control horizon, respectively, and $P \geq M$, $\mathbf{y}_i^d$ is the set-point of subsystem $\mathcal{S}_i$, and $\Delta\mathbf{u}_i(k) = \mathbf{u}_i(k) - \Delta\mathbf{u}_i(k-1)$ is the input increment vector of subsystem $\mathcal{S}_i$.

Moreover, in many situations, the communication resources are not unlimited for the safety reason and communication bandwidth limitation, or the global information is unavailable to every subsystem due to the physical or man-made reasons. Those require a simple structure of a local controller. Thus, as pointed out in [100], how to improve the performance of the entire system is still a challenge for this class of system under the distributed control framework with a simple control structure.

In the previous chapters, two classes of distributed MPCs for the large-scale systems as described above were presented.

1. Distributed algorithms where each local controller minimizes following the local optimization objective

$$J_i(k) = \sum_{l=1}^{P} \left\| \mathbf{y}_i(k+l) - \mathbf{y}_i^d(k+l) \right\|_{\mathbf{Q}_i}^2 + \sum_{l=1}^{M} \left\| \Delta \mathbf{u}_i(k+l-1) \right\|_{\mathbf{R}_i}^2 \tag{7.4}$$

   When computing the optimal solution, each local controller exchanges estimation states with its neighbors; therefore, it improves the performance of the closed-loop subsystem.
2. Distributed algorithms where each local controller minimizes a global cost function

$$J(k) = \sum_{i=1}^{m} J_i(k) \tag{7.5}$$

   This strategy could achieve a good performance closing to the centralized MPC under the condition that the global information is available to each subsystem-based MPC and there are more communication resources.

In this chapter, a method based on the neighborhood optimization is proposed for the large-scale system in which each subsystem interacts with each other by both inputs and states. The goal of it is to achieve a significantly improving performance of the entire system with little increase of the network resources comparing to the (1) strategy when the global information is not available and the communication resources are limited.

7.2.2 DMPC Formulation

The proposed control architecture is based on a set of independent MPC controllers $\mathcal{C}_i$, $i = 1, 2, \ldots, m$, for each subsystem $\mathcal{S}_i$. Each controller could exchange information with its neighbors through a network. To discuss the control methodology proposed in this chapter, the simplifying hypothesis of accessible local states $\mathbf{x}_i(k)$ and Assumption 1 are considered. Moreover, Definition 1 and notations listed in Table 7.1 are defined to describe the proposed methodology clearly.

Assumption 7.1

(a) Controllers are synchronous.
(b) Controllers communicate only once within a sampling time interval.
(c) Communication cannel introduces a delay of a single sampling time interval.

This set of assumptions is not restrictive. The controllers are synchronous is not so strong because the sampling interval is usually rather long compared to the computational time in process control. The assumption (b) of single information exchange with a sampling time interval is due to the necessity of minimizing the amount of data exchange through the network. In real situations, an instantaneous data transfer is not possible; therefore, assumption (c) of unit delay is required.

Table 7.1 Notations used in this chapter

Notations	Explanations
$\hat{\mathbf{x}}_i(l\|h)$, $\hat{\mathbf{y}}_i(l\|h)$	The predictions of $\mathbf{x}_i(l)$ and $\mathbf{y}_i(l)$ computed at time h, and $l, h \in \mathcal{P}$, $h < l$
$\Delta\mathbf{u}_i(l\|h)$, $\mathbf{u}_i(l\|h)$	The input $\mathbf{u}_i(l)$ and the input increment $\Delta\mathbf{u}_i(l)$ computed by controller $\mathcal{C}_i$ at time h, $l, h \in \mathcal{P}$, and $h < l$
$\mathbf{y}_i^d(l\|h)$	The set-point of $\mathbf{y}_i(l\|h)$
$\widehat{\mathbf{x}}_i(k)$, $\widehat{\mathbf{y}}_i(k)$	The state and output vectors of the downstream neighbors of $\mathcal{S}_i$. $\widehat{\mathbf{x}}_i(k) = \left[\mathbf{x}_i^{\mathrm{T}}(k) \quad \mathbf{x}_{i_1}^{\mathrm{T}}(k) \cdots \mathbf{x}_{i_n}^{\mathrm{T}}(k)\right]^{\mathrm{T}}$ and $\widehat{\mathbf{y}}_i(k) = \left[\mathbf{y}_i^{\mathrm{T}}(k) \quad \mathbf{y}_{i_1}^{\mathrm{T}}(k) \quad \cdots \quad \mathbf{y}_{i_n}^{\mathrm{T}}(k)\right]^{\mathrm{T}}$, n is the number of the downstream neighbors of $\mathcal{S}_i$
$\widehat{\mathbf{w}}_i(k)$, $\widehat{\mathbf{v}}_i(k)$	The interactions act on the state and output of downstream neighbors of $\mathcal{S}_i$. See (7.13) and (7.14)
$\hat{\widehat{\mathbf{x}}}_i(l\|h)$, $\hat{\widehat{\mathbf{y}}}_i(l\|h)$	The predictions of $\widehat{\mathbf{x}}_i(l)$ and $\widehat{\mathbf{y}}_i(l)$ computed at time h, $l, h \in \mathcal{P}$ and $h < l$
$\hat{\widehat{\mathbf{w}}}_i(l\|h)$, $\hat{\widehat{\mathbf{v}}}_i(l\|h)$	The estimations of $\widehat{\mathbf{w}}_i(l)$ and $\widehat{\mathbf{v}}_i(l)$ computed at time h, $l, h \in \mathcal{P}$ and $h < l$
$\widehat{\mathbf{y}}_i^d(l\|h)$	The set-point of $\widehat{\mathbf{y}}_i(l\|h)$
$\mathbf{U}_i(l,p\|h)$	A complete input vector, $\mathbf{U}_i(l,p\|h) = \left[\mathbf{u}_i^{\mathrm{T}}(l\|h) \quad \mathbf{u}_i^{\mathrm{T}}(l+1\|h) \quad \cdots \quad \mathbf{u}_i^{\mathrm{T}}(l+p\|h)\right]^{\mathrm{T}}$, $p, l, h \in \mathbb{N}$ and $h < l$;
$\Delta\mathbf{U}_i(l,p\|h)$	Input increment sequence vector, $\Delta\mathbf{U}_i(l,p\|h) = \left[\Delta\mathbf{u}_i^{\mathrm{T}}(l\|h) \quad \Delta\mathbf{u}_i^{\mathrm{T}}(l+1\|h) \quad \cdots \quad \Delta\mathbf{u}_i^{\mathrm{T}}(l+p\|h)\right]^{\mathrm{T}}$, $h < l$;
$\mathbf{U}(l,p\|h)$	A complete stacked input vector, $\mathbf{U}(l,p\|h) = [\mathbf{u}_1^{\mathrm{T}}(l\|h) \quad \cdots \quad \mathbf{u}_m^{\mathrm{T}}(l\|h) \quad \cdots \quad \mathbf{u}_1^{\mathrm{T}}(l+p\|h) \quad \cdots \quad \mathbf{u}_m^{\mathrm{T}}(l+p\|h)]^{\mathrm{T}}$;
$\hat{\mathbf{X}}_i(l,p\|h)$	A stacked distributed state vector, $\hat{\mathbf{X}}_i(l,p\|h) = \left[\hat{\mathbf{x}}_i^{\mathrm{T}}(l\|h) \quad \hat{\mathbf{x}}_i^{\mathrm{T}}(l+1\|h) \quad \cdots \quad \hat{\mathbf{x}}_i^{\mathrm{T}}(l+p\|h)\right]^{\mathrm{T}}$;
$\hat{\mathbf{X}}(l,p\|h)$	A complete stacked state vector, $\hat{\mathbf{X}}(l,p\|h) = [\hat{\mathbf{x}}_1^{\mathrm{T}}(l\|h) \quad \cdots \quad \hat{\mathbf{x}}_m^{\mathrm{T}}(l\|h) \quad \cdots \quad \hat{\mathbf{x}}_1^{\mathrm{T}}(l+p\|h) \quad \cdots \quad \hat{\mathbf{x}}_m^{\mathrm{T}}(l+p\|h)$;
$\hat{\widehat{\mathbf{X}}}_i(l,p\|h)$	A stacked state vector, $\hat{\widehat{\mathbf{X}}}_i(l,p\|h) = \left[\hat{\widehat{\mathbf{x}}}_i^{\mathrm{T}}(l\|h) \quad \hat{\widehat{\mathbf{x}}}_i^{\mathrm{T}}(l+1\|h) \quad \cdots \quad \hat{\widehat{\mathbf{x}}}_i^{\mathrm{T}}(l+p\|h)\right]^{\mathrm{T}}$, $p, l, h \in \mathcal{P}$ and $h < l$
$\hat{\widehat{\mathbf{Y}}}_i(l,p\|h)$	A stacked output vector, $\hat{\widehat{\mathbf{Y}}}_i(l,p\|h) = \left[\hat{\widehat{\mathbf{y}}}_i^{\mathrm{T}}(l\|h) \quad \hat{\widehat{\mathbf{y}}}_i^{\mathrm{T}}(l+1\|h) \quad \cdots \quad \hat{\widehat{\mathbf{y}}}_i^{\mathrm{T}}(l+p\|h)\right]^{\mathrm{T}}$, $p, l, h \in \mathcal{P}$ and $h < l$
$\widehat{\mathbf{Y}}_i^d(l,p\|h)$	The set-point of $\hat{\widehat{\mathbf{Y}}}_i(l,p\|h)$
$\hat{\widehat{\mathbf{W}}}_i(l,p\|h)$	A stacked interaction vector, $\left[\hat{\widehat{\mathbf{w}}}_i^{\mathrm{T}}(l\|h) \quad \hat{\widehat{\mathbf{w}}}_i^{\mathrm{T}}(l+1\|h) \quad \cdots \quad \hat{\widehat{\mathbf{w}}}_i^{\mathrm{T}}(l+p\|h)\right]^{\mathrm{T}}$, $p, l, h \in \mathcal{P}$ and $h < l$
$\hat{\widehat{\mathbf{V}}}_i(l,P\|h)$	A stacked interaction vector, $\left[\hat{\widehat{\mathbf{v}}}_i^{\mathrm{T}}(l\|h) \quad \hat{\widehat{\mathbf{v}}}_i^{\mathrm{T}}(l+1\|h) \quad \cdots \quad \hat{\widehat{\mathbf{v}}}_i^{\mathrm{T}}(l+p\|h)\right]^{\mathrm{T}}$, $p, l, h \in \mathcal{P}$ and $h < l$
$\hat{\mathbb{X}} = (l,p\|h)$	A complete stacked state vector, $\hat{\mathbb{X}}(l,p\|h) = \left[\hat{\mathbf{X}}_1^{\mathrm{T}}(l,p\|h) \quad \cdots \quad \hat{\mathbf{X}}_m^{\mathrm{T}}(l,p\|h)\right]^{\mathrm{T}}$
$\mathbb{U} = (l,p\|h)$	A complete stacked state vector, $\mathbb{U}(l,p\|h) = \left[\mathbf{U}_1^{\mathrm{T}}(l,p\|h) \quad \cdots \quad \mathbf{U}_m^{\mathrm{T}}(l,p\|h)\right]^{\mathrm{T}}$

Definition 7.1 Neighboring subsystem. *Subsystem $\mathcal{S}_i$ interacts with $\mathcal{S}_j$, and the state outputs of subsystem $\mathcal{S}_i$ are affected by subsystem $\mathcal{S}_j$. In this case, $\mathcal{S}_i$ is called the upstream neighboring subsystem of $\mathcal{S}_j$ and $\mathcal{S}_i$ is called the downstream neighboring subsystem of $\mathcal{S}_j$. $\mathcal{S}_i$ and $\mathcal{S}_j$ are the neighboring subsystems or neighbors.*

Neighborhood of a subsystem. *The upstream (downstream)-neighborhood $\mathcal{P}_{+i}(\mathcal{P}_{-i})$ of subsystem $\mathcal{S}_i$ is the set of the subscripts of all its upstream (downstream) neighbors:*

$$\mathcal{P}_{+i} = \{\mathcal{S}_i, \mathcal{S}_j | \mathcal{S}_j \text{ is an upstream neighbor of } \mathcal{S}_i\}$$

$$\mathcal{P}_{-i} = \{\mathcal{S}_i, \mathcal{S}_j | \mathcal{S}_j \text{ is a downstream neighbor of } \mathcal{S}_i\}$$

The neighborhood $\mathcal{P}_i$ of subsystem $\mathcal{S}_i$ is the set of all its neighbors:

$$\mathcal{P}_i = \mathcal{P}_{+i} \cup \mathcal{P}_{-i}$$

7.2.2.1 Performance Index

For the large-scale system considered here, the global performance index (5.4) can be decomposed in terms of the local index J_i for each subsystem $\mathcal{S}_i, i = 1, 2, \ldots, m$ [101].

$$J_i(k) = \sum_{l=1}^{P} \left\| \hat{\mathbf{y}}_i(k+l|k) - \mathbf{y}_i^d(k+l|k) \right\|_{\mathbf{Q}_i}^2 + \sum_{l=1}^{M} \left\| \Delta \mathbf{u}_i(k+l-1|k) \right\|_{\mathbf{R}_i}^2 \tag{7.6}$$

The local control decision of $\mathcal{S}_i$ is computed by solving the optimization problem $\min\limits_{\Delta \mathbf{U}(k,M|k)} J_i(k)$ with local input/output variables and constraints in the distributed MPC based on the state (or input) estimations of neighbors at time $k-1$ [46] or Nash optimality [29].

However, since the state evolution of downstream neighbors of subsystem $\mathcal{S}_i$ is affected by the control decision of subsystem $\mathcal{S}_i$, see Equation (5.1) (1), the performance of these neighbors may be destroyed by the improper control decision of $\mathcal{S}_i$ in some cases. To solve this problem, the so-called *neighborhood optimization* [19, 54] is adopted and the performance index is expressed as

$$\begin{aligned} \overline{J}_i(k) &= \sum_{j \in \mathrm{N}_i^{\text{out}}} J_i(k) \\ &= \sum_{j \in \mathrm{N}_i^{\text{out}}} \left[\sum_{l=1}^{P} \left\| \hat{\mathbf{y}}_j(k+l|k) - \mathbf{y}_j^d(k+l|k) \right\|_{\mathbf{Q}_j}^2 + \sum_{l=1}^{M} \left\| \Delta \mathbf{u}_j(k+l-1|k) \right\|_{\mathbf{R}_j}^2 \right] \end{aligned} \tag{7.7}$$

Since $\Delta \mathbf{u}_j(k+l-1|k)$ $(j \in \mathcal{P}_{-i}, j \neq i, l = 1, \ldots, M)$ is unknown and independent of the control decision of $\mathcal{S}_i$, $\Delta \mathbf{u}_j(k+l-1|k-1)$ is used to approximate $\Delta \mathbf{u}_j(k+l-1|k)$. Then, Equation (7.7) becomes

$$\begin{aligned} \overline{J}_i(k) &= \sum_{j \in \mathcal{P}_{-i}} \sum_{l=1}^{P} \left\| \hat{\mathbf{y}}_j(k+l|k) - \mathbf{y}_j^d(k+l|k) \right\|_{\mathbf{Q}_j}^2 + \sum_{l=1}^{M} \left\| \Delta \mathbf{u}_i(k+l-1|k) \right\|_{\mathbf{R}_i}^2 \\ &\quad + \sum_{j \in \mathcal{P}_{-i}, j \neq i} \sum_{l=1}^{M} \left\| \Delta \mathbf{u}_i(k+l-1|k-1) \right\|_{\mathbf{R}_j}^2 \\ &= \sum_{j \in \mathcal{P}_{-i}} \sum_{l=1}^{P} \left\| \hat{\mathbf{y}}_j(k+l|k) - \mathbf{y}_j^d(k+l|k) \right\|_{\mathbf{Q}_j}^2 + \sum_{l=1}^{M} \left\| \Delta \mathbf{u}_i(k+l-1|k) \right\|_{\mathbf{R}_i}^2 + \text{Constant} \end{aligned}$$

For a simplifying reason, redefine $\overline{J}_i(k)$ as

$$\overline{J}_i(k) = \sum_{l=1}^{P} \left\| \hat{\mathbf{y}}_i\left(k+l\,|k\right) - \mathbf{y}_i^d(k+l|k) \right\|_{\hat{\mathbf{Q}}_i}^2 + \sum_{l=1}^{M} \left\| \Delta \mathbf{u}_i\left(k+l-1\,|k\right) \right\|_{\mathbf{R}_i}^2 \tag{7.8}$$

where $\hat{\mathbf{Q}}_i = \mathrm{diag}(\mathbf{Q}_i, \mathbf{Q}_{i_1}, \ldots, \mathbf{Q}_{i_b})$.

The optimization index $\overline{J}_i(k)$ considers not only the performance of subsystem $\mathcal{S}_i$ but also that of the downstream neighbors of $\mathcal{S}_i$. The impact of the control decision of $\mathcal{S}_i$ to $\mathcal{S}_j \in \mathcal{P}_{-i}$ are fully considered in the neighborhood optimization, and, therefore, the global performance improvement is guaranteed. It should be noticed that the global performance may further be improved by using the optimization objective (7.5) in each subsystem, but it requires a high-quality and complicated network communication and introduces more complex computation.

7.2.2.2 Prediction Model

Since the state evolution of $\mathcal{S}_j \in \mathcal{P}_{-i}$ is affected by $\mathbf{u}_i(k)$, to improve the predictive precision, subsystem $\mathcal{S}_i$, its downstream neighbors should be considered as one relatively large integral subsystem when predicting the future states of $\mathcal{S}_i$ and its downstream neighbors. Assume that the number of downstream neighbors of $\mathcal{S}_i$ is n; then the state evolution model of the downstream neighbors of $\mathcal{S}_i$ can be easily deduced by (5.1) and expressed as

$$\begin{cases} \hat{\mathbf{x}}_i\left(k+1\right) = \hat{\mathbf{A}}_i\hat{\mathbf{x}}_i(k) + \hat{\mathbf{B}}_i\mathbf{u}_i(k) + \hat{\mathbf{w}}_i(k) \\ \hat{\mathbf{y}}_i(k) = \hat{\mathbf{C}}_i\hat{\mathbf{x}}_i(k) + \hat{\mathbf{v}}_i(k) \end{cases} \tag{7.9}$$

where

$$\hat{\mathbf{A}}_i = \begin{bmatrix} \hat{\mathbf{A}}_i^{(1)} & \hat{\mathbf{A}}_i^{(2)} \end{bmatrix} = \begin{bmatrix} \mathbf{A}_{ii} & \mathbf{A}_{ii_1} & \cdots & \mathbf{A}_{ii_n} \\ \mathbf{A}_{i_1 i} & \mathbf{A}_{i_1 i_1} & \cdots & \mathbf{A}_{i_n i_n} \\ \vdots & \vdots & \ddots & \vdots \\ \mathbf{A}_{i_n i} & \mathbf{A}_{i_n i_1} & \cdots & \mathbf{A}_{i_n i_n} \end{bmatrix} \tag{7.10}$$

$$\hat{\mathbf{B}}_i = \begin{bmatrix} \mathbf{B}_{ii} \\ \mathbf{B}_{i_1 i} \\ \vdots \\ \mathbf{B}_{i_n i} \end{bmatrix} \tag{7.11}$$

$$\hat{\mathbf{C}}_i = \begin{bmatrix} \mathbf{C}_{ii} & \mathbf{C}_{ii_1} & \cdots & \mathbf{C}_{ii_m} \\ \mathbf{C}_{i_1 i} & \mathbf{C}_{i_1 i_1} & \cdots & \mathbf{C}_{i_1 i_m} \\ \vdots & \vdots & \ddots & \vdots \\ \mathbf{C}_{i_m i} & \mathbf{C}_{i_m i_1} & \cdots & \mathbf{C}_{i_m i_m} \end{bmatrix} \tag{7.12}$$

$$\hat{\mathbf{w}}_i(k) = \begin{bmatrix} \sum\limits_{j\in\mathcal{P}_{+i}, j\neq i} \mathbf{B}_{ij}u_j\left(k\right) + \mathbf{0} \\ \sum\limits_{j\in\mathcal{P}_{+i_1}, j\neq i} \mathbf{B}_{i_1 j}u_j(k) + \sum\limits_{j\in\mathcal{P}_{+i_1}, j\notin j\in\mathcal{P}_{-i},} \mathbf{A}_{i_1 j}x_j(k) \\ \vdots \\ \sum\limits_{j\in\mathcal{P}_{+i_n}, j\neq i} \mathbf{B}_{i_n j}u_j(k) + \sum\limits_{j\in\mathcal{P}_{+i_n}, j\notin j\in\mathcal{P}_{-i},} \mathbf{A}_{i_n j}x_j(k) \end{bmatrix} \tag{7.13}$$

$$\hat{\mathbf{v}}_i(k) = \begin{bmatrix} \mathbf{0} \\ \sum\limits_{j\in\mathcal{P}_{+i_1}, j\notin j\in\mathcal{P}_{-i},} \mathbf{C}_{i_1 j} x_j(k) \\ \vdots \\ \sum\limits_{j\in\mathcal{P}_{+i_n}, j\notin j\in\mathcal{P}_{-i},} \mathbf{C}_{i_n j} x_j(k) \end{bmatrix} \tag{7.14}$$

It should be noticed that the input of this neighborhood model is still the input of $\mathcal{S}_i$, and the inputs of $\mathcal{S}_j \in \mathcal{P}_{-i}$, $j \neq i$ are regarded as disturbances. It is because each local MPC can only determine the manipulated variables of the corresponding subsystem.

Due to the unit delay introduced by the network (see Assumption 1), the information of other subsystems is available only after one sampling time interval. Therefore, controller $\mathcal{C}_i$ uses $\hat{\hat{\mathbf{w}}}_i(k+l-s|k-1))$ and $\hat{\mathbf{v}}_i(k+l|k-1)$ computed on the basis of information related to time $k-1$ to estimate the interactions, and the initial states of downstream neighbors are substituted with $\hat{\mathbf{x}}^{\mathrm{T}}_{i_h}(k|k-1)(h = 1, \dots, n)$. For $i = 1, \dots, m$, define

$$\hat{\mathbf{x}}_i(k|k) = \begin{bmatrix} \mathbf{x}_i^{\mathrm{T}}(k\,|k) & \hat{\mathbf{x}}^{\mathrm{T}}_{i_1}(k|k-1) & \cdots & \hat{\mathbf{x}}^{\mathrm{T}}_{i_n}(k|k-1) \end{bmatrix}^{\mathrm{T}} \tag{7.15}$$

Then the states and outputs of the downstream neighbors in l-step ahead can be predicted by

$$\begin{cases} \hat{\hat{\mathbf{x}}}_i(k+l\,|k) = \hat{\mathbf{A}}_i^l \hat{\mathbf{x}}_i(k|k) + \sum\limits_{s=1}^{l} \hat{\mathbf{A}}_i^{s-1} \hat{\mathbf{B}}_i \mathbf{u}_i(k+l-s|k) \\ \qquad\qquad + \sum\limits_{s=1}^{l} \hat{\mathbf{A}}_i^{s-1} \hat{\hat{\mathbf{w}}}_i(k+l-s|k-1)) \\ \hat{\hat{\mathbf{y}}}_i(k+l|k) = \hat{\mathbf{C}}_i \hat{\hat{\mathbf{x}}}_i(k+l|k) + \hat{\mathbf{v}}_i(k+l|k-1) \end{cases} \tag{7.16}$$

7.2.2.3 Optimization Problem

For each independent controller $\mathcal{C}_i$, $i = 1, \dots, m$, the unconstrained ND-MPC problem with the prediction horizon P and control horizon M, $M < P$, at time k solves the following optimization problem:

$$\min_{\Delta \mathbf{U}_i(k,M|k)} \overline{J}_i(k) = \sum_{l=1}^{P} \left\| \hat{\hat{\mathbf{y}}}_i(k+l\,|k) - \mathbf{y}_i^d(k+l|k) \right\|^2_{\hat{\mathbf{Q}}_i} + \sum_{l=1}^{M} \left\| \Delta\mathbf{u}_i(k+l-1\,|k) \right\|^2_{\mathbf{R}_i}$$
$$s.t. \quad Eq.\ (7.16) \tag{7.17}$$

At time k, the interaction predictions $\hat{\hat{\mathbf{w}}}_i(k+l-1|k-1)$ and $\hat{\mathbf{v}}_i(k+l|k-1)$, together with $\hat{\mathbf{x}}_i(k|k)$, are used to resolve the optimization problem (7.17) in each $\mathcal{C}_i$, $(i = 1, \dots, m)$. The first element of the optimal solution $\Delta\mathbf{U}_i^*(k)$ is selected and $\mathbf{u}_i(k) = \mathbf{u}_i(k-1) + \Delta\mathbf{u}(k|k)$ is applied to $\mathcal{S}_i$. Then, by (7.15), each local controller estimates the future state trajectory over the prediction horizon and broadcasts it in a network together with the optimal control sequence over the control horizon. At time $k+1$, each local controller uses this information for evaluating the interaction predictions and the whole procedure is repeated.

The only information that each $\mathcal{C}_i$, $i = 1, \ldots, m$, needs is the future behavior of $\mathcal{S}_j \in \mathcal{P}_i$ and $\mathcal{S}_g \in \mathcal{P}_j$. Similarly, $\mathcal{C}_i$ broadcasts the future behavior of $\mathcal{S}_i$ to the controller of $\mathcal{S}_j \in \mathcal{P}_i$ and that of $\mathcal{S}_g \in \mathcal{P}_j$.

7.2.3 Closed-Form Solution

The main result of this subsection is the computation of the closed-form solution to the ND-MPC proposed. For this purpose, expressions of the interaction prediction and the state prediction are provided first. Define that

$$\tilde{\mathbf{A}}_i^{(1)} = \text{diag}_P \left\{ \times \underbrace{\begin{bmatrix} \mathbf{A}_{i,1} & \cdots & \mathbf{A}_{i,i-1} & \mathbf{0}_{n_{x_i}\times n_{x_i}} & \mathbf{A}_{i,i+1} & \cdots & \mathbf{A}_{i,i_1-1} & \mathbf{0}_{n_{x_i}\times n_{x_{i1}}} & \mathbf{A}_{i,i_1+1} & \cdots & \mathbf{A}_{i,i_n-1} & \mathbf{0}_{n_{x_i}\times n_{x_{in}}} & \mathbf{A}_{i,i_n+1} & \cdots & \mathbf{A}_{i,m} \\ \mathbf{A}_{i_1,1} & \cdots & \mathbf{A}_{i_1,i-1} & \mathbf{0}_{n_{x_{i1}}\times n_{x_i}} & \mathbf{A}_{i_1,i+1} & \cdots & \mathbf{A}_{i_1,i_1-1} & \mathbf{0}_{n_{x_{i1}}\times n_{x_{i1}}} & \mathbf{A}_{i_1,i_1+1} & \cdots & \mathbf{A}_{i_1,i_n-1} & \mathbf{0}_{n_{x_{i1}}\times n_{x_{in}}} & \mathbf{A}_{i_1,i_n+1} & \cdots & \mathbf{A}_{i_1,m} \\ \vdots & \cdots & \vdots & \vdots & \vdots & \cdots & \vdots & \vdots & \vdots & \cdots & \vdots & \vdots & \vdots & \cdots & \vdots \\ \mathbf{A}_{i_n,1} & \cdots & \mathbf{A}_{i_n,i-1} & \mathbf{0}_{n_{x_{in}}\times n_{x_i}} & \mathbf{A}_{i_n,i+1} & \cdots & \mathbf{A}_{i_n,i_1-1} & \mathbf{0}_{n_{x_{in}}\times n_{x_{i1}}} & \mathbf{A}_{i_n,i_1+1} & \cdots & \mathbf{A}_{i_n,i_n-1} & \mathbf{0}_{n_{x_{im}}\times n_{x_{in}}} & \mathbf{A}_{i_m,i_n+1} & \cdots & \mathbf{A}_{i_m,m} \end{bmatrix}}_{n \text{ columns}} \right\} \tag{7.18}$$

$$\tilde{\mathbf{A}}_i^{(2)} = \text{diag} \left(\underbrace{\begin{bmatrix} \mathbf{0}_{n_{x_i}\times n_{x_1}} & \cdots & \mathbf{0}_{n_{x_i}\times n_{x_{i1-1}}} & \mathbf{A}_{i,i_1} & \mathbf{0}_{n_{x_i}\times n_{x_{i1+1}}} & \cdots & \mathbf{0}_{n_{x_i}\times n_{x_{in-1}}} & \mathbf{A}_{i,i_n} & \mathbf{0}_{n_{x_i}\times n_{x_{in+1}}} & \cdots & \mathbf{0}_{n_{x_i}\times n_{x_m}} \\ \mathbf{0}_{n_{x_{i1}}\times n_{x1}} & \cdots & \mathbf{0}_{n_{x_{i1}}\times n_{x_{i1-1}}} & \mathbf{A}_{i_1,i_1} & \mathbf{0}_{n_{x_{i1}}\times n_{x_{i1+1}}} & \cdots & \mathbf{0}_{n_{x_{i1}}\times n_{x_{in-1}}} & \mathbf{A}_{i_1,i_n} & \mathbf{0}_{n_{x_{i1}}\times n_{x_{in+1}}} & \cdots & \mathbf{0}_{n_{x_i}\times n_{x_m}} \\ \vdots & \cdots & \vdots & \vdots & \vdots & \cdots & \vdots & \vdots & \vdots & \cdots & \vdots \\ \mathbf{0}_{n_{x_{in}}\times n_{x_1}} & \cdots & \mathbf{0}_{n_{x_{i1}}\times n_{x_{i1-1}}} & \mathbf{A}_{i_n,i_1} & \mathbf{0}_{n_{x_{i1}}\times n_{x_{i1+1}}} & \cdots & \mathbf{0}_{n_{x_{i1}}\times n_{x_{in-1}}} & \mathbf{A}_{i_n,i_n} & \mathbf{0}_{n_{x_{i1}}\times n_{x_{in+1}}} & \cdots & \mathbf{0}_{n_{x_i}\times n_{x_m}} \end{bmatrix}}_{n \text{ columns}}, \right.$$

$$\left. \text{diag}_{P-1} \left\{ \mathbf{0}_{\sum_{l\in\mathcal{P}_{+i}} n_{x_l} \times \sum_{l=1}^{m} n_{x_l}} \right\} \right) \tag{7.19}$$

$$\tilde{\tilde{\mathbf{B}}}_i = \text{diag}_P \left\{ \begin{bmatrix} \mathbf{B}_{i,1} & \cdots & \mathbf{B}_{i,i-1} & \mathbf{0}_{n_{x_i}\times n_{u_i}} & \mathbf{B}_{i,i+1} & \cdots & \mathbf{B}_{i,m} \\ \mathbf{B}_{i_1,1} & \cdots & \mathbf{B}_{i_1,i-1} & \mathbf{0}_{n_{x_{i1}}\times n_{u_i}} & \mathbf{B}_{i_1,i+1} & \cdots & \mathbf{B}_{i_1,m} \\ \vdots & \ddots & \vdots & \vdots & \vdots & \cdots & \vdots \\ \mathbf{B}_{i_n,1} & \cdots & \mathbf{B}_{i_m,i-1} & \mathbf{0}_{n_{x_{in}}\times n_{u_i}} & \mathbf{B}_{i_n,i+1} & \cdots & \mathbf{B}_{i_n,m} \end{bmatrix} \right\} \tag{7.20}$$

$$\tilde{\mathbf{C}}_i = \text{diag}_P \left\{ \underbrace{\begin{bmatrix} \mathbf{C}_{i,1} & \cdots & \mathbf{C}_{i,i-1} & \mathbf{0}_{n_{y_i}\times n_{x_i}} & \mathbf{C}_{i,i+1} & \cdots & \mathbf{C}_{i,i_1-1} & \mathbf{0}_{n_{y_i}\times n_{x_{i1}}} & \mathbf{C}_{i,i_1+1} & \cdots & \mathbf{C}_{i,i_n-1} & \mathbf{0}_{n_{y_i}\times n_{x_{in}}} & \mathbf{C}_{i,i_n+1} & \cdots & \mathbf{C}_{i,m} \\ \mathbf{C}_{i_1,1} & \cdots & \mathbf{C}_{i_1,i-1} & \mathbf{0}_{n_{y_{i1}}\times n_{x_i}} & \mathbf{C}_{i_1,i+1} & \cdots & \mathbf{C}_{i_1,i_1-1} & \mathbf{0}_{n_{y_{i1}}\times n_{x_{i1}}} & \mathbf{C}_{i_1,i_1+1} & \cdots & \mathbf{C}_{i_1,i_n-1} & \mathbf{0}_{n_{y_{i1}}\times n_{x_{in}}} & \mathbf{C}_{i_1,i_n+1} & \cdots & \mathbf{C}_{i_1,m} \\ \vdots & \cdots & \vdots & \vdots & \vdots & \cdots & \vdots & \vdots & \vdots & \cdots & \vdots & \vdots & \vdots & \cdots & \vdots \\ \mathbf{C}_{i_n,1} & \cdots & \mathbf{C}_{i_n,i-1} & \mathbf{0}_{n_{y_{in}}\times n_{x_i}} & \mathbf{C}_{i_n,i+1} & \cdots & \mathbf{C}_{i_n,i_1-1} & \mathbf{0}_{n_{y_{in}}\times n_{x_{i1}}} & \mathbf{C}_{i_n,i_1+1} & \cdots & \mathbf{C}_{i_n,i_n-1} & \mathbf{0}_{n_{y_{in}}\times n_{x_{in}}} & \mathbf{C}_{i_n,i_n+1} & \cdots & \mathbf{C}_{i_n,m} \end{bmatrix}}_{n \text{ columns}} \right\} \tag{7.21}$$

where $\mathbf{A}_{i,j}$, $\mathbf{B}_{i,j}$, and $\mathbf{C}_{i,j}$ are the zero blocks of congruent dimensions if $\mathcal{S}_j \notin P_{+h}$ ($S_h \in \mathcal{P}_{-i}$).

Moreover, define

$$\tilde{\mathbf{\Gamma}} = \begin{bmatrix} \mathbf{0}_{(M-1)n_u \times n_u} & \mathbf{I}_{(M-1)n_u} \\ \mathbf{0}_{n_u \times (M-1)n_u} & \mathbf{I}_{n_u} \\ \vdots & \vdots \\ \mathbf{0}_{n_u \times (M-1)n_u} & \mathbf{I}_{n_u} \end{bmatrix}, n_u = \sum_{l=1}^{m} n_{u_l}, \tilde{\mathbf{B}}_i = \tilde{\tilde{\mathbf{B}}}_i \tilde{\mathbf{\Gamma}} \tag{7.22}$$

Then, the following lemmas can be deduced based on definitions (7.18)–(7.22). Proofs of the lemmas can be found in appendixes.

Lemma 7.1 ***(Interaction prediction)*** *Under Assumption 7.1, for each controller $\mathcal{C}_i$, $i = 1, \ldots, n$, the stacked predictions of the interaction vectors at time k, based on the information computed at time $k-1$, are given by*

$$\begin{cases} \hat{\hat{\mathbf{W}}}_i(k, P\,|k-1) = \tilde{\mathbf{A}}_{i1}\hat{\mathbf{X}}(k, P|k-1) + \tilde{\mathbf{B}}_i\mathbf{U}(k-1, M|k-1), \\ \hat{\hat{\mathbf{V}}}_i(k, P|k-1) = \tilde{\mathbf{C}}_i\hat{\mathbf{X}}(k, P|k-1) \end{cases} \tag{7.23}$$

Lemma 7.2 ***(State prediction)*** *Under Assumption 7.1, for each controller $\mathcal{C}_i$, $i = 1, \ldots, n$, the stacked predictions of the state and output of the downstream neighbors of subsystem $\mathcal{S}_i$ at time k are expressed by*

$$\begin{cases} \hat{\hat{\mathbf{X}}}_i(k+1, P\,|k) = \overline{\mathbf{S}}_i[\overline{\mathbf{A}}_i^{(1)}\hat{\mathbf{x}}(k|k) + \overline{\mathbf{B}}_i\mathbf{U}_i(k, M|k) \\ \qquad + \tilde{\mathbf{A}}_i\hat{\mathbf{X}}(k, P|k-1) + \tilde{\mathbf{B}}_i\mathbf{U}(k-1, M|k-1)], \\ \hat{\hat{\mathbf{Y}}}_i(k+1, P|k) = \overline{\mathbf{C}}_i\hat{\hat{\mathbf{X}}}_i(k+1, P|k) + \mathbf{T}_i\tilde{\mathbf{C}}_i\hat{\mathbf{X}}(k+1, P|k-1) \end{cases} \tag{7.24}$$

where

$$\overline{\mathbf{A}}_i = \begin{bmatrix} \overline{\mathbf{A}}_i^{(1)} & \overline{\mathbf{A}}_i^{(2)} \end{bmatrix} = \begin{bmatrix} \hat{\mathbf{A}}_i^{(1)} & \hat{\mathbf{A}}_i^{(2)} \\ \mathbf{0}_{Pn_{\hat{x}_i} \times n_{x_i}} & \mathbf{0}_{Pn_{\hat{x}_i} \times (n_{\hat{x}_i} - n_{x_i})} \end{bmatrix}, n_x = \sum_{l=1}^{n} n_{x_l}, \tag{7.25}$$

$$\mathbf{T}_i = \begin{bmatrix} \mathbf{0}_{(P-1)n_{\hat{y}} \times n_{\hat{y}}} & \mathbf{I}_{(P-1)n_{\hat{y}}} \\ \mathbf{0}_{n_{\hat{y}} \times (P-1)n_{\hat{y}}} & \mathbf{I}_{n_{\hat{y}}} \end{bmatrix} \tag{7.26}$$

$$\overline{\mathbf{B}}_i = \begin{bmatrix} \multicolumn{2}{c}{\mathrm{diag}_M\left(\hat{\mathbf{B}}_i\right)} \\ \mathbf{0}_{n_{\hat{x}_i} \times (M-1)n_{u_i}} & \hat{\mathbf{B}}_i \\ \vdots & \vdots \\ \mathbf{0}_{n_{\hat{x}_i} \times (M-1)n_{u_i}} & \hat{\mathbf{B}}_i \end{bmatrix} \tag{7.27}$$

$$\overline{\mathbf{S}}_i = \begin{bmatrix} \hat{\mathbf{A}}_i^{0} & \cdots & \mathbf{0} \\ \vdots & \ddots & \vdots \\ \hat{\mathbf{A}}_i^{P-1} & \cdots & \hat{\mathbf{A}}_i^{0} \end{bmatrix} \tag{7.28}$$

$$\overline{\mathbf{C}}_i = \mathrm{diag}_P\{\hat{\mathbf{C}}_i\} \tag{7.29}$$

The proposed networked DMPC problem stated in Equation (7.17) now can be formulated as a QP by the introduction of the following matrices:

$$
\begin{aligned}
\mathbf{S}_i &= \overline{\mathbf{C}}_i\overline{\mathbf{S}}_i, \\
\mathbf{N}_i &= \mathbf{S}_i\overline{\mathbf{B}}_i\overline{\boldsymbol{\Gamma}}_i \\
\underset{(M \text{ blocks})}{\boldsymbol{\Gamma}'_i} &= \begin{bmatrix} \mathbf{I}_{n_{u_i}} \\ \vdots \\ \mathbf{I}_{n_{u_i}} \end{bmatrix} \\
\underset{(M\times M \text{ blocks})}{\overline{\boldsymbol{\Gamma}}_i} &= \begin{bmatrix} \mathbf{I}_{n_{u_i}} & \cdots & 0 \\ \vdots & \ddots & \vdots \\ \mathbf{I}_{n_{u_i}} & \cdots & \mathbf{I}_{n_{u_i}} \end{bmatrix} \\
\overline{\mathbf{Q}}_i &= \text{diag}_P\{\hat{\mathbf{Q}}_i\} \\
\overline{\mathbf{R}}_i &= \text{diag}_P\{\mathbf{R}_i\}
\end{aligned}
\tag{7.30}
$$

Lemma 7.3 ***(Quadratic program)*** *Under Assumption 7.1, each controller $\mathcal{C}_i$, $i = 1, \ldots, m$, has to solve at time k the following optimization problem:*

$$
\min_{\Delta \mathrm{U}(k,M|k)} [\Delta \mathrm{U}_i^T k, M|k)\mathbf{H}_i\Delta \mathrm{U}_i(k, M|k) - \mathbf{G}(k+1, P)\Delta \mathrm{U}_i(k, M|k)] \tag{7.31}
$$

where the positive definite matrix $\mathbf{H}_i$ has the form

$$
\mathbf{H}_i = \mathbf{N}_i^T\overline{\mathbf{Q}}_i\mathbf{N}_i + \overline{\mathbf{R}}_i \tag{7.32}
$$

and

$$
\mathbf{G}_i(k+1, P|k) = 2\mathbf{N}_i^T\overline{\mathbf{Q}}_i[\mathbf{Y}_i^d(k+1, P|k) - \hat{\mathbf{Z}}(k+1, P|k)] \tag{7.33}
$$

with

$$
\begin{aligned}
\hat{\mathbf{Z}}_i(k+1, P|k) = {} & \mathbf{S}_i[\overline{\mathbf{B}}_i\boldsymbol{\Gamma}'_i\mathbf{u}_i(k-1) + \overline{\mathbf{A}}_i^{(1)}\hat{x}(k|k) + \tilde{\mathbf{A}}_i\hat{\mathbf{X}}(k, P|k-1) \\
& +\tilde{\mathbf{B}}_i\mathbf{U}(k-1, M|k-1)] + \mathbf{T}_i\tilde{\mathbf{C}}_i\hat{\mathbf{X}}(k+1, P|k-1)
\end{aligned}
\tag{7.34}
$$

Making use of these definitions:

$$
\mathbf{K}_i = \boldsymbol{\Gamma}_i\overline{\mathbf{K}}_i, \boldsymbol{\Gamma}_i = \begin{bmatrix} \mathbf{I}_{n_{u_i}} & \mathbf{0}_{n_{u_i}\times Mn_{u_i}} \end{bmatrix}, \overline{\mathbf{K}}_i = \mathbf{H}_i^{-1}\mathbf{N}_i^{\mathrm{T}}\overline{\mathbf{Q}}_i \tag{7.35}
$$

The proof can be found in Appendix C. Based on Lemma 7.3, the following theorem can be deduced.

Theorem 7.1 ***(Closed-form solution)*** *Under Assumption 7.1, for each controller C_i, $i = 1, \ldots, n$, the closed form of the control law applied at time k at controller C_i to subsystem S_i is given by*

$$
\mathbf{u}_i(k) = \mathbf{u}(k-1) + \mathbf{K}_i[\mathbf{Y}_i^d(k+1, P|k) - \hat{\mathbf{Z}}_i(k+1, P|k)] \tag{7.36}
$$

The proof can be found in Appendix D.

Remark 7.1 *The resulting complexity to obtain the closed-form solution for the local subsystem* S_i *is mainly given by the inversion of matrix* $\mathbf{H}_i$*. Considering that the size of matrix* $\mathbf{H}_i$ *equals* $M \cdot n_{u_i}$*, the complexity of the inversion algorithm is* $O(M^3, n_{u_i}^3)$ *if using the Gauss–Jordan algorithm. Therefore, the total computational complexity of the distributed solution is* $O\left(M^3, \sum_{i=1}^{n} n_{u_i}^3\right)$*, while the computational complexity of the centralized MPC is* $O\left(M^3, \left(\sum_{i=1}^{n} n_{u_i}\right)^3\right)$.

7.2.4 Stability Analysis

On the basis of the closed-form solution stated in Theorem 7.1, the closed-loop dynamics can be specified and the stability condition can be verified by analyzing the closed-loop dynamic matrix. Thus, the following theorem is obtained.

Theorem 7.2 ***(Networked DMPC stability)*** *The closed-loop system given by the feedback connection of plant S with the set of independent controller* C_i*,* $i = 1, \ldots, n$*, whose closed-form control laws are given by Equation (7.36), is asymptotically stable if and only if*

$$|\lambda_j\{\mathbf{A}_N\}| < 1, \forall j = 1, \ldots, n_N \tag{7.37}$$

where $n_N = Pn_x + n_x + 2Mn_u$ *is the order of the global closed-loop system.*

$$\mathbf{A}_N = \begin{bmatrix} \mathbf{A} & \mathbf{0} & \mathbf{B\Gamma} & \mathbf{0} \\ \mathbf{L\bar{S}\bar{A}} & \mathbf{L\bar{S}\tilde{A}\Omega} & \mathbf{L\bar{S}\bar{B}} & \mathbf{L\bar{S}\tilde{B}\Pi} \\ \mathbf{\Theta A} + \mathbf{\Phi L\bar{S}\bar{A}} & \mathbf{\Phi L\bar{S}\tilde{A}\Omega} & \mathbf{\Theta B\Gamma} + \mathbf{\Phi L\bar{S}\bar{B}} + \mathbf{\Psi} & \mathbf{\Phi L\bar{S}\tilde{B}\Pi} \\ \mathbf{0} & \mathbf{0} & \mathbf{I}_{Mn_u} & \mathbf{0} \end{bmatrix} \tag{7.38}$$

The proof can be found in Appendix. E.

Remark 7.2 *It should be noticed that the first two block rows of the dynamic matrix* $\mathbf{A}_N$ *depend on element of matrix* $\mathbf{A}$ *(the first two block columns) and the element of matrix* $\mathbf{B}$ *(in the last two block columns), while the third block row depends on process matrices* $\mathbf{A}$*,* $\mathbf{B}$*, and* $\mathbf{C}$*, weight matrices* $\mathbf{Q}_i$*,* $\mathbf{R}_i$ *and horizons* $\mathbf{P}$ *and* $\mathbf{M}$*. This fact suggests a key for the design of networked-DMPC. The degree of freedom available to the designer is on the choices of weight matrices* $\mathbf{Q}_i$*,* $\mathbf{R}_i$*, and horizons* $\mathbf{P}$ *and* $\mathbf{M}$*, which introduce significant modifications on the third block row of matrix* $\mathbf{A}_N$*.*

7.2.5 Analysis of Performance

To explain the essential differences between the optimization problem with the neighborhood optimization index and the optimization problem with the local performance index, for each controller C_i, $i = 1, \ldots, n$, the optimization problem (7.17) of the proposed networked DMPC is rewritten into the following form:

$$\min_{\Delta \mathbf{U}_i(k,M|k)} \sum_{i=1}^{m} \left[\sum_{l=1}^{P} \left\| \hat{\mathbf{y}}_i(k+l|k) - \mathbf{y}_i^d(k+l|k) \right\|_{\mathbf{Q}_i}^2 + \sum_{l=1}^{M} \| \Delta \mathbf{u}_i(k+l-1|k) \|_{\mathbf{R}_i}^2 \right]$$

$$\text{s.t.} \quad \begin{bmatrix} \hat{\mathbf{x}}_i(k+l+1|k) \\ \hat{\mathbf{x}}_{i_1}(k+l+1|k) \\ \vdots \\ \hat{\mathbf{x}}_{i_n}(k+l+1|k) \end{bmatrix} = \begin{bmatrix} \mathbf{A}_{ii} & \mathbf{A}_{ii_1} & \cdots & \mathbf{A}_{ii_n} \\ \mathbf{A}_{i_1 i} & \mathbf{A}_{i_1 i_1} & \cdots & \mathbf{A}_{i_n i_n} \\ \vdots & \vdots & \ddots & \vdots \\ \mathbf{A}_{i_n i} & \mathbf{A}_{i_n i_1} & \cdots & \mathbf{A}_{i_n i_n} \end{bmatrix} \begin{bmatrix} \hat{\mathbf{x}}_i(k+l|k) \\ \hat{\mathbf{x}}_{i_1}(k+l|k) \\ \vdots \\ \hat{\mathbf{x}}_{i_n}(k+l|k) \end{bmatrix} + \begin{bmatrix} \mathbf{B}_{ii} \\ \mathbf{B}_{i_1 i} \\ \vdots \\ \mathbf{B}_{i_n i} \end{bmatrix} \mathbf{u}_i(k+l|k) + \hat{\mathbf{w}}_i(k+l|k-1)) \tag{7.39}$$

$$\hat{\mathbf{x}}_j(k+l+1|k) = \hat{\mathbf{x}}_j(k+l+1|k-1), (j \notin \mathcal{P}_{-i})$$
$$\hat{\mathbf{y}}_i(k+l|k) = \mathbf{C}_i \hat{\mathbf{x}}_i(k+l|k) + \hat{\mathbf{v}}_i(k+l|k-1), (i = 1, \ldots, m)$$
$$\Delta \mathbf{u}_j(k+l-1|k) = \Delta \mathbf{u}_j(k+l-1|k-1), (j \neq i)$$

If using the local performance index (7.4), as in [46], for each controller $\mathcal{C}_i$, $i = 1, \ldots, m$, the optimization problem of the distributed MPC can be written in the following form:

$$\min_{\Delta \mathbf{U}_i(k,M|k)} \sum_{j=1}^{m} \left[\sum_{l=1}^{P} \left\| \hat{\mathbf{y}}_j(k+l|k) - \mathbf{y}_j^d(k+l|k) \right\|_{\mathbf{Q}_j}^2 + \sum_{l=1}^{M} \left\| \Delta \mathbf{u}_j(k+l-1|k) \right\|_{\mathbf{R}_j}^2 \right]$$

$$\text{s.t.} \quad \hat{x}_i(k+l+1|k) = \mathbf{A}_{ii} x_i(k+l|k) + \mathbf{B}_{ii} u_i(k+l|k) + \hat{\mathbf{w}}_i(k+l|k-1));$$
$$\hat{\mathbf{y}}_i(k+l|k) = \hat{\mathbf{C}}_i \hat{x}_i(k+l|k) + \hat{\mathbf{v}}_i(k+l|k-1)$$
$$\hat{\mathbf{y}}_j(k+l|k) = \hat{\mathbf{y}}_j(k+l|k-1) \quad (j \neq i)$$
$$\Delta \mathbf{u}_j(k+l-1|k) = \Delta \mathbf{u}_j(k+l-1|k-1)(j \neq i) \tag{7.40}$$

It can be seen that the performance indices of problems (7.39) and (7.40) are similar to each other. In neighborhood optimization, the state evolutions of subsystem S_i and its downstream neighbors are solved together, and the impact of control decision $\Delta \mathbf{U}_i(k, M|k)$ on the states of $\mathcal{S}_i$ and its downstream neighbors is fully considered. However, in problem (7.40), only the state evolution of subsystem S_i is determined by $\Delta \mathbf{U}_i(k, M|k)$ and the states of other subsystems are substituted by the estimations at time $k - 1$. It is clear that the predictive model in problem (7.39) is closer to the system model (5.3). Thus, it is more reasonable to adopt neighborhood performance index rather than to use local performance index.

In fact, after several control periods, the control decision $\Delta \mathbf{U}_i(k, M|k)$ affects not only the downstream neighbors of S_i but also other subsystems (e.g., the downstream neighbors of the downstream neighbors of $\mathcal{S}_i$). Here, the interactions with other subsystems except downstream neighbors are neglected. If there is enough network band-width for employing iterative algorithm, these interactions can also be taken into account.

It should be noticed that each controller only communicates with its neighbors and its neighbors' neighbors in ND-MPC. Moreover, if each controller communicates with its neighbors twice within a sampling time interval, the information of its neighbors' neighbors can be obtained from its neighbors. That means only the information exchanging among neighborhood is required using this method. Thus, if one subsystem fails, the other subsystem unrelated to $\mathcal{S}_i$ can be run normally. The communication loads related to $\mathcal{S}_i$ are that $\mathcal{S}_i$ get its future states to its neighbors and sent its neighbor's states and inputs to its neighbors. This means that the maximum communication loads is

$$\sum_{i=1}^{m} n_i \left(Pn_{xi} + \sum_{j \in \mathcal{N}_i} Pn_{xj} \right) \tag{7.41}$$

In Equation (7.41), some information is calculated repeatedly; thus, Equation (7.41) is the maximum communication burden using ND-MPC. The required physical connections equal the number of nonzero elements in **A** subtract the rank of **A**.

Since the computational burden mainly comes from the complexity of the inversion algorithm, the computational burden is similar to (a little more than) that of the method proposed in [46]. The memory required is a little larger than that in [46] since the system matrice's dimension of each subsystem's states evolution equation is larger than that in [46]. However, the memory is not a problem with the modern computer technology. The efficiency of this method will be validated in the next section.

7.2.6 *Numerical Validation*

To illustrate the performance of the proposed method, application of this method to the accelerated cooling process (ACC) test rig is performed in one steel company in Shanghai, China.

7.2.6.1 Process Description

The ACC process, simplified in Figure 7.1, is used to cool a metal plate from initial temperature around 750–800°C down to final temperature in the range of 450–560°C. A constant cooling curve of plate is required in ACC, which helps a lot to strongly improve the mechanical characteristics of the corresponding products. The cooling area is partitioned into three sections: air cooling section, water cooling section, and re-reddening section, labeled A, B, and C, respectively. Fifteen cooling header units are uniformly spaced along section B. The number of cooling header units in operation (N) and the water flux of each cooling unit (F) can be adjusted separately. The temperature drop is caused by the heat radiation in sections A and C, and caused by both radiation and water cooling in section B [102, 103]. Four pyrometers $T_{P1} - T_{P4}$ are located in the positions of 13.5 m, 58.6 m, 89 m, and 109.5 m, respectively. The temperatures of the plate inside cooling section are measured by soft sensors.

The control objective is to control the location-dependent temperatures at location $l_1, l_2, \ldots, l_n$ to be consistent with the reference temperature denoted with $\mathbf{y}^d = \begin{bmatrix} y_1^d & y_2^d & \cdots & y_m^d \end{bmatrix}^{\mathrm{T}}$ through adjusting the flux of each water cooling header unit and plate velocity. (l_0 is the location of T_{P1}; l_i, $i = 1, 2, \ldots, 15$, is the entry of the ith cooling header unit; l_{m-1} is the exit of section B; and l_m is the location of T_{P3}.)

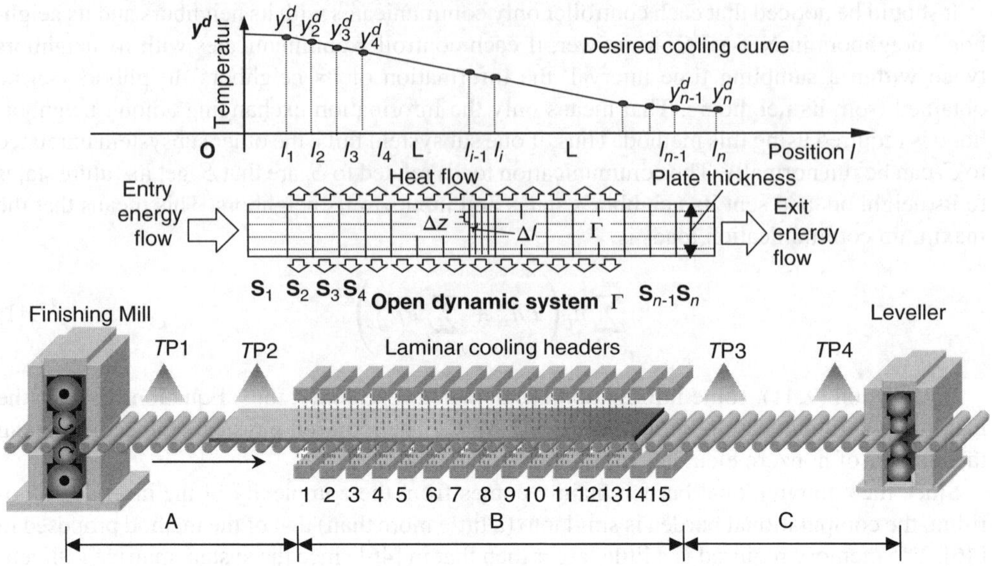

Figure 7.1 ACC process for middle and heavy plates

7.2.6.2 System State-Space Model

Consider the overall system of the cooling area from the point of view of geometrically distributed setting system, the geometrical locations of temperature sensor T_{P2} and T_{P3} as well as the plate top and bottom sides represent an open thermodynamic system $\mathbf{\Gamma}$. The overall system, therefore, can be dismembered into n subsystems in accordance with the location coordinate l. The sth subsystem ranges from l_{s-1} to l_s ($s = 1, 2, \ldots, m$) as shown in Figure 7.5. The input of sth subsystem is the water flux of corresponding header and the output is the plate temperature at location l_s.

For the numeric treatment of heat conductivity through plate thickness, the system $\mathbf{\Gamma}$ is broken into L layers, and each subsystem is divided into n_s volumes in the l-direction. Denote the temperature of ith in the z-direction and jth in the l-direction volume with $x_s^{(i,j)}$ and set the sampling time be Δt. Then the following linear state-space representation of subsystem $\mathcal{S}_s$ can be deduced [19]:

$$\begin{cases} \mathbf{x}_s(k+1) = \mathbf{A}_{ss} \cdot \mathbf{x}_s(k) + \mathbf{B}_{ss} \cdot u_s(k) + \mathbf{D}_{s,s-1} \cdot \mathbf{x}_{s-1}(k) \\ y_s(k) = \mathbf{C}_{ss} \cdot \mathbf{x}_s(k) \\ s = 1, 2, \ldots, m \end{cases} \tag{7.42}$$

where

$$\mathbf{x}_s = \left[(\mathbf{x}_{s,1})^{\mathrm{T}} \quad (\mathbf{x}_{s,2})^{\mathrm{T}} \quad \cdots \quad (\mathbf{x}_{s,n_s})^{\mathrm{T}} \right]^{\mathrm{T}}$$

$$\mathbf{x}_{s,j} = \left[x_s^{(1,j)} \quad x_s^{(2,j)} \quad \cdots \quad x_s^{(L,j)} \right]^{\mathrm{T}}$$

$j = 1, 2, \ldots, n_s$ is the state vector of subsystem $\mathcal{S}_s$, y_s is the average temperature of the last column volumes of subsystem $\mathcal{S}_s$, u_s is the input of subsystem $\mathcal{S}_s$, and there is a fixed relationship between u_s and the water flux in subsystem $\mathcal{S}_s$. $\mathbf{A}_{ss}$, $\mathbf{B}_{ss}$, $\mathbf{D}_{s,s-1}$, and $\mathbf{C}_{ss}$ are coefficient matrices of subsystem $\mathcal{S}_s$ with

$$\mathbf{A}_{ss} = \begin{bmatrix} \boldsymbol{\Phi}_s^{(1)} \cdot \boldsymbol{\Lambda} & \mathbf{0} & \cdots & \mathbf{0} \\ \mathbf{0} & \boldsymbol{\Phi}_s^{(2)} \cdot \boldsymbol{\Lambda} & & \vdots \\ \vdots & & \ddots & \mathbf{0} \\ \mathbf{0} & \cdots & \mathbf{0} & \boldsymbol{\Phi}_s^{(n_s)} \cdot \boldsymbol{\Lambda} \end{bmatrix}$$

$$+ \begin{bmatrix} (1-\gamma)\mathbf{I}_L & \mathbf{0} & \cdots & \mathbf{0} \\ \gamma \mathbf{I}_L & (1-\gamma)\mathbf{I}_L & \ddots & \vdots \\ \vdots & \ddots & \ddots & \mathbf{0} \\ \mathbf{0} & \cdots & \gamma \mathbf{I}_L & (1-\gamma)\mathbf{I}_L \end{bmatrix}$$

$$\mathbf{B}_{ss} = \begin{bmatrix} \boldsymbol{\Psi}_s^{(1)} \\ \vdots \\ \boldsymbol{\Psi}_s^{(n_s)} \end{bmatrix}$$

$$\mathbf{C}_{ss} = m^{-1} \cdot \left[\mathbf{0}^{1\times L\cdot(n_s-1)} \quad \mathbf{1}^{1\times L}\right]$$

$$\mathbf{D}_{s,s-1} = \begin{bmatrix} \mathbf{0}^{L\times L\cdot(n_s-1)} & \gamma \mathbf{I}_L \\ \mathbf{0}^{L(n_s-1)\times L\cdot(n_s-1)} & \mathbf{0}^{L\cdot(n_s-1)\times L} \end{bmatrix}$$

$$\boldsymbol{\Phi}_s^{(j)} = \begin{bmatrix} a\left(\breve{x}_s^{(1,j)}\right) & \cdots & 0 \\ \vdots & \ddots & \vdots \\ 0 & \cdots & a(\breve{x}_s^{(L,j)}) \end{bmatrix}$$

$$\boldsymbol{\Psi}_s^{(j)}(\mathbf{x}_s) = \begin{bmatrix} \theta_s^{(1,j)}(\breve{x}_s^{(1,j)} - x_\infty) \cdot \beta(\breve{x}_s^{(1,j)}) \\ \mathbf{0}^{(L-2)\times 1} \\ \theta_s^{(L,j)}(\breve{x}_s^{(L,j)} - x_\infty) \cdot \beta(\breve{x}_s^{(L,j)}) \end{bmatrix}$$

$$\boldsymbol{\Lambda} = \begin{bmatrix} -1 & 1 & 0 & \cdots & 0 \\ 1 & -2 & 1 & \ddots & \vdots \\ 0 & \ddots & \ddots & \ddots & 0 \\ \vdots & \ddots & 1 & -2 & 1 \\ 0 & \cdots & 0 & 1 & -1 \end{bmatrix}$$

$$\mathbf{I}_L \in \mathbb{R}^{L\times L}$$

$$\begin{cases} \theta_s^{(i,j)} = (\breve{x}_s^{(i,j)}/x)^a, & s \in \mathcal{C}_\mathcal{W} \\ \theta_s^{(i,j)} = h_{air}(\breve{x}_s^{(i,j)}), & s \in \mathcal{C}_\mathcal{A} \end{cases}$$

$$\begin{cases} u_s = 2186.7 \times 10^{-6} \times a \cdot \left(v/v_0\right)^b \times (F_s/F_0)^c, & s \in \mathcal{C}_\mathcal{W} \\ u_s = 1, s \in \mathcal{C}_\mathcal{A} \end{cases}$$

$$a(x_s^{(i,j)}) = -\Delta t \cdot \lambda(x_s^{(i,j)})/(\Delta z^2 \rho(x_s^{(i,j)}))c_p(x_s^{(i,j)}))$$

$$\beta(x_s^{(i,j)}) = \Delta t \cdot a(x_s^{(i,j)})/\lambda(x_s^{(i,j)})$$

$$\gamma = \Delta t \cdot v/\Delta l, \quad i = 1, 2, \dots, L,\ j = 1, 2, \dots, n_s$$

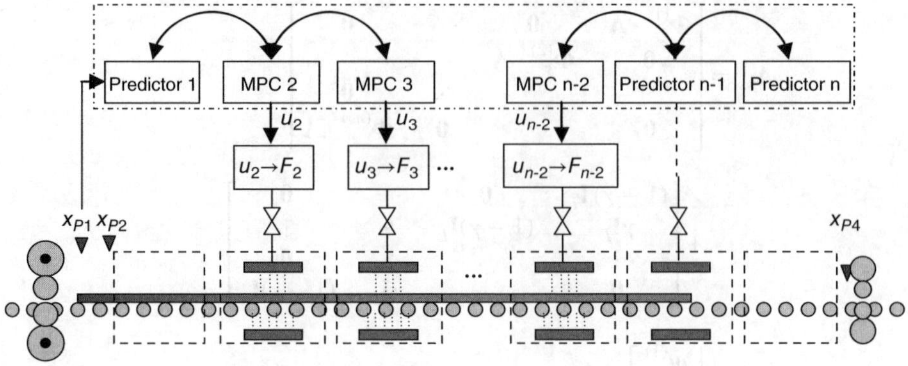

Figure 7.2 Control strategy of ACC

where Δl and Δz are the longitude and thickness of each volume, respectively, ρ is the plate density, c_p is the specific heat capacity, λ is the heat conductivity, v is the plate velocity, $\breve{x}_s^{(i,j)}$ is the equilibrium state of $\mathcal{S}_s$, $\mathcal{C}_{\mathcal{W}}$ is the set of subsystems, in which the plate is cooled by water, C_A is the set of subsystems in which the plate is cooled major through radiation, F_s is the water flux of the header unit in S_s, F_0, v_0, a, b, and c are constants, and their detailed definitions are available in [19, 104]. The different value of ρ, c_p, and λ with different steel temperature can also be found in [104].

7.2.6.3 Control Strategy

With the development of computer technologies and control theory, many advanced control methodologies have been successfully applied to complex metallurgical processes [105–108]. As for the ACC process, the proposed N-DMPC is adopted in this work. As shown in Figure 7.2, each subsystem is controlled by a local MPC. As for the subsystems in which the corresponding cooling water header unit is closed, the local MPC is substituted with a predictor. The predictor estimates the future states of corresponding subsystem and broadcasts the estimations to its neighbors.

7.2.6.4 Performance of System

One X70 pipe steel plate is taken as an example. The parameters of this steel plate are listed in Table 7.2. And the equilibrium temperature of this plate is shown in Figure 7.3.

Set both the prediction horizon (P) and control horizon (M) of each local MPC equal to 10. And set the weights of outputs and inputs equal to 1 in the optimization index. Let the starting cooling temperature (T_{P2}) in the whole process be 780°C. The resulting performance of the closed-loop system using centralized MPC, N-DMPC, and LCO-MPC described in [46] are presented in Figure 7.4, and the corresponding manipulated variables (unit: 1 m^{-2} min^{-1}) are shown in Figure 7.5.

It can be observed from Figures 7.4 and 7.5 that the performance of the closed-loop system using N-DMPC can be improved significantly comparing to that using the LCO-MPC

Table 7.2 The plate parameters and the operating points

Item	Value
Thickness of plate	19.28 mm
Length of plate	25 m
Environment temperature	25°C
Starting temperature	750–800°C
Desired final temperature	510–550°C
Average velocity of plate	1.6 m/s
Number of header opened	12
Sampling period	0.37 s

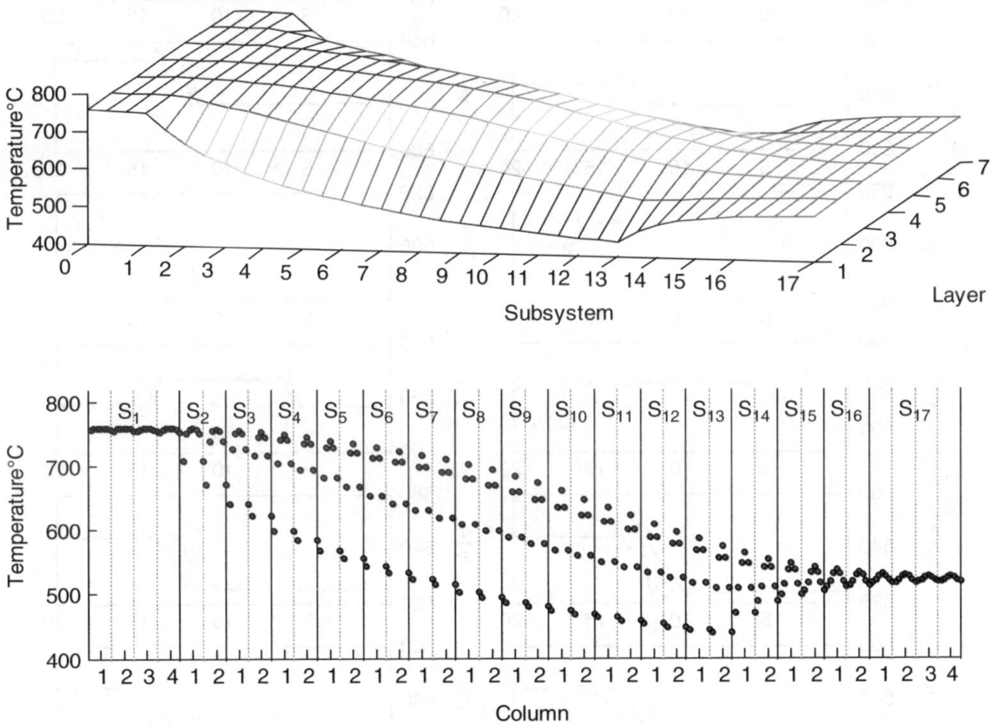

Figure 7.3 Equilibriums of states of entire system

proposed in [46]. Both the control decision and performance of the closed-loop system using N-DMPC are very close to those using the centralized MPC. Furthermore, there is less computation demand using the N-DMPC than using the centralized MPC. Thus, the N-DMPC is an effective method which could guarantee global performance improvement with fast computational speed and less communication burden.

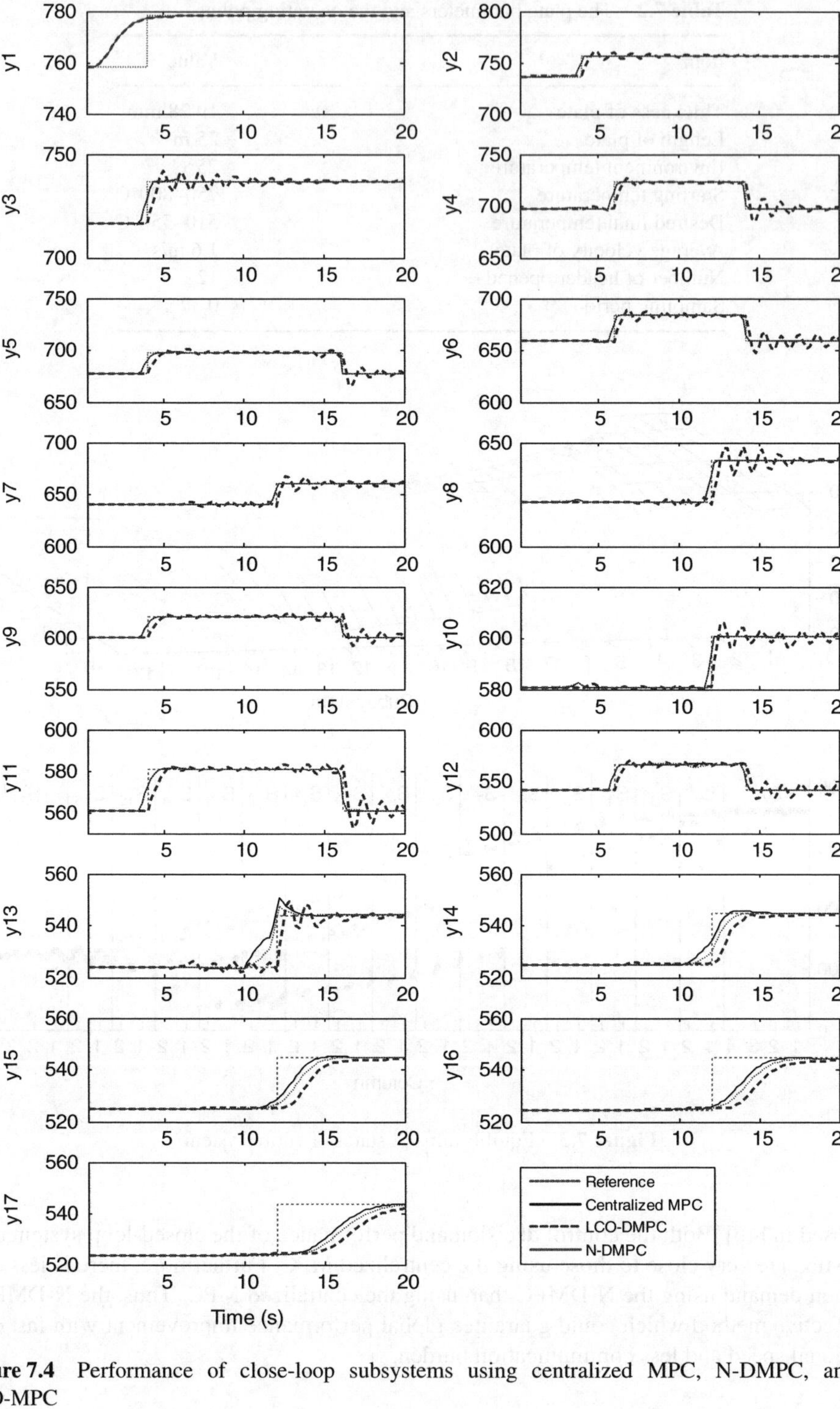

Figure 7.4 Performance of close-loop subsystems using centralized MPC, N-DMPC, and the LCO-MPC

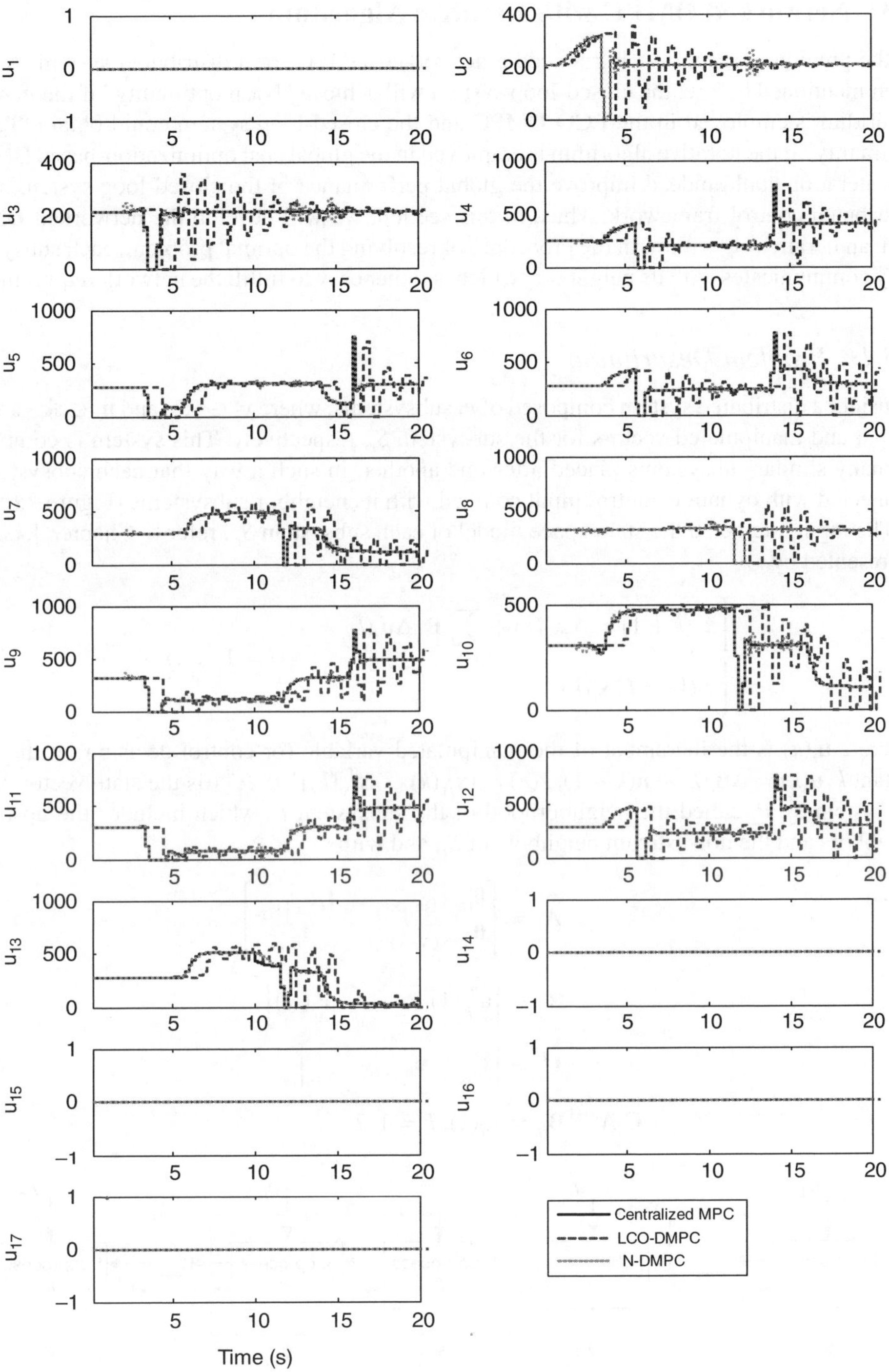

Figure 7.5 Flux of each header unit using centralized MPC, N-DMPC, and LCO-MPC

7.3 Networked DMPC with Iterative Algorithm

In the previous section, we introduced a networked DMPC for a distributed system. As has been mentioned before, the closed-loop system will achieve "Nash optimality" if the iterative algorithm is employed in the LCO-DMPC and the closed-loop system could obtain "Pareto optimality" if the iterative algorithm is employed in the global cost optimization-based DMPC. The iteration could indeed improve the global performance of the closed-loop system in the distributed control framework. Thus, in this section, we will develop the networked DMPC with an iterative algorithm. In the procedure of resolving the optimal problem, each subsystem only communicates with its neighbors, which is rather easy to fulfill the network requirements.

7.3.1 Problem Description

Consider a distributed system composed of m subsystems, where $y_i \in \mathbb{R}^{n_n}$ and $\mathbf{u}_i \in \mathbb{R}^{n_i}$ are the output and manipulated vectors for the subsystem S_i, respectively. This system is composed of many similar subsystems placed after one another, in such a way that each subsystem is connected with dynamic control input coupled with its neighbor subsystems (Figure 7.6).

The input interacted the state-space model of each subsystem S_i , refer to Chapter 4, can be represented by

$$\begin{cases} \mathbf{x}_i(k+1) = \mathbf{A}_i \boldsymbol{x}_i(k) + \sum\limits_{j\in\mathcal{P}_i} \mathbf{B}_{ij}\Delta\mathbf{u}_j(k) \\ \mathbf{y}_i(k) = \mathbf{C}_i\mathbf{x}_i(k) \end{cases} \quad (i = 1, \ldots, m) \tag{7.43}$$

where $\Delta\mathbf{u}_i(k)$ is the increment of the manipulated variable (or control decision) at the time instant k, $\mathbf{u}(k) = \Delta\mathbf{u}_i(k) + \mathbf{u}(k-1)$, $\mathbf{x}(k) = [\mathbf{x}_{i1}^{\mathrm{T}}(k)\cdots\mathbf{x}_{iN}^{\mathrm{T}}(k)]^{\mathrm{T}} \in \mathbb{R}^{n_{yi}}$ is the state vector of the subsystem S_i, P_i called the neighborhood of the subsystem i , which includes the upstream neighbors and the downstream neighbors of S_i, and with

$$\mathbf{A}_i = \begin{bmatrix} \mathbf{0}_{(N-1)n_{yi}\times n_{yi}} & \mathbf{I}_{(N-1)n_{yi}} \\ \mathbf{0}_{n_{yi}\times(N-1)n_{yi}} & \mathbf{I}_{n_{yi}} \end{bmatrix}$$

$$\mathbf{B}_{ij} = \begin{bmatrix} \mathbf{a}_{ij}^T(1) & \cdots & \mathbf{a}_{ij}^T(N) \end{bmatrix}^T$$

$$\mathbf{C}_i = \begin{bmatrix} \mathbf{I}_{n_{yi}} & \mathbf{0}_{n_{yi}\times(N-1)n_{yi}} \end{bmatrix}$$

$$\mathbf{C}_i\mathbf{A}_i^{k-1}\mathbf{B}_{ij} = \mathbf{a}_{ij}(k),\ k = 1, 2, \ldots$$

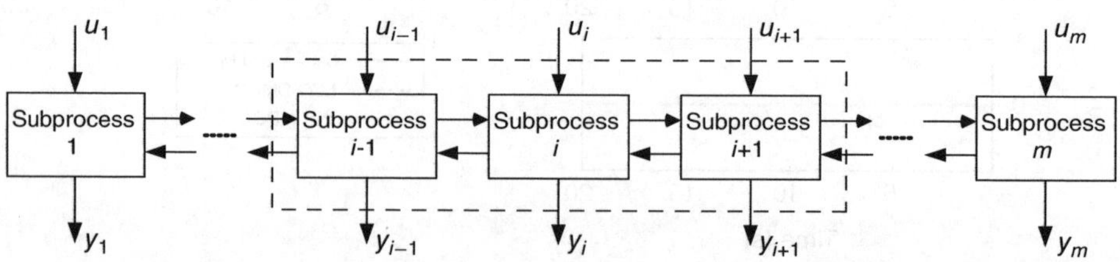

Figure 7.6 Diagram of a serially connected process

$\{\mathbf{C}_i, \mathbf{A}_i\}$ is observable, $\mathbf{a}_{ij}(k)$ $(k = 1, \dots, N; j \in \mathcal{P}_i)$ is the sampled output matrix array of the subsystem $\mathcal{S}_i$ for the subsystem $\mathcal{S}_j$ unit step input at the sampling time instant k, N is the modeling horizon. It can be seen that the output of each subsystem is dependent on control inputs of its own and its neighbor subsystems. It can be seen that the state-space model described in Equation (7.43) for each subsystem is developed directly from step responses without further identification. It is not necessary to identify the structure of the subsystem during the modeling. The optimization and/or control strategy can be employed with the usual state-space techniques.

The objective of the whole system is to regulate the system output to the expected values while keeping the performance index minimal and satisfying the above constraints. In other words, the control objective is minimizing the performance index $J(k)$ and can be expressed as

$$\min_{\Delta U_M(k)} J(k) = \sum_{s=1}^{P} \|\mathbf{r}\,(k+s) - \hat{\mathbf{y}}(k+s|k)\|_{\mathbf{Q}}^2 + \sum_{h=1}^{M} \|\Delta\mathbf{u}\,(k+h-1\,|k)\|_{\mathbf{R}}^2 \tag{7.44}$$

where

$$\begin{aligned}
\mathbf{r}(k+s) &= \begin{bmatrix}\mathbf{r}_1^T\,(k+s) & \cdots & \mathbf{r}_m^T(k+s)\end{bmatrix}^T\\
\hat{\mathbf{y}}(k+s|k) &= \begin{bmatrix}\hat{\mathbf{y}}_1^T\,(k+s\,|k) & \cdots & \hat{\mathbf{y}}_m^T(k+s|k)\end{bmatrix}^T\\
\Delta\mathbf{u}(k+h-1|k) &= [\Delta\mathbf{u}_1^T(k+h-1|k)\cdots\Delta\mathbf{u}_m^{\mathrm{T}}(k+h-1|k)]^{\mathrm{T}}\\
\Delta\mathbf{U}_M(k) &= \begin{bmatrix}\Delta\mathbf{U}_{1,M}^T\,(k) & \cdots & \Delta\mathbf{U}_{m,M}^T(k)\end{bmatrix}^T\\
\Delta\mathbf{U}_{i,M}(k) &= \begin{bmatrix}\Delta\mathbf{u}_i^T\,(k\,|k) & \cdots & \Delta\mathbf{u}_i^T\,(k+M-1|k)\end{bmatrix}^T
\end{aligned}$$

$\mathbf{r}(k+s)$ is the output reference signals of the whole system, $\hat{\mathbf{y}}(k+s|k) \in \mathbb{R}^{n_y}$ is the predictive output at the time instant $k+s$ based on the output at the time instant k, $\sum_{i=1}^{m} n_{yi} = n_y$, $\Delta\mathbf{u}(k+h-1|k) \in \mathbb{R}^{n_u}$ is the increment of the future manipulated vector, $\sum_{i=1}^{m} n_{ui} = n_u$, $\Delta\mathbf{u}_i(k|k) = \Delta\mathbf{u}_i(k)$; $\mathbf{Q} \in \mathbb{R}^{n_y \times n_y}$ and $\mathbf{R} \in \mathbb{R}^{n_u \times n_u}$ are output and control weighting matrices, which are positive definite and symmetrical, $\|\mathbf{z}\|_{\mathbf{S}}^2 = \mathbf{z}^T\mathbf{S}\mathbf{z}$, P is the prediction horizon, M is the control horizon, and $P > M$. Each manipulated vector $\mathbf{u}_i(k+h-1|k)$, $\Delta\mathbf{u}_i(k+h-1|k)$ and predictive output $\hat{y}_i(k+s|k)$ are subject to the physical constraints $\mathbf{u}_i^{\max}$, $\mathbf{u}_i^{\min}$, $\Delta\mathbf{u}_i^{\max}$, $\Delta\mathbf{u}_i^{\min}$, and $\mathbf{y}_i^{\max}$, $\mathbf{y}_i^{\min}$.

7.3.2 DMPC Formulation

For serially connected systems, supposing that the weighting matrices $\mathbf{Q}$ and $\mathbf{R}$ have block-diagonal forms in Equation (7.44), the global performance index can be decomposed in terms of the local indexes for each subsystem:

$$\min J(k) = \sum_{i=1}^{m} J_i(k) \tag{7.45}$$

$$J_i(k) = \sum_{s=1}^{P} \left\|\mathbf{r}_i\,(k+s) - \hat{\mathbf{y}}_i(k+s|k)\right\|_{\mathbf{Q}_i}^2 + \sum_{h=1}^{M} \left\|\Delta\mathbf{u}_i\,(k+h-1\,|k)\right\|_{\mathbf{R}_i}^2 \tag{7.46}$$

where

$$\mathbf{Q} = \text{block-diag}(\mathbf{Q}_1, \dots, \mathbf{Q}_m)$$

$$\mathbf{R} = \text{block-diag}(\mathbf{R}_1, \dots, \mathbf{R}_m)$$

$$\mathbf{Q}_i \in \mathbb{R}^{n_{yi} \times n_{yi}}, \mathbf{R}_i \in \mathbb{R}^{n_{ui} \times n_{ui}}$$

The local control decision is computed by solving the local optimization problem $J_i(k)$ with local input/output variables and constraints in the distributed MPC based on Nash optimality presented in [29]. The whole system will arrive at Nash equilibrium if the convergent condition of the algorithm is satisfied. However, the Nash optimal solution to the local optimization problem collectively is not equal to the global optimal control decision of the whole system, that is, there exists loss in control performance.

The new performance index for each subsystem can be improved by

$$\min_{\Delta U_{i,M}(k)} \overline{J}_i(k) = \sum_{j \in \mathcal{P}_i} J_j(k) \quad (i = 1, \dots, m) \tag{7.47}$$

It is noted that the new performance index for the subsystem $\mathcal{S}_i$ includes cost functions not only by the subsystem $\mathcal{S}_i$ but also by its neighbors, which is called neighborhood optimization. Cooperation between subsystems can be incorporated in the performance index of the local optimization problem by including cost functions of neighbor subsystems. At the time instant k, the future predictive output of the ith subsystem can be expressed as

$$\hat{\mathbf{y}}_i(k+s|k) = \mathbf{C}_i\mathbf{A}_i^s\mathbf{x}_i(k) + \sum_{j \in \mathcal{P}_i} \sum_{h=1}^{\min(M,s)} \mathbf{C}_i\mathbf{A}_i^{s-h}\mathbf{B}_{ij}\Delta\mathbf{u}_j(k+h-1|k)$$

$$s = 1, \dots, P \tag{7.48}$$

However, all the elements in $\mathbf{x}_i(k)$ are not directly measurable except $\mathbf{x}_{i1}(k) = \mathbf{y}_i(k)$, a predictive state observer for each subsystem can be constructed to estimate them as follows:

$$\begin{cases} \hat{\mathbf{x}}_i(k+1) = \mathbf{A}_i\hat{\mathbf{x}}_i(k) + \sum_{j \in \mathcal{P}_i} \mathbf{B}_{ij}\Delta\mathbf{u}_j(k) + \mathbf{V}_i(\mathbf{y}_i(k+1) - \hat{\mathbf{y}}_i(k+1)) \\ \hat{\mathbf{y}}_i(k+1) = \mathbf{C}_i\mathbf{A}_i\hat{\mathbf{x}}_i(k) + \sum_{j \in \mathcal{P}_i} \mathbf{C}_i\mathbf{B}_{ij}\Delta\mathbf{u}_j(k) \\ i = 1, \dots, m \end{cases} \tag{7.49}$$

where $\mathbf{V}_i$ is the observer feedback matrix.

With the notation $\tilde{\mathbf{x}}_i(k) = \mathbf{x}_i(k) - \hat{\mathbf{x}}_i(k)$, the observer dynamics for each subsystem can be described as

$$\tilde{\mathbf{x}}_i(k+1) = (\mathbf{I} - \mathbf{V}_i\mathbf{C}_i)\mathbf{A}_i\tilde{\mathbf{x}}_i(k) \tag{7.50}$$

It can be seen that will $\hat{\boldsymbol{x}}_i(k)$ converge to $\boldsymbol{x}_i(k)$ under a proper selection of $\mathbf{V}_i$. A more flexible error correction is naturally derived by employing the predictive state observer, and the stability for the prediction error correction can be guaranteed by analyzing the observer dynamics. Furthermore, the selection of the observer feedback matrix for each subsystem is independent of that of other subsystems.

The local optimization problem for each subsystem at the sampling time instant k is

$$\min_{\Delta U_{i,M}(k)} \overline{J}_i(k) = \sum_{j\in\mathcal{P}_i}\left[\sum_{s=1}^{P}\left\|\mathbf{r}_j(k+s)-\hat{\mathbf{y}}_j(k+s|k)\right\|_{\mathbf{Q}_j}^2+\sum_{h=1}^{M}\left\|\Delta\mathbf{u}_j(k+h-1|k)\right\|_{\mathbf{R}_j}^2\right]$$

s.t.

$$\hat{\mathbf{y}}_i(k+s|k) = \mathbf{C}_i\mathbf{A}_i^s\hat{\mathbf{x}}_i(k)+\sum_{j\in\mathcal{P}_i}\sum_{h=1}^{\min(M,s)}\mathbf{C}_i\mathbf{A}_i^{s-h}\mathbf{B}_{ij}\Delta\mathbf{u}_j(k+h-1|k), s=1,\ldots,P \quad (7.51)$$

$$\begin{aligned}
&\mathbf{u}_i^{\min}\le\mathbf{u}_i(k+h-1|k)\le\mathbf{u}_i^{\max}, h=1,\ldots,M\\
&\Delta\mathbf{u}_i^{\min}\le\Delta\mathbf{u}_i(k+h-1|k)\le\Delta\mathbf{u}_i^{\max}, h=1,\ldots,M\\
&\mathbf{y}_i^{\min}\le\hat{\mathbf{y}}_i(k+s|k)\le\mathbf{y}_i^{\max}, s=1,\ldots,P
\end{aligned}$$

The above optimization problem can be cast as a quadratic programming (QP) with the optimized vector (derivation process can be seen in Appendix A), which is solved online at each sampling instant and can be solved by the QP algorithm available in the MATLAB toolbox. The resulting online QP is

$$\begin{aligned}
&\min_{\Delta\mathbf{U}_{i,M}(k)}\frac{1}{2}\Delta\mathbf{U}_{i,M}^T(k)\mathbf{\Pi}_i\Delta\mathbf{U}_{i,M}(k)+\mathbf{f}_i^T(k)\Delta\mathbf{U}_{i,M}(k)\\
&\text{s.t.}\quad \mathbf{\Omega}_i\Delta\mathbf{U}_{i,M}(k)\le\mathbf{b}_i(k)
\end{aligned}\quad (7.52)$$

where $\mathbf{\Pi}_i=\mathbf{\Pi}_i^T>\mathbf{0}$, $\mathbf{f}_i(k)$, $\mathbf{\Omega}_i$, and $\mathbf{b}_i(k)$ are given in Appendix A.

The diagram of the MPC unit for each subsystem in a networked MPC scheme is shown in Figure 7.7, including the subsystem model represented by Equation (7.48), the predictive state observer expressed in Equation (7.49), and the QP optimizer described in Equation (7.52). Based on the "predictive state" model representation, the control decision for each subsystem can be derived by using the usual state-space techniques (e.g., state observer and moving horizon optimization) with the assumption that the local control decisions of its neighbors are available.

It is assumed that the connectivity of the communication network is sufficient for each subsystem to obtain information from its neighbors. Regarding the network capacity, we consider that it is possible for each subsystem to exchange information several times during it solves its local optimization problem at each sampling time instant, which is an ideal information model for network communication.

7.3.3 Networked MPC Algorithm

According to neighborhood optimization, the local optimal control decision for each subsystem can be obtained by solving the local problem (7.51) if the local optimal control decision of its neighbors is available, that is,

$$\Delta\mathbf{U}_{i,M}^*(k)=\arg\left\{\min_{\Delta\mathbf{U}_{i,M}(k)}\overline{J}_i(k)|_{\Delta\mathbf{U}_{j,M}^*(k)(j\in\mathcal{P}_i,j\neq i)}\right\}\quad(i=1,\ldots,m) \quad (7.53)$$

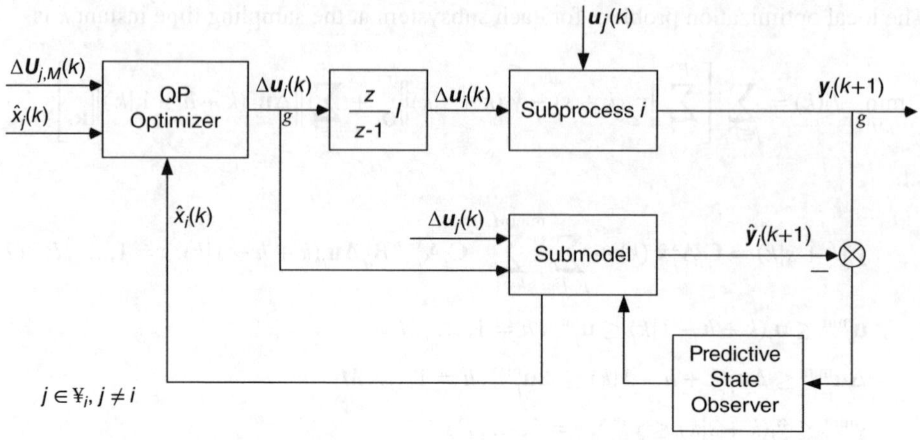

Figure 7.7 Diagram of the MPC unit for each subsystem

It can be seen that the local optimal control decision for each subsystem depends on the local optimal control decision of its neighbors at the same time. However, the local optimal control decisions of neighbors are not necessarily available to each subsystem, and hence each subsystem must use an estimate of the local optimal control decision by its neighbors. Since the mutual communication and the information exchange are adequately taken into account, each subsystem resolves its local optimization problem with the estimate of the local optimal control decision of its neighbors available. It is possible for each subsystem to exchange information several times during it solves its local optimization problem. The objective of the communication among subsystems is to achieve the control decisions of local optimization problems convergent to the global optimal control decision of the whole optimization problem. To realize this objective, the following assumptions are provided for all subsystems:

- There is no disturbance.
- $\bar{J}_i(k)$ is bounded from below in the feasible region.
- The starting point is in the interior of the feasible region.

An iterative algorithm for the networked MPC (shown in Figure 7.8) is developed to seek the local optimal control decision for each subsystem at each sampling time. Each subsystem resolves its local optimization problem with the estimate of the local optimal control decision of its neighbors available. Then each subsystem compares the newly computed control decision with that obtained in last iteration and checks if the terminal condition is satisfied. If the algorithm is convergent, all the terminal conditions of the m subsystems will be satisfied and the iteration will be terminated; otherwise, each subsystem will then announce the newly computed control decision to its neighbors and resolve its local problem with the updated values for neighbors. This optimization process will be repeated at the next sampling time.

Algorithm 7.1 Networked DMPC iterative algorithm

Step 1. Initialization and communication: At the sampling time instant k, each subsystem exchanges $\hat{x}_i(k)$ with its neighbors, makes the initial estimate of its local optimal

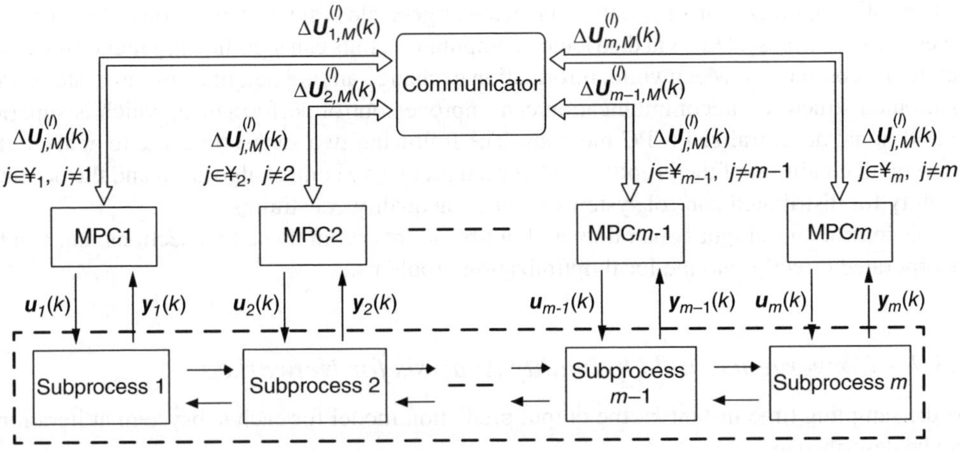

Figure 7.8 Diagram of networked MPC algorithm

control decision, and transmits it to its neighbors by communicator, and lets the iterative index $l = 0$:

$$\Delta \mathbf{U}_{i,M}^{(l)}(k) = \Delta \hat{\mathbf{U}}_{i,M}(k)(i = 1, \dots, m)$$

Step 2. Subsystem optimization: Each subsystem resolves its local optimization problem described in

$$\min_{\Delta U_{i,M}(k)} \bar{J}_i(k)\Big|_{\Delta \mathbf{U}_{j,M}^{(l)}(k)(j\in\mathbb{N}_i, j\neq i)}$$

simultaneously to derive its control decision $\Delta \mathbf{U}_{i,M}^{(l+1)}(k)$.

Step 3. Checking and updating: Each subsystem checks if its terminal iteration condition is satisfied, that is, for the given error accuracy $\varepsilon_i \in \mathbb{R}$, $(i = 1, \dots, m)$, if there exist

$$\left\|\Delta \mathbf{U}_{i,M}^{(l+1)}(k) - \Delta \mathbf{U}_{i,M}^{(l)}(k)\right\| \leq \varepsilon_i, (i = 1, \dots, m)$$

If all the terminal conditions are satisfied at iteration l^*, then to end the iteration, set the local optimal control decision for each subsystem $\Delta \mathbf{U}_{i,M}^{*}(k) = \Delta \mathbf{U}_{i,M}^{(l^*)}(k)$, and go to Step 4; otherwise, let $l = l + 1$, and each subsystem communicate to exchange the new information $\Delta \mathbf{U}_{i,M}^{(l)}(k)$ with its neighbors, and go to Step 2.

Step 4. Assignment and implementation: Compute the instant control law

$$\Delta \mathbf{u}_i^*(k) = \begin{bmatrix}\mathbf{I}_{n_{ui}} & \mathbf{0} & \cdots & \mathbf{0}\end{bmatrix} \Delta \mathbf{U}_{i,M}^{*}(k), (i = 1, \dots, m)$$

and apply $u_i^*(k) = \Delta u_i^*(k) + u_i^*(k-1)$ to each subsystem.

Step 5. Reassigning the initial estimate: Set the initial estimate of the local optimal control decision for the next sampling time

$$\Delta \hat{U}_{i,M}(k+1) = \Delta U_{i,M}^{*}(k)(i = 1, \dots, m)$$

Step 6. Receding horizon: Move horizon to the next sampling time, that is, $k + 1 \to k$, go to Step 1, and repeat the above steps.

The online optimization of serially connected large-scale systems can be converted to that of several small-scale systems via distributed computation, thus can significantly reduce the computational complexity. Meanwhile, information exchange among neighboring subsystems in a distributed structure via communication can improve control performance, which is superior to traditional decentralized MPC methods. The following two subsections are to analyze the convergent condition of the proposed networked predictive control algorithm and the nominal stability for distributed control systems without inequality constraints.

The control and output constraints will not be addressed in these two sections, but can be incorporated directly into the local optimization problems.

7.3.4 Convergence and Optimality Analysis for Networked

At the sampling time instant k , the output prediction model for each subsystem at iteration l can be described as

$$\hat{Y}^{(l)}_{i,P}(k) = \mathbf{G}_i\hat{x}_i(k) + \mathbf{H}_{ii}\Delta U_{i,M}(k) + \sum_{j\in\mathbb{N}_i, j\neq i} \mathbf{H}_{ij}\Delta U^{(l)}_{j,M}(k) \quad (i = 1, \dots, m) \tag{7.54}$$

where $\mathbf{G}_i$, $\mathbf{H}_{ij}$, and $\hat{Y}^{(l)}_{i,P}(k)$ are given in Equation (F1) in Appendix F.

The relationship of control decision for the subsystem S_i between iteration l and iteration $l+1$ can be derived by solving the local QP problem (7.52) without inequality constraints

$$\begin{aligned}\Delta\mathbf{U}^{(l+1)}_{i,M}(k) = {} & \left(\sum_{j\in\mathcal{P}_i} \mathbf{H}^T_{ji}\overline{\mathbf{Q}}_j\mathbf{H}_{ji} + \overline{\mathbf{R}}_i\right)^{-1} \\ & \sum_{j\in\mathcal{P}_i} \mathbf{H}^T_{ji}\overline{\mathbf{Q}}_j\left[\mathbf{R}_{j,P}(k) - \mathbf{G}_j\hat{\boldsymbol{x}}_j(k) - \sum_{h\in\mathcal{P}_j, h\neq i} \mathbf{H}_{jh}\Delta\mathbf{U}^{(l)}_{h,M}(k)\right]\end{aligned} \tag{7.55}$$

The integral control decision of the whole system can be written as

$$\Delta\mathbf{U}^{(l+1)}_{M}(k) = -(\boldsymbol{\Phi}_d + \overline{\mathbf{R}})^{-1}\Phi_{nd}\Delta\mathbf{U}^{(l)}_{M}(k) + (\boldsymbol{\Phi}_d + \overline{\mathbf{R}})^{-1}\mathbf{H}^{\mathrm{T}}\overline{\mathbf{Q}}[\mathbf{R}_P(k) - \mathbf{G}\hat{\boldsymbol{x}}(k)] \tag{7.56}$$

where

$$\begin{aligned}\mathbf{R}_P(k) &= \left[\mathbf{R}^{\mathrm{T}}_{1,P}(k) \quad \cdots \quad \mathbf{R}^{\mathrm{T}}_{m,P}(k)\right]^T \\ \mathbf{x}(k) &= \left[\mathbf{x}^T_1(k) \quad \cdots \quad \mathbf{x}^T_m(k)\right]^T \\ \mathbf{G} &= \text{block-diag}(\mathbf{G}_1, \dots, \mathbf{G}_m) \\ \overline{\mathbf{Q}} &= \text{block-diag}(\overline{\mathbf{Q}}_1, \dots, \overline{\mathbf{Q}}_m) \\ \overline{\mathbf{R}} &= \text{block-diag}(\overline{\mathbf{R}}_1, \dots, \overline{\mathbf{R}}_m) \\ \boldsymbol{\Phi}_d &= \text{block-diag}(\mathbf{S}_1, \dots, \mathbf{S}_m) \\ \mathbf{S}_i &= \sum_{j\in\mathbb{N}_i} \mathbf{H}^T_{ji}\overline{\mathbf{Q}}_j\mathbf{H}_{ji}\end{aligned}$$

$$\mathbf{H}^T\overline{\mathbf{Q}}\mathbf{H} = \Phi_d + \Phi_{nd}$$

$$\mathbf{H}(i,j) = \begin{cases} \mathbf{H}_{ij}, j \in \mathcal{P}_i \\ \mathbf{0}, j \notin \mathcal{P}_i \end{cases}$$

$\mathbf{H}(i, j)$ represents the block element matrix at the ith row and jth column of $\mathbf{H}$.

At the time instant k, $\mathbf{R}_P(k)$ and $\mathbf{G}\hat{\mathbf{x}}(k)$ are known in advance; hence, $(\mathbf{\Phi}_d + \overline{\mathbf{R}})^{-1}\mathbf{H}^T\overline{\mathbf{Q}}$ $[\mathbf{R}_P(k) - \mathbf{G}\hat{\mathbf{x}}(k)]$ is the constant term irrelevant to the iteration. The convergence of expression (7.55) is equivalent to that of the following equation:

$$\Delta\mathbf{U}_M^{(l+1)}(k) = -(\mathbf{\Phi}_d + \overline{\mathbf{R}})^{-1}\mathbf{\Phi}_{nd}\Delta\mathbf{U}_M^{(l)}(k) \tag{7.57}$$

From the above analysis, the convergent condition for the algorithm in application to networked linear MPC is

$$|\rho((\mathbf{\Phi}_d + \overline{\mathbf{R}})^{-1}\mathbf{\Phi}_{nd})| < 1 \tag{7.58}$$

That is, the spectrum radius must be less than 1 to guarantee a convergent computation.

If the convergent condition (7.58) is satisfied, the integral optimal control decision of the whole system at the time instant k is

$$\begin{aligned}\Delta\mathbf{U}_M^*(k) = &-(\mathbf{\Phi}_d + \overline{\mathbf{R}})^{-1}\mathbf{\Phi}_{nd}\Delta\mathbf{U}_M^*(k)\\ &+ (\mathbf{\Phi}_d + \overline{\mathbf{R}})^{-1}\mathbf{H}^T\overline{\mathbf{Q}}[\mathbf{R}_P(k) - \mathbf{G}\hat{\mathbf{x}}(k)]\end{aligned} \tag{7.59}$$

which can be rewritten as

$$\begin{aligned}\Delta\mathbf{U}_M^*(k) &= \left[\mathbf{I} + \left(\mathbf{\Phi}_d + \overline{\mathbf{R}}\right)^{-1}\mathbf{\Phi}_{nd}\right]^{-1}\left(\mathbf{\Phi}_d + \overline{\mathbf{R}}\right)^{-1}\mathbf{H}^T\overline{\mathbf{Q}}\left[\mathbf{R}_P(k) - \mathbf{G}\hat{\mathbf{x}}(k)\right]\\ &= \left\{\left(\mathbf{\Phi}_d + \overline{\mathbf{R}}\right)\cdot\left[\mathbf{I} + \left(\mathbf{\Phi}_d + \overline{\mathbf{R}}\right)^{-1}\mathbf{\Phi}_{nd}\right]\right\}^{-1}\mathbf{H}^T\overline{\mathbf{Q}}\left[\mathbf{R}_P(k) - \mathbf{G}\hat{\mathbf{x}}(k)\right]\\ &= \left(\mathbf{\Phi}_d + \overline{\mathbf{R}} + \mathbf{\Phi}_{nd}\right)^{-1}\mathbf{H}^T\overline{\mathbf{Q}}\left[\mathbf{R}_P(k) - \mathbf{G}\hat{\mathbf{x}}(k)\right]\\ &= \left(\mathbf{H}^T\overline{\mathbf{Q}}\mathbf{H} + \overline{\mathbf{R}}\right)^{-1}\mathbf{H}^T\overline{\mathbf{Q}}\left[\mathbf{R}_P(k) - \mathbf{G}\hat{\mathbf{x}}(k)\right]\end{aligned} \tag{7.60}$$

At the sampling time instant k, the output prediction model for the whole system can be described as

$$\hat{\mathbf{Y}}_P(k) = \mathbf{G}\hat{\mathbf{x}}(k) + \mathbf{H}\Delta\mathbf{U}_M^c(k) \tag{7.61}$$

where $\Delta\mathbf{U}_M^c(k)$ represents control decision derived by centralized MPC,

$$\Delta\mathbf{U}_M^c(k) = \left[\Delta\mathbf{U}_{1,M}^{\mathrm{T}}(k) \quad \cdots \quad \Delta\mathbf{U}_{m,M}^{\mathrm{T}}(k)\right]^T$$

$$\hat{\mathbf{Y}}_P(k) = \left[\hat{\mathbf{Y}}_{1,P}^{\mathrm{T}}(k) \quad \cdots \quad \hat{\mathbf{Y}}_{m,P}^{\mathrm{T}}(k)\right]^T$$

The performance index of the whole system is

$$\begin{aligned} J(k) &= [\mathbf{R}_P(k) - \mathbf{G}\hat{x}(k) - \mathbf{H}\Delta\mathbf{U}_M^c(k)]^T\,\overline{\mathbf{Q}}[\mathbf{R}_P(k) - \mathbf{G}\hat{x}(k) - \mathbf{H}\Delta\mathbf{U}_M^c(k)] \\ &\quad + (\Delta\mathbf{U}_M^c(k))^T\,\overline{\mathbf{R}}\Delta\mathbf{U}_M^c(k) \\ &= (\Delta\mathbf{U}_M^c(k))^T(\mathbf{H}^T\overline{\mathbf{Q}}\mathbf{H} + \overline{\mathbf{R}})\Delta\mathbf{U}_M^c(k) \\ &\quad - 2[\mathbf{H}^T\overline{\mathbf{Q}}(\mathbf{R}_P(k) - \mathbf{G}\hat{\mathbf{x}}(k))]^T\Delta\mathbf{U}_M^c(k) + \text{constant} \end{aligned} \tag{7.62}$$

The integral optimal control decision of the whole system derived from the centralized optimization is

$$\Delta\mathbf{U}_M^c(k) = (\mathbf{H}^T\overline{\mathbf{Q}}\mathbf{H} + \overline{\mathbf{R}})^{-1}\mathbf{H}^T\overline{\mathbf{Q}}[\mathbf{R}_P(k) - \mathbf{G}\hat{\mathbf{x}}(k)] \tag{7.63}$$

which is equal to the result by distributed optimization.

Under the network environment, the capacity of the communication network is assumed to be sufficient for each subsystem to obtain information from its neighbors, so it is possible for each subsystem to exchange information several times during it solves its local optimization problem at the sampling time instant. Furthermore, when the convergent condition is satisfied, the solution to the local optimization problems collectively will be the global optimal control decision of the whole system, that is, the coordinated distributed computations solve an equivalent centralized MPC problem. An illustrative example will be provided in Section 7.3.6 to test the effectiveness of the networked MPC algorithm with neighborhood optimization.

7.3.5 Nominal Stability Analysis for Distributed Control Systems

The state-space model of the whole system can be written as

$$\begin{cases} \mathbf{x}(k+1) = \mathbf{A}\mathbf{x}(k) + \mathbf{B}\Delta\mathbf{u}(k) \\ \mathbf{y}(k) = \mathbf{C}\mathbf{x}(k) \end{cases} \tag{7.64}$$

where

$$\Delta\mathbf{u}(k) = \begin{bmatrix}\Delta\mathbf{u}_1^T(k) & \cdots & \Delta\mathbf{u}_m^T(k)\end{bmatrix}^{\mathrm{T}}$$

$$\mathbf{y}(k) = \begin{bmatrix}\mathbf{y}_1^T(k) & \cdots & \mathbf{y}_m^T(k)\end{bmatrix}^{\mathrm{T}}$$

$$\mathbf{A} = \text{block-diag}(\mathbf{A}_1, \ldots, \mathbf{A}_m)$$

$$\mathbf{B}(i,j) = \begin{cases} \mathbf{B}_{ij},\ j \in \mathbb{N}_i \\ \mathbf{0},\ j \notin \mathbb{N}_i \end{cases}$$

$$\mathbf{C} = \text{block-diag}(\mathbf{C}_1, \ldots, \mathbf{C}_m)$$

The observer dynamics for the whole system can be described as

$$\tilde{\mathbf{x}}(k+1) = (\mathbf{I} - \mathbf{V}\mathbf{C})\mathbf{A}\tilde{\mathbf{x}}(k) \tag{7.65}$$

where

$$\tilde{\mathbf{x}}(k) = \begin{bmatrix}\tilde{\mathbf{x}}_1^T(k) & \cdots & \tilde{\mathbf{x}}_m^T(k)\end{bmatrix}^{\mathrm{T}},$$

$$\mathbf{V} = \text{block-diag}(\mathbf{V}_1, \ldots, \mathbf{V}_m).$$

If the convergent condition (7.58) is satisfied, the optimal instant control law of the whole system at the time instant k is

$$\Delta \mathbf{u}^*(k) = \mathbf{L}\Delta \mathbf{U}_M^*(k) = \mathbf{W}[\mathbf{R}_P(k) - \mathbf{G}\hat{\mathbf{x}}(k)] \tag{7.66}$$

where

$$\mathbf{L} = \text{block-diag}\left(\underbrace{\mathbf{L}_0 \quad \cdots \quad \mathbf{L}_0}_{m} \right)$$

$$\mathbf{L}_0 = \begin{bmatrix} \mathbf{I}_{n_{ui}} & \mathbf{0}_{n_{ui}\times(M-1)n_{ui}} \end{bmatrix}$$

$$\mathbf{W} = \mathbf{L}(\mathbf{H}^T\overline{\mathbf{Q}}\mathbf{H} + \overline{\mathbf{R}})^{-1}\mathbf{H}^T\overline{\mathbf{Q}}$$

The whole closed-loop system with the networked MPC control and predictive state observer can be described as

$$\begin{bmatrix} \mathbf{x}(k+1) \\ \tilde{\mathbf{x}}(k+1) \end{bmatrix} = \begin{bmatrix} \mathbf{A} - \mathbf{BWG} & \mathbf{BWG} \\ \mathbf{0} & (\mathbf{I} - \mathbf{VC})\,\mathbf{A} \end{bmatrix} \begin{bmatrix} \mathbf{x}(k) \\ \tilde{\mathbf{x}}(k) \end{bmatrix} + \begin{bmatrix} \mathbf{BW} \\ \mathbf{0} \end{bmatrix} \mathbf{R}_P(k) \tag{7.67}$$

The nominal stability of the whole closed-loop system can be guaranteed if only if

$$\left| \lambda_j \left\{ \begin{bmatrix} \mathbf{A} - \mathbf{BWG} & \mathbf{BWG} \\ \mathbf{0} & (\mathbf{I} - \mathbf{VC})\,\mathbf{A} \end{bmatrix} \right\} \right| < 1, \forall j = 1, \ldots, 2Nn_y \tag{7.68}$$

That is, the eigenvalues of the above matrix are all in the unit circle.

Remark 7.3 *It has been noticed that the convergence of N-MPC Algorithm 7.1 is local, that is, whether the distributed computation is convergent is only related to the current sampling time instant. While the stability analysis in this section is global, the convergence of the distributed computation and stability for distributed control systems are concerned during the whole receding horizon.*

7.3.6 Simulation Study

A simple simulation example is illustrated to verify the optimality of the N-DMPC interative Algorithm 7.1, and the N-DMPC Algorithm 7.1 is applied to the fuel feed flow control for the walking beam reheating furnace in this section.

7.3.6.1 Illustrative Example

Consider the following two-input two-output process:

$$\begin{pmatrix} y_1 \\ y_2 \end{pmatrix} = \begin{pmatrix} \dfrac{1}{10s+1} & \dfrac{2}{11s+1} \\ -\dfrac{1}{8s+1} & \dfrac{1.5}{9s+1} \end{pmatrix} \begin{pmatrix} u_1 \\ u_2 \end{pmatrix}$$

with the proposed networked MPC scheme, first of all divide the whole process into two subsystems, they are

$$\text{Subsystem 1}: y_1 = \frac{1}{10\,s+1}u_1$$
$$\text{Subsystem 2}: y_2 = \frac{1.5}{9\,s+1}u_2$$

The local performance index for each subsystem is

$$J_i(k) = \sum_{s=1}^{P} \|r_i\,(k+s) - \hat{y}_i(k+s|k)\|^2_{Q_i} + \sum_{h=1}^{M} \|\Delta u_i\,(k+h-1|k)\|^2_{R_i}$$

With regard to neighborhood optimization, the new performance index for each subsystem is

$$\overline{J}_i(k) = J_1(k) + J_2(k)$$

The tuning parameters for each subsystem are set with $P = 8, M = 2, Q_i = 10, R_i = 1(i = 1, 2)$ using sampling time of 1 s, and $\varepsilon_i = 0.01(i = 1, 2)$. MATLAB-based simulation results are carried out to evaluate the proposed networked MPC interactive algorithm (shown in Figure 7.11) through performance comparisons with decentralized MPC (shown in Figure 7.9) and distributed MPC with Nash optimization (shown in Figure 7.10). Set points step from 0 to 1 for the first subsystem and from 1 to 0 for the second subsystem at the time 20 s shown as the dotted lines in figures. As can be seen, system outputs are divergent and fluctuant under the decentralized MPC, while system outputs under the networked MPC Algorithm 7.1 with neighborhood optimization reach the set-points in a very shorter time exhibiting a very smaller overshoots than that of the distributed MPC with Nash optimization under similar conditions. In addition, the iteration based networked MPC scheme with neighborhood optimization can work as well as the centralized MPC method.

Performance index comparisons are shown for these three schemes in Figure 7.12. Simulation results demonstrate that the control performance with decentralized MPC (shown as dotted line) is worse than the distributed MPC with Nash optimization (shown as dashed line) and the proposed networked MPC with neighborhood optimization (shown as solid line), and the networked MPC with neighborhood optimization can efficiently improve the control performance.

7.3.6.2 Walking Beam Reheating Furnace System

Walking beam reheating furnace is one of the most important equipment in steel rolling industry. The structure of a walking beam reheating furnace is depicted in Figure 7.13. As seen in the figure, the reheating furnace consists of three chambers: preheating, heating, and soaking zones. Billets are fed to the furnace and move forward a step at every interval, it will be heated to the specified temperature at the exit of the furnace for metallurgical quality and for hot rolling [109]. If furnace temperature is too high, the billets in the furnace will be overheated. Otherwise, it cannot be heated to the desired temperature. So it is important to control the fuel feed flow appropriately for each zone in order to heat billets to the desired temperature with minimum energy consumption.

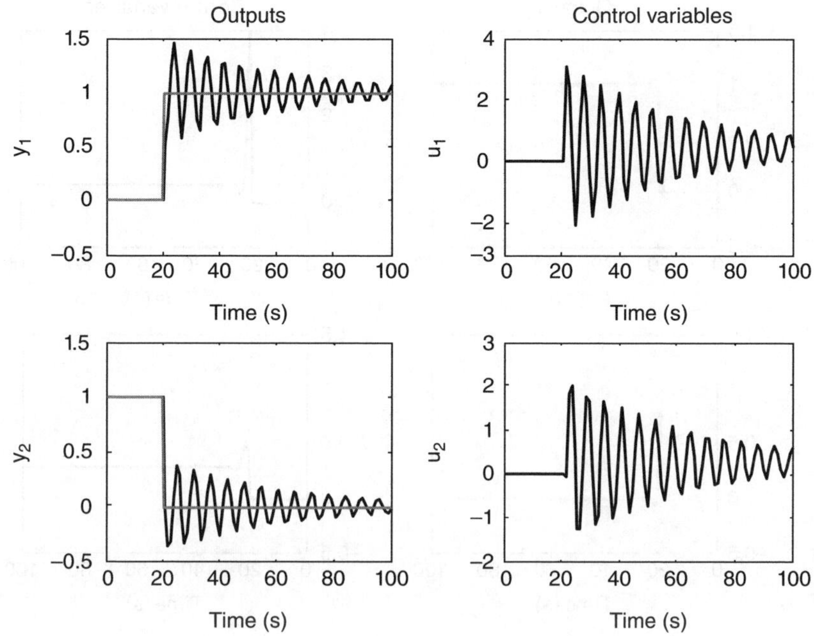

Figure 7.9 Outputs and control signals under the decentralized MPC

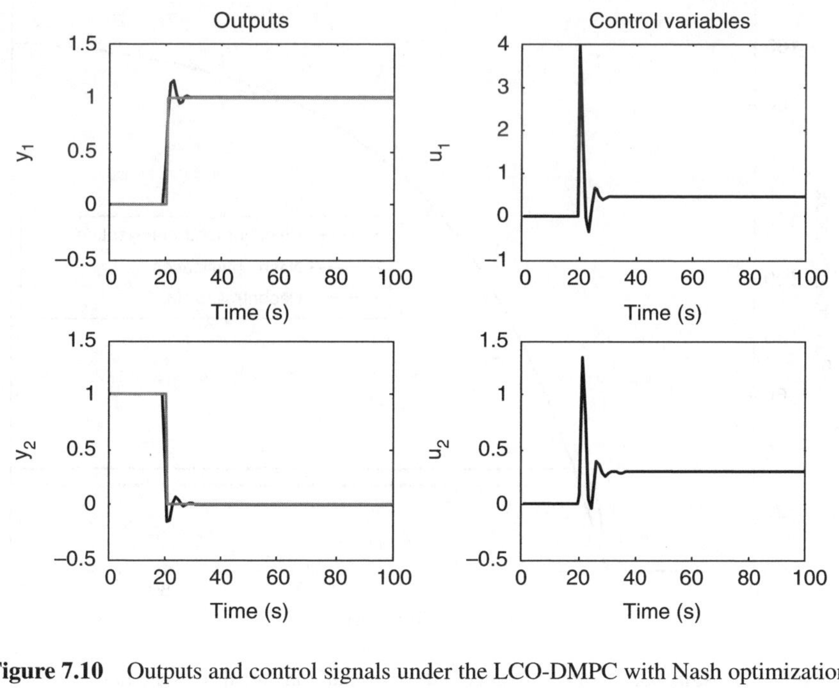

Figure 7.10 Outputs and control signals under the LCO-DMPC with Nash optimization

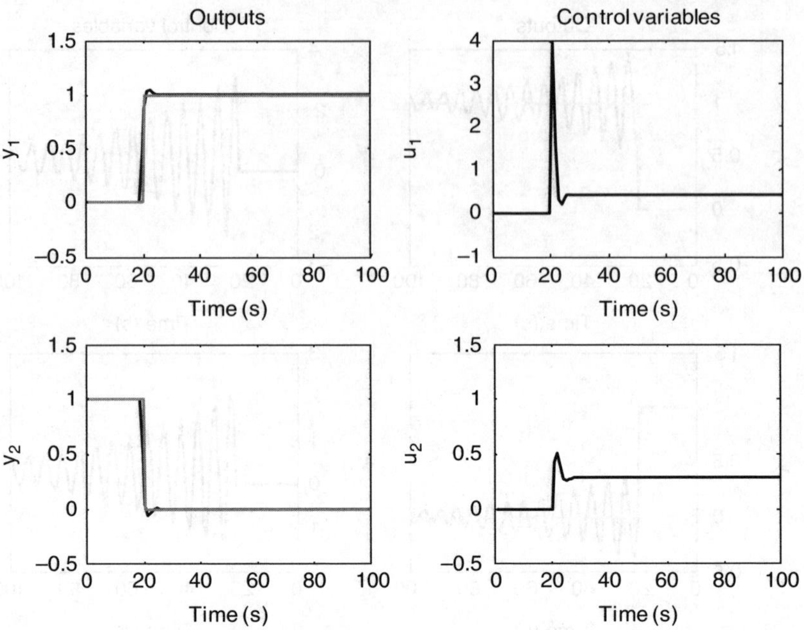

Figure 7.11 Outputs and control signals under the N-DMPC iterative algorithm

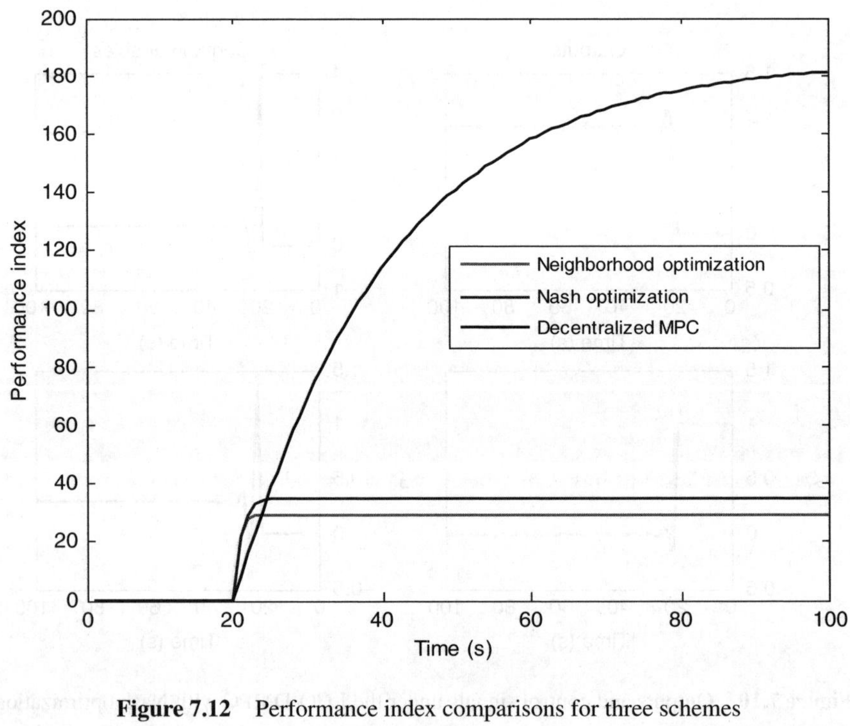

Figure 7.12 Performance index comparisons for three schemes

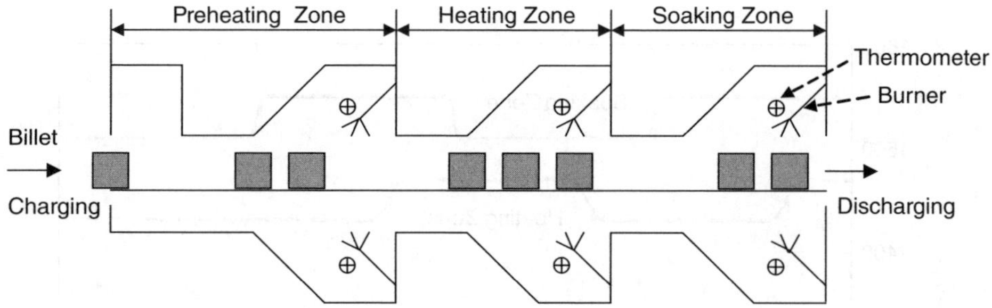

Figure 7.13 Structure of a walking beam reheating furnace

Thermal transmission is carried through only from the end to the head of the furnace, the dynamic model of the reheating furnace is derived using material and energy balances [110]. The linear model of the reheating furnace for networked MPC design can be described as

$$\begin{cases} y_1(s) = \dfrac{0.002757}{s+0.00572}u_1(s) \\ y_2(s) = \dfrac{0.003983}{s+0.001891}\mathrm{e}^{-32.8s}u_2(s) + \dfrac{0.001009}{s+0.001907}\mathrm{e}^{-52.5s}y_1(s) \\ y_3(s) = \dfrac{0.001088}{s+0.003857}\mathrm{e}^{-33.2s}u_3(s) + \dfrac{0.003224}{s+0.004634}y_2(s) \end{cases}$$

where y_1, y_2 and y_3 are the furnace temperatures of the soaking, heating and preheating zones, and u_1, u_2 and u_3 are fuel feed flows of the soaking, heating and preheating zones, respectively.

With the proposed networked MPC scheme, the reheating furnace process can be decomposed into three subsystems, they are subsystems of soaking zone (the first subsystem), heating zone (the second subsystem) and preheating zone (the third subsystem), respectively, i.e.

$$\begin{cases} y_1(s) = \dfrac{0.002757}{s+0.00572}u_1(s) \\ y_2(s) = \dfrac{0.003983}{s+0.001891}\mathrm{e}^{-32.8s}u_2(s) \\ y_3(s) = \dfrac{0.001088}{s+0.003857}\mathrm{e}^{-33.2s}u_3(s) \end{cases}$$

The local performance index for each subsystem employs the expression in (36), With regards to neighborhood optimization, the new performance index for each subsystem is as follows:

$$\overline{J}_i(k) = J_1(k) + J_2(k) + J_3(k)$$

each subsystem are set with $P = 30, M = 10, Q_i = 10, R_i = 2 (i = 1, 2, 3)$, using sampling time of 1 min. MATLAB-based simulation results (Figure 7.14) show the furnace temperatures and the fuel feed flows of three zones running for 4 h. It can be seen that the furnace temperatures of three zones in this distributed structure follow their set-points (shown as the dotted lines in the figure) satisfactorily and that the interaction between the channels is effectively reduced.

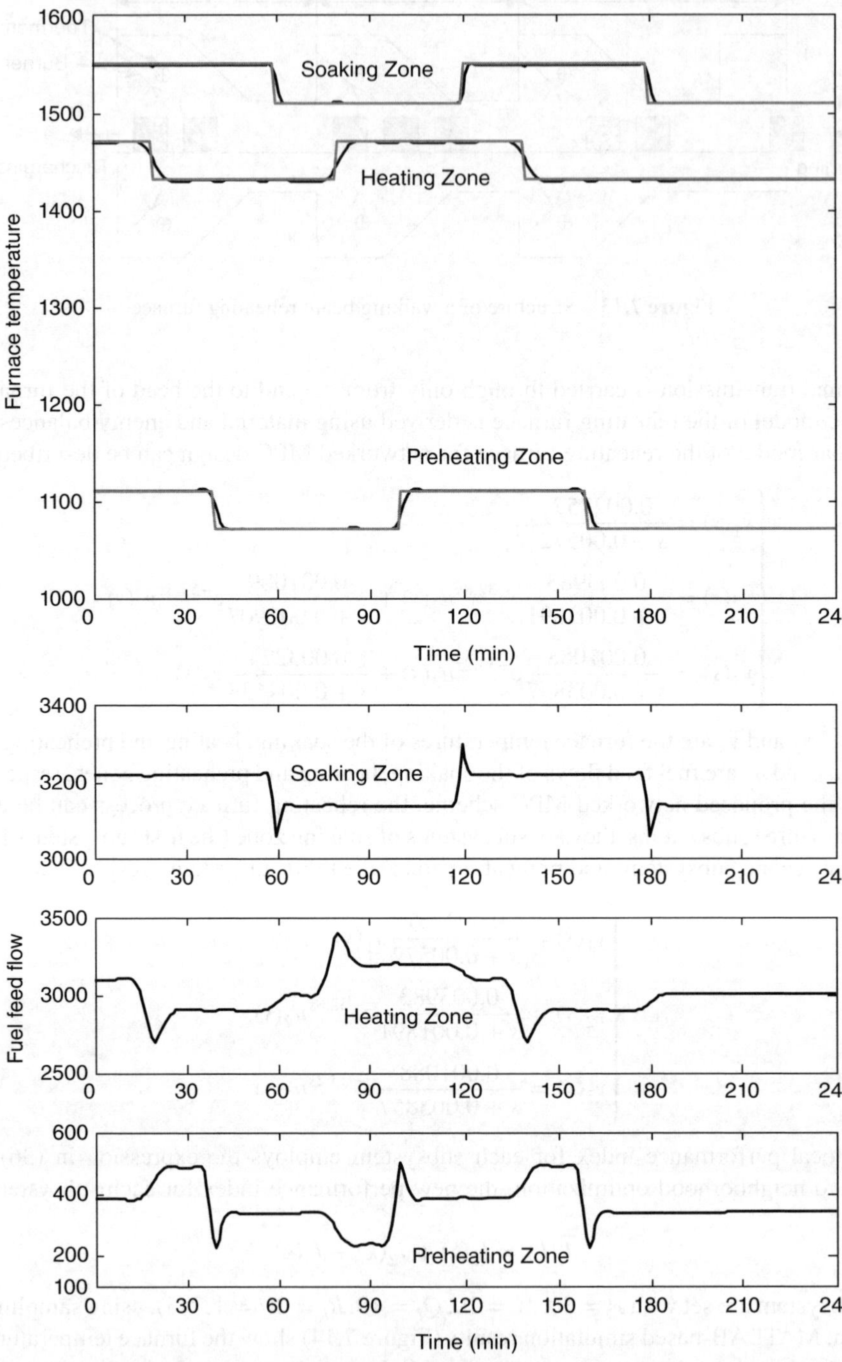

Figure 7.14 Furnace temperature and fuel feed flow for each zone

From the simulation, the networked MPC scheme can work as well as the centralized MPC method. In addition, the design parameters for each subsystem such as prediction horizon, control horizon, weighting matrices, etc. can all be designed and tuned separately, which provides more flexibility for the analysis and applications. Notice that each subsystem is not necessary limited to SISO case and also it can be MIMO subsystem, whose dimension is still much lower than that of the whole system.

7.4 Conclusion

In this chapter, the noniterative distributed MPC algorithm based on neighborhood optimization is present and the condition of closed-loop stability is given for local MPCs tuning. In the procedure of resolving optimal solution, each subsystem only communicates with its neighbors, which is rather easy to fulfill the network requirements. Moreover, the discussion of the performance of proposed methodology and the application of N-DMPC to ACC test rig prove that the proposed method guarantees an improving performance of entire system with relative relaxed communication requirements. Further investigation will focus on designing stable distributed MPC with constraints and global performance improvement for this class of large-scale systems.

In addition, regarding the network capacity, an iterative networked DMPC method based on neighborhood optimization is developed for a class of serially connected processes. The state-space model for each subsystem is developed directly from step responses without further identification. It is not necessary to identify the structure of the subsystem during the modeling, and the more flexible error correction is naturally derived by employing the predictive state observer. Neighborhood optimization employs a cooperative strategy so that the whole control performance can be efficiently improved. Regarding the network capacity, an iterative algorithm for networked MPC is presented with one situation that it is possible for each subsystem to exchange information several times during it solves its local optimization problem. The computational convergence of the iteration based N-MPC algorithm and the nominal stability are derived for distributed control systems without the control and output constraints. In addition, when convergent condition is satisfied, the solution to the local optimization problems collectively is proved to equal the Nash optimality. An illustrative example and the simulation study of the fuel feed flow control for the walking beam reheating furnace if presented to verify the efficiency of the proposed networked MPC algorithms.

Appendix

Appendix A. Proof of Lemma 7.1

The proof is stated by writing, for $i = 1, \dots, n$, the h-ahead predictions at time k based on the information computed at time $k-1$ of the interaction vectors (9) (10) and by representing them in a stacked form for $h = 1, \dots, P$. The last $P-M-1$ samples of the stacked control action predictions $\mathbf{U}_j(k, P|k-1)(j = 1, 2, \dots, n)$, that are not contained in $\mathbf{U}_j(k-1, M|k-1)$, are assumed equal to the last element of $\mathbf{U}_j(k-1, M|k-1)$. By definitions (14–18) and Table 7.1, this implies that relations (7.23) hold.

Appendix B. Proof of Lemma 7.2

By (11), (12), and imposing that

$$\mathbf{u}_i(k+P-1|k) = \mathbf{u}_i(k+P-2|k) = \cdots = \mathbf{u}_i(k+M|k) = \mathbf{u}_i(k+M-1|k)$$

and

$$\hat{\mathbf{v}}_i(k+P|k-1) = \hat{\mathbf{v}}_i(k+P-1|k-1)$$

also substituting $\hat{\hat{\mathbf{W}}}_i(k,p|k-1)$ and $\hat{\hat{\mathbf{V}}}_i(k,p|k-1)$ with their explicit expressions (7.23), it results the following stacked state prediction for controller C_i

$$\begin{aligned}\hat{\hat{\mathbf{X}}}_i(k+1,P|k) = \overline{\mathbf{S}}_i[\overline{\mathbf{A}}_i\hat{x}_i(k|k) + \overline{\mathbf{B}}_i\mathbf{U}_i(k,M|k) \\ +\tilde{\mathbf{A}}_{i1}\hat{\mathbf{X}}(k,p|k-1) + \tilde{\mathbf{B}}_i\mathbf{U}(k-1,M|k-1)]\end{aligned}$$

Let $\hat{\mathbf{x}}_i'(k|k-1) = \begin{bmatrix}\hat{\mathbf{x}}_{i_1}^{\mathrm{T}}(k|k-1) & \cdots & \hat{\mathbf{x}}_{i_{mi}}^{\mathrm{T}}(k|k-1)\end{bmatrix}^{\mathrm{T}}$, and by definitions (7.12), (7.18) (7.19), the above equation becomes

$$\begin{aligned}\hat{\hat{\mathbf{X}}}_i(k+1,P|k) &= \overline{\mathbf{S}}_i[\overline{\mathbf{A}}_i^{(1)}\hat{x}(k|k) + \overline{\mathbf{A}}_i^{(2)}\hat{\hat{x}}_i'(k|k-1) + \overline{\mathbf{B}}_i\mathbf{U}_i(k,M|k) \\ &\quad + \tilde{\mathbf{A}}_i^{(1)}\hat{\mathbf{X}}(k,P|k-1) + \tilde{\mathbf{B}}_i\mathbf{U}(k-1,M|k-1)] \\ &= \overline{\mathbf{S}}_i[\overline{\mathbf{A}}_i^{(1)}\hat{x}(k|k) + \overline{\mathbf{B}}_i\mathbf{U}_i(k,M|k) + (\tilde{\mathbf{A}}_i^{(1)} + \tilde{\mathbf{A}}_i^{(2)})\hat{\mathbf{X}}(k,P|k-1) \\ &\quad + \tilde{\mathbf{B}}_i\mathbf{U}(k-1,M|k-1)] \\ &= \overline{\mathbf{S}}_i[\overline{\mathbf{A}}_i^{(1)}\hat{x}(k|k) + \overline{\mathbf{B}}_i\mathbf{U}_i(k,M|k) \\ &\quad + \tilde{\mathbf{A}}_i\hat{\mathbf{X}}(k,P|k-1) + \tilde{\mathbf{B}}_i\mathbf{U}(k-1,M|k-1)]\end{aligned}$$

By model (7.9) and the coefficient defined in (7.24), the stacked output prediction for controller C_i can be expressed as

$$\hat{\hat{\mathbf{Y}}}_i(k+1,P|k) = \overline{\mathbf{C}}_i\hat{\hat{\mathbf{X}}}_i(k+1,P|k) + \mathbf{T}_i\tilde{\mathbf{C}}_i\hat{\mathbf{X}}(k+1,P|k-1).$$

This proves Lemma 7.2.

Appendix C. Proof of Lemma 7.3

Making use of stacked vectors and definitions (7.30), the cost function (7.8) to be minimized by controller C_i can be expressed in the equivalent form

$$\overline{J}_i = ||\hat{\hat{\mathbf{Y}}}_i(k+1,P|k) - \hat{\mathbf{Y}}_i^d(k+1,P|k)||^2_{\mathbf{Q}_i} + ||\Delta\mathbf{U}_i(k,M|k)||^2_{\mathbf{R}_i}$$

The stacked local output prediction $\hat{\hat{\mathbf{Y}}}_i(k+1,P|k)$ is a function of the control action, therefore, in order to express $\overline{J}_i$ as a function of the control sequence $\Delta\mathbf{U}_i(k,M|k)$, an explicit expression for such a prediction is needed. Considering that $\mathbf{u}_i(k+h|k) = \mathbf{u}_i(k-1) + \sum_{r=0}^{h}\Delta\mathbf{u}_i(k+r|k), h = 1,2,\ldots,M$, the local stacked control sequence $\mathbf{U}_i(k,M|k)$

is used together with Equations (7.24) and (7.30) to obtain the output prediction in the form $\hat{\hat{\mathbf{Y}}}_i(k+1,P|k) = \mathbf{N}_i\Delta\mathbf{U}_i(k,M|k) + \hat{\mathbf{Z}}(k+1,P|k)$. By substituting this expression, the local cost function $\overline{J}_i$ takes the form (7.31). The positive definiteness of matrices $\overline{\mathbf{Q}}_i$ and $\overline{\mathbf{R}}_i$ implies the same property for matrix $\mathbf{H}_i$.

In this way, the ND-MPC problem has been transformed into an equivalent unconstrained QP problem which has to be locally solved online at each sampling instant.

Appendix D. Proof of Theorem 7.1

States that a solution to the ND-MPC problem minimizes the cost function (7.31) with respect to the control sequence $\Delta\mathbf{U}_i(k,M|k)$. This solution has the form $\Delta\mathbf{U}_i(k,M|k) = ((1/2)\mathbf{H}_i^{-1}\mathbf{G}_i(k+1,P|k))$. Following the receding horizon strategy, only the first element of the optimal sequence is actually applied to the process and the control action is expressed as $\mathbf{u}_i(k) = \mathbf{u}_i(k-1) + \mathbf{\Gamma}_i\Delta\mathbf{U}_i(k,M|k)$ which gives the final closed form (7.36).

Appendix E. Proof of Theorem 7.2

To simplify the process of stability proof, define that

$$\begin{aligned}
&\mathbf{\Omega} = \begin{bmatrix}\mathbf{\Omega}_1^{\mathrm{T}} & \cdots & \mathbf{\Omega}_P^{\mathrm{T}}\end{bmatrix}^{\mathrm{T}}\\
&\mathbf{\Omega}_j = \mathrm{diag}\{\mathbf{\Omega}_{1j}, \ldots, \mathbf{\Omega}_{nj}\}\\
&\mathbf{\Omega}_{ij} = \begin{bmatrix}\mathbf{0}_{n_{x_i}\times(j-1)n_{x_i}} & \mathbf{I}_{n_{x_i}} & \mathbf{0}_{n_{x_i}\times(P-j)n_{x_i}}\end{bmatrix}\\
&(i = 1, \ldots, n,\ j = 1, \ldots, P);
\end{aligned} \tag{E.1}$$

$$\begin{aligned}
&\mathbf{\Pi} = \begin{bmatrix}\mathbf{\Pi}_1^{\mathrm{T}} & \cdots & \mathbf{\Pi}_M^{\mathrm{T}}\end{bmatrix}^{\mathrm{T}}\\
&\mathbf{\Pi}_j = \mathrm{diag}\{\mathbf{\Pi}_{1j}, \ldots, \mathbf{\Pi}_{nj}\}\\
&\mathbf{\Pi}_{ij} = \begin{bmatrix}\mathbf{0}_{n_{u_i}\times(j-1)n_{u_i}} & \mathbf{I}_{n_{u_i}} & \mathbf{0}_{n_{u_i}\times(M-j)n_{u_i}}\end{bmatrix}\\
&(i = 1, \ldots, n,\ j = 1, \ldots, M)
\end{aligned} \tag{E.2}$$

The following equations are achieved

$$\hat{\mathbf{X}}(k,P|k-1) = \mathbf{\Omega}\hat{\mathbf{X}}(k,P|k-1) \tag{E.3}$$

$$\mathbf{U}(k,M|k-1) = \mathbf{\Pi}\mathbf{U}(k,M|k-1) \tag{E.4}$$

Define

$$\begin{aligned}
&\overline{\mathbf{A}} = \mathrm{diag}\{\overline{\mathbf{A}}_{11}, \ldots, \overline{\mathbf{A}}_{n1}\};\\
&\tilde{\mathbf{A}} = \begin{bmatrix}\tilde{\mathbf{A}}_1^{\mathrm{T}} & \cdots & \tilde{\mathbf{A}}_n^{\mathrm{T}}\end{bmatrix}^{\mathrm{T}};\\
&\overline{\mathbf{B}} = \mathrm{diag}\{\overline{\mathbf{B}}_1, \ldots, \overline{\mathbf{B}}_n\};
\end{aligned}$$

$$\begin{aligned}
\tilde{\mathbf{B}} &= \left[\tilde{\mathbf{B}}_1^{\mathrm{T}} \quad \cdots \quad \tilde{\mathbf{B}}_n^{\mathrm{T}}\right]^{\mathrm{T}} \\
\mathbf{L} &= \mathrm{diag}\{\mathbf{L}_1, \ldots, \mathbf{L}_n\} \\
\mathbf{L}_i &= \mathrm{diag}_P\left\{\left[\mathbf{I}_{n_{x_i}} \quad \mathbf{0}_{n_{x_i}\times(n_{\hat{x}_i}-n_{x_i})}\right]\right\} \\
\overline{\mathbf{S}} &= \mathrm{diag}\{\overline{\mathbf{S}}_1, \ldots, \overline{\mathbf{S}}_n\}
\end{aligned} \tag{E.5}$$

Then, for each controller C_i, $i = 1, \ldots, n$, by Lemma 2 and definitions (E.5), the stacked distributed state prediction at time k are expressed by

$$\begin{aligned}
\hat{\mathbf{X}}_i(k+1, P|k) &= \mathbf{L}_i\hat{\hat{\mathbf{X}}}_i(k+1, P|k) \\
&= \mathbf{L}_i\overline{\mathbf{S}}_i[\overline{\mathbf{A}}_{i1}\hat{x}(k|k) + \overline{\mathbf{B}}_i\mathbf{U}_i(k, M|k) \\
&\quad + \tilde{\mathbf{A}}_i\hat{\mathbf{X}}(k, P|k-1) + \tilde{\mathbf{B}}_i\mathbf{U}(k-1, M|k-1)]
\end{aligned} \tag{E.6}$$

By definitions (E.5), the completed stacked distributed prediction can be expressed as

$$\begin{aligned}
\hat{\mathbb{X}}(k+1, P|k) &= \mathbf{L}\overline{\mathbf{S}}[\overline{\mathbf{A}}\hat{x}(k|k) + \overline{\mathbf{B}}\mathbb{U}(k, M|k) \\
&\quad + \tilde{\mathbf{A}}\hat{\mathbf{X}}(k, P|k-1) + \tilde{\mathbf{B}}\mathbf{U}(k-1, M|k-1)]
\end{aligned} \tag{E.7}$$

Substituting (E.3) and (E.2) into (E.7), the following complete version of the stacked distributed prediction can be deduced:

$$\begin{aligned}
\hat{\mathbb{X}}(k+1, P|k) &= \mathbf{L}\overline{\mathbf{S}}[\overline{\mathbf{A}}\hat{x}(k|k) + \overline{\mathbf{B}}\mathbb{U}(k, M|k) \\
&\quad + \tilde{\mathbf{A}}\boldsymbol{\Omega}\hat{\mathbb{X}}(k, P|k-1) + \tilde{\mathbf{B}}\boldsymbol{\Pi}\hat{\mathbb{U}}(k-1, M|k-1)]
\end{aligned} \tag{E.8}$$

Considering that the local control action applied at time $k-1$ is given by $\mathbf{u}_i(k-1) = \boldsymbol{\Gamma}_i\mathbf{U}_i(k-1, m|k-1)$, the open-loop optimal sequence $\mathbf{U}_i(k, M|k)$ of controller C_i at time k can be expressed as

$$\mathbf{U}_i(k, M|k) = \boldsymbol{\Gamma}'_i\boldsymbol{\Gamma}_i\mathbf{U}_i(k-1, M|k-1) + \overline{\boldsymbol{\Gamma}}_i\Delta\mathbf{U}_i(k, M|k).$$

Then by Equations (7.35) and (7.37), the stacked open-loop optimal control sequence at time k can be directly expressed as

$$\begin{aligned}
\mathbf{U}_i(k, M|k) &= \boldsymbol{\Gamma}'_i\mathbf{u}_i(k-1) + \overline{\boldsymbol{\Gamma}}_i\overline{\mathbf{K}}_i[\mathbf{Y}_i^d(k+1, P|k) - \hat{\mathbf{Z}}_i(k+1, P|k)] \\
&= \boldsymbol{\Gamma}'_i\mathbf{u}_i(k-1) + \overline{\boldsymbol{\Gamma}}_i\overline{\mathbf{K}}_i\{\mathbf{Y}_i^d(k+1, P|k) \\
&\quad - \mathbf{S}_i[\overline{\mathbf{B}}_i\boldsymbol{\Gamma}'_i\mathbf{u}_i(k-1) + \overline{\mathbf{A}}_i^{(1)}\hat{x}(k|k) + \tilde{\mathbf{A}}_i\hat{\mathbf{X}}(k, P|k-1) \\
&\quad + \tilde{\mathbf{B}}_i\mathbf{U}(k-1, M|k-1)] - \mathbf{T}_i\tilde{\mathbf{C}}_i\hat{\mathbf{X}}(k, P|k-1)\}
\end{aligned} \tag{E.9}$$

Define that

$$\begin{aligned}
\boldsymbol{\Gamma}' &= \mathrm{diag}\{\boldsymbol{\Gamma}'_1, \ldots, \boldsymbol{\Gamma}'_n\} \\
\boldsymbol{\Gamma} &= \mathrm{diag}\{\boldsymbol{\Gamma}_1, \ldots, \boldsymbol{\Gamma}_n\} \\
\mathbf{S} &= \mathrm{diag}\{\mathbf{S}_1, \ldots, \mathbf{S}_n\}
\end{aligned}$$

$$\mathbf{T} = \text{diag}\{\mathbf{T}_1, \dots, \mathbf{T}_n\}$$
$$\boldsymbol{\Xi} = \text{diag}\{\overline{\boldsymbol{\Gamma}}_1\overline{\mathbf{K}}_1, \dots, \overline{\boldsymbol{\Gamma}}_n\overline{\mathbf{K}}_n\} \tag{E.10}$$

By definitions (E.5) and (E.10), and substituting (E.3) and (E.4) into (E.9), the completed stacked open-loop optimal sequence can be expressed as

$$\begin{aligned}\mathbb{U}(k,M|k) &= \boldsymbol{\Gamma}'\boldsymbol{\Gamma}\hat{\mathbb{X}}(k-1,M|k-1) + \boldsymbol{\Xi}\{\mathbf{Y}^d(k+1,P|k)\\ &\quad - \mathbf{S}[\overline{\mathbf{B}}\boldsymbol{\Gamma}'\boldsymbol{\Gamma}\mathbb{U}(k-1,M|k-1) + \overline{\mathbf{A}}\hat{x}(k|k)\\ &\quad + \tilde{\mathbf{A}}\boldsymbol{\Omega}\hat{\mathbb{X}}(k,P|k-1) + \tilde{\mathbf{B}}\boldsymbol{\Pi}\mathbb{U}(k-1,M|k-1)] - \mathbf{T}\tilde{\mathbf{C}}\boldsymbol{\Omega}\hat{\mathbb{X}}(k,P|k-1)\}\end{aligned} \tag{E.11}$$

Define

$$\begin{aligned}\boldsymbol{\Theta} &= -\boldsymbol{\Xi}\mathbf{S}\overline{\mathbf{A}}\\ \boldsymbol{\Phi} &= -\boldsymbol{\Xi}(\mathbf{S}\tilde{\mathbf{A}}\boldsymbol{\Omega} + \mathbf{T}\tilde{\mathbf{C}}\boldsymbol{\Omega})\\ \boldsymbol{\Psi} &= \boldsymbol{\Gamma}'\boldsymbol{\Gamma} - \boldsymbol{\Xi}\mathbf{S}(\overline{\mathbf{B}}\boldsymbol{\Gamma}'\boldsymbol{\Gamma} + \tilde{\mathbf{B}}\boldsymbol{\Pi})\end{aligned} \tag{E.12}$$

Then the completed stacked open-loop optimal sequence (D.11) has the form

$$\begin{aligned}\mathbb{U}(k,M|k) &= \boldsymbol{\Psi}\mathbb{U}(k-1,M|k-1) + \boldsymbol{\Theta}\hat{x}(k|k)\\ &\quad + \boldsymbol{\Phi}\hat{\mathbb{X}}(k,P|k-1) + \boldsymbol{\Xi}\mathbf{Y}^d(k+1,P|k)\end{aligned} \tag{E.13}$$

Therefore, the complete feedback control law computed by all controllers can be expressed as

$$\mathbf{u}(k) = \boldsymbol{\Gamma}\mathbb{U}(k,M|k) \tag{E.14}$$

Merging the process model (5.3), the feedback control law (E.14), the global prediction equation given by (E.8) and the controller equation given by (E.13), the closed-loop state-space representation for the distributed case is derived

$$\begin{cases}\mathbf{x}(k) = \mathbf{A}\mathbf{x}(k-1) + \mathbf{B}\boldsymbol{\Gamma}\mathbb{U}(k-1,M|k-1)\\ \hat{\mathbb{X}}(k,P|k-1) = \mathbf{L}\overline{\mathbf{S}}[\overline{\mathbf{A}}\hat{x}(k-1) + \tilde{\mathbf{A}}\boldsymbol{\Omega}\hat{\mathbb{X}}(k-1,P|k-2)\\ \qquad\qquad + \overline{\mathbf{B}}\mathbb{U}(k-1,M|k-1) + \tilde{\mathbf{B}}\boldsymbol{\Pi}v(k-2,M|k-2)]\\ \mathbb{U}(k,M|k) = \boldsymbol{\Theta}\hat{x}(k) + \boldsymbol{\Phi}\hat{\mathbb{X}}(k,P|k-1) + \boldsymbol{\Psi}\mathbb{U}(k-1,M|k-1) + \boldsymbol{\Xi}\mathbf{Y}^d(k+1,P|k)\\ \qquad = \boldsymbol{\Theta}[\mathbf{A}\mathbf{x}(k-1) + \mathbf{B}\boldsymbol{\Gamma}\mathbb{U}(k-1,M|k-1)]\\ \qquad\quad + \boldsymbol{\Phi}\mathbf{L}\overline{\mathbf{S}}[\overline{\mathbf{A}}\hat{x}(k-1) + \tilde{\mathbf{A}}\boldsymbol{\Omega}\hat{\mathbb{X}}(k-1,P|k-2)\\ \qquad\quad + \overline{\mathbf{B}}\mathbb{U}(k-1,M|k-1) + \tilde{\mathbf{B}}\boldsymbol{\Pi}\mathbb{U}(k-2,M|k-2)]\\ \qquad\quad + \boldsymbol{\Psi}\mathbb{U}(k-1,M|k-1) + \boldsymbol{\Xi}\mathbf{Y}^d(k+1,P|k)\\ \mathbf{y}(k) = \mathbf{C}\mathbf{x}(k)\end{cases} \tag{E.15}$$

where $\hat{x}(k|k)$ in Equations (E.8) and (E.13) has been substituted with $x(k)$ due to the assumption of fully accessible state. Defining the extended state

$$\mathbf{X}_N(k) = \left[\mathbf{x}^{\mathrm{T}}(k) \quad \hat{\mathbb{X}}^{\mathrm{T}}(k, P|k-1) \quad \hat{\mathbb{X}}^{\mathrm{T}}(k, M|k) \quad \hat{\mathbb{X}}^{\mathrm{T}}(k-1, M|k-1)\right]^{\mathrm{T}},$$

the closed-loop state-space representation has the form

$$\begin{cases} \mathbf{X}_N(k) = \mathbf{A}_N \mathbf{X}_N(k-1) + \mathbf{B}_N \mathbf{Y}^d(k+1, p|k) \\ \mathbf{y}(k) = \mathbf{C}_N \mathbf{X}_N(k) \end{cases} \tag{E.16}$$

where

$$\mathbf{A}_N = \begin{bmatrix} \mathbf{A} & \mathbf{0} & \mathbf{B\Gamma} & \mathbf{0} \\ \mathbf{L\overline{S}\,\overline{A}} & \mathbf{L\overline{S}\tilde{A}\Omega} & \mathbf{L\overline{S}\,\overline{B}} & \mathbf{L\overline{S}\tilde{B}\Pi} \\ \mathbf{\Theta A} + \mathbf{\Phi L\overline{S}\,\overline{A}} & \mathbf{\Phi L\overline{S}\tilde{A}\Omega} & \mathbf{\Theta B\Gamma} + \mathbf{\Phi L\overline{S}\,\overline{B}} + \mathbf{\Psi} & \mathbf{\Phi L\overline{S}\tilde{B}\Pi} \\ \mathbf{0} & \mathbf{0} & \mathbf{I}_{Mn_u} & \mathbf{0} \end{bmatrix} \tag{E.17}$$

Thus, Theorem 7.2 is obtained.

Appendix F. Derivation of the QP problem (7.52)

At the sampling time instant k, the output prediction model for each subsystem can be derived from (7.48)

$$\hat{\mathbf{Y}}_{i,P}(k) = \mathbf{G}_i \hat{\mathbf{x}}_i(k) + \sum_{j \in \mathbb{N}_i} \mathbf{H}_{ij} \Delta \mathbf{U}_{j,M}(k) \tag{F.1}$$

where

$$\hat{\mathbf{Y}}_{i,P}(k) = \left[\hat{\mathbf{y}}_i^T(k+1|k) \quad \cdots \quad \hat{\mathbf{y}}_i^T(k+P|k)\right]^{\mathrm{T}}$$
$$\mathbf{G}_i = \left[(\mathbf{C}_i\mathbf{A}_i)^T \quad \cdots \quad (\mathbf{C}_i\mathbf{A}_i^P)^T\right]$$
$$\mathbf{H}_{ij} = \begin{bmatrix} \mathbf{C}_i\mathbf{B}_{ij} & \cdots & \mathbf{0} \\ \vdots & \ddots & \vdots \\ \mathbf{C}_i\mathbf{A}_i^{M-1}\mathbf{B}_{ij} & \cdots & \mathbf{C}_i\mathbf{B}_{ij} \\ \vdots & \vdots & \vdots \\ \mathbf{C}_i\mathbf{A}_i^{P-1}\mathbf{B}_{ij} & \cdots & \mathbf{C}_i\mathbf{A}_i^{P-M}\mathbf{B}_{ij} \end{bmatrix}$$

The local performance index for each subsystem in (7.51) can be rewritten as

$$\begin{aligned} \overline{J}_i(k) = {} & \left[\overline{\mathbf{R}}_{i,P}(k) - \mathbf{H}_{ii}\Delta\mathbf{U}_{i,M}(k)\right]^{\mathrm{T}} \overline{\mathbf{Q}}_i \left[\overline{\mathbf{R}}_{i,P}(k) - \mathbf{H}_{ii}\Delta\mathbf{U}_{i,M}(k)\right] \\ & + \Delta\mathbf{U}_{i,M}^{\mathrm{T}}(k)\overline{\mathbf{R}}_i\Delta\mathbf{U}_{i,M}(k) \\ & + \sum_{j \in \mathcal{P}_i, j \neq i} \left\{ \left[\overline{\mathbf{R}}_{j,P}(k) - \mathbf{H}_{ji}\Delta\mathbf{U}_{i,M}(k)\right]^{\mathrm{T}} \overline{\mathbf{Q}}_j \left[\overline{\mathbf{R}}_{j,P}(k) - \mathbf{H}_{ji}\Delta\mathbf{U}_{i,M}(k)\right] \right. \\ & \left. + \Delta\mathbf{U}_{j,M}^{\mathrm{T}}(k)\overline{\mathbf{R}}_j\Delta\mathbf{U}_{j,M}(k) \right\} \end{aligned}$$

$$= \Delta\mathbf{U}_{i,M}^{\mathrm{T}}(k)\left(\sum_{j\in\mathcal{P}_i}\mathbf{H}_{ji}^{T}\overline{\mathbf{Q}}_j\mathbf{H}_{ji}+\overline{\mathbf{R}}_i\right)\Delta\mathbf{U}_{i,M}(k)$$

$$-2\left(\sum_{j\in\mathcal{P}_i}\mathbf{H}_{ji}^{T}\overline{\mathbf{Q}}_j\overline{\mathbf{R}}_{j,P}(k)\right)^{\mathrm{T}}\Delta\mathbf{U}_{i,M}(k)+\text{constant} \tag{F.2}$$

where

$$\overline{\mathbf{R}}_{j,P}(k)=\mathbf{R}_{j,P}(k)-\mathbf{G}_j\hat{\mathbf{x}}_j(k)-\sum_{h\in\mathcal{P}_j,h\neq i}\mathbf{H}_{jh}\Delta\mathbf{U}_{h,M}(k)$$

$$\mathbf{R}_{i,P}(k)=\left[\mathbf{r}_i^{\mathrm{T}}(k+1)\quad\cdots\quad\mathbf{r}_i^{\mathrm{T}}(k+P)\right]^{T}$$

$$\overline{\mathbf{Q}}_i=\text{block-diag}(\underbrace{\mathbf{Q}_i,\ldots,\mathbf{Q}_i}_{P})$$

$$\overline{\mathbf{R}}_i=\text{block-diag}(\underbrace{\mathbf{R}_i,\ldots,\mathbf{R}_i}_{M}).$$

By removing the constant terms, the local optimization problem (7.51) can be written in the following quadratic form:

$$\min_{\Delta\mathbf{U}_{i,M}(k)}\overline{J}_i(k)\Longleftrightarrow\min_{\Delta\mathbf{U}_{i,M}(k)}\frac{1}{2}\Delta\mathbf{U}_{i,M}^{\mathrm{T}}(k)\Pi_i\Delta\mathbf{U}_{i,M}(k)+\mathbf{f}_i^{\mathrm{T}}(k)\Delta\mathbf{U}_{i,M}(k) \tag{F.3}$$

where

$$\mathbf{\Pi}_i=\sum_{j\in\mathbb{N}_i}\mathbf{H}_{ji}^{\mathrm{T}}\overline{\mathbf{Q}}_j\mathbf{H}_{ji}+\overline{\mathbf{R}}_i>0$$

$$\mathbf{f}_i(k)=-\sum_{j\in\mathbb{N}_i}\mathbf{H}_{ji}^{\mathrm{T}}\overline{\mathbf{Q}}_j\overline{\mathbf{R}}_{j,P}(k)$$

The inequality constraints in (7.51) can be converted into

$$\mathbf{\Omega}_i\Delta\mathbf{U}_{i,M}(k)\leq\mathbf{b}_i(k) \tag{F.4}$$

where

$$\mathbf{\Omega}_i=\begin{bmatrix}\mathbf{T}_i\\-\mathbf{T}_i\\\mathbf{I}_{Mn_{ui}}\\-\mathbf{I}_{Mn_{ui}}\\\mathbf{H}_{ii}\\-\mathbf{H}_{ii}\end{bmatrix},\ \mathbf{b}_i(k)=\begin{bmatrix}\overline{\mathbf{U}}_i\\-\underline{\mathbf{U}}_i\\\Delta\overline{\mathbf{U}}_i\\-\Delta\underline{\mathbf{U}}_i\\\overline{\mathbf{Y}}_i-\mathbf{G}_i\hat{\mathbf{x}}_i(k)-\sum\limits_{\substack{j\in\mathbb{N}_i\\j\neq i}}\mathbf{H}_{ij}\Delta\mathbf{U}_{j,M}(k)\\\mathbf{G}_i\hat{\mathbf{x}}_i(k)+\sum\limits_{\substack{j\in\mathbb{N}_i\\j\neq i}}\mathbf{H}_{ij}\Delta\mathbf{U}_{j,M}(k)-\underline{\mathbf{Y}}_i\end{bmatrix}$$

$$\mathbf{T}_i = \begin{bmatrix} \mathbf{I}_{n_{ui}} & & & \mathbf{0} \\ \mathbf{I}_{n_{ui}} & \mathbf{I}_{n_{ui}} & & \\ \vdots & \ddots & \ddots & \\ \mathbf{I}_{n_{ui}} & \cdots & \mathbf{I}_{n_{ui}} & \mathbf{I}_{n_{ui}} \end{bmatrix}_{Mn_{ui}\times Mn_{ui}}$$

$$\overline{\mathbf{U}}_i = \begin{bmatrix} \mathbf{u}_i^{\max} - \mathbf{u}_i(k-1) \\ \vdots \\ \mathbf{u}_i^{\max} - \mathbf{u}_i(k-1) \end{bmatrix}_{Mn_{ui}\times 1}, \quad \underline{\mathbf{U}}_i = \begin{bmatrix} \mathbf{u}_i^{\min} - \mathbf{u}_i(k-1) \\ \vdots \\ \mathbf{u}_i^{\min} - \mathbf{u}_i(k-1) \end{bmatrix}_{Mn_{ui}\times 1}$$

$$\Delta\overline{\mathbf{U}}_i = \begin{bmatrix} \Delta\mathbf{u}_i^{\max} \\ \vdots \\ \Delta\mathbf{u}_i^{\max} \end{bmatrix}_{Mn_{ui}\times 1}, \quad \Delta\underline{\mathbf{U}}_i = \begin{bmatrix} \Delta\mathbf{u}_i^{\min} \\ \vdots \\ \Delta\mathbf{u}_i^{\min} \end{bmatrix}_{Mn_{ui}\times 1}$$

$$\overline{\mathbf{Y}}_i = \begin{bmatrix} \mathbf{y}_i^{\max} \\ \vdots \\ \mathbf{y}_i^{\max} \end{bmatrix}_{Pn_{yi}\times 1}, \quad \underline{\mathbf{Y}}_i = \begin{bmatrix} \mathbf{y}_i^{\min} \\ \vdots \\ \mathbf{y}_i^{\min} \end{bmatrix}_{Pn_{yi}\times 1}.$$

Part Three

Constraint Distributed Predictive Control

8

Local Cost Optimization Based Distributed Predictive Control with Constraints

8.1 Introduction

In the second part, the unconstrained distributed model predictive control (DMPC) is introduced given the concept of each DPMC coordination strategy and helps the readers to comprehensively understand the essential characteristics of each kind of coordination strategies of DMPC. There are three kinds of DMPC strategies presented in Part Two. Let us first briefly review these three kinds of coordination strategies.

- *The local cost optimization (LCO)-based DMPC*, where each local controller minimizes its own subsystem's cost and uses the state prediction of the previous time instant to approximate the state sequence at the current time instant in computing the optimal solution. If the iterative algorithm is employed, the Nash optimality of closed-loop system can be achieved.
- *Cooperative-based DMPC*, where each subsystem-based MPC optimizes the cost of overall system to improve the global performance. While computing the optimal solution, it also uses the state prediction of the previous time instant to approximate the state sequence at the current time instant. This strategy could achieve a good global performance in some cases, but it reduces the flexibility and increases the communication load. We refer it as global cost optimization based DMPC here, and the *Pareto optimality* of the closed-loop system is obtained by this method.
- *Networked DMPC with information structure constraint*. In an effort to achieve a trade-off between the global performance of the entire system and the computational burden, an intuitively appealing strategy is provided in Chapter 7, where each subsystem-based MPC only considers the cost of its own subsystem and those of the subsystems it directly impacts on.

The application areas of all these approaches are complementary. Each method possesses its own strengths and weaknesses. The practitioner, using knowledge and experience, must choose the control algorithm that is more appropriate for the problem at hand.

Distributed Model Predictive Control for Plant-Wide Systems, First Edition. Shaoyuan Li and Yi Zheng.

It is well known that the explicitly handling constraint is an important advantage of MPC. Designing stabilized DMPC with constraints is definitely an important and difficult task. In this part, we will develop the stabilized DMPC with constraints. And in this chapter the basic and most used local optimization based DMPC is designed first.

Control design that takes state and/or input constraints into account, whether or not under the MPC framework, is an important and challenging problem. Many methods can be found in the literature [78, 111–113]. Under the MPC framework, closed-loop stability is ensured by judiciously integrating designs of the terminal cost, the terminal constraint set, and the local controllers [78]. In DMPC, the future state sequences of upstream neighbors, which are calculated based on the solution in the previous time instant, may not be equal to the predictive states calculated by the corresponding subsystem at the current time instant, and the errors between them are hard to estimate. In addition, in the presence of constraints, the feasibility of each subsystem-based MPC cannot be guaranteed. The remaining part of the optimal control sequence calculated at the previous time instant may not be a feasible solution at the current time instant. It is difficult to construct a feasible solution in the current time instant. All these make it difficult to design a stabilizing LCO-DMPC that takes constraints into consideration.

In this chapter, a stabilized LCO-DMPC algorithm is developed, which uses a constraint to limit the error between the future state sequences (or called presumed sequences) of upstream neighbors, which are calculated based on the solution in the previous time instant, and the predictive states calculated by the corresponding subsystem in the current time instant. Then the stability is ensured by judiciously integrating designs of the bound of the error between presumed state sequence and predictive state sequence, the terminal cost, the constraint set, and the local controllers.

The remainder of this chapter is organized as follows: Section 8.2 describes the problem to be solved in this chapter. Section 8.3 presents the design of the stabilizing LCO-DMPC with communicating once a sampling time. The feasibility of the proposed LCO-DMPC and the stability of the resulting closed-loop system are analyzed in Section 8.4. Section 8.5 presents the simulation results to demonstrate the effectiveness of the proposed LCO-DMPC algorithm. Finally, a brief conclusion to the chapter is drawn in Section 8.6.

8.2 Problem Description

Consider a spatially distributed system, as illustrated in Figure 4.1, which is composed of many physically partitioned interacted subsystems, and each subsystem is controlled by a subsystem-based controller which in turn is able to exchange information with its neighboring subsystem-based controllers.

Suppose the distributed system $\mathcal{S}$ is composed of m discrete-time linear subsystems $\mathcal{S}_i$, $i \in P = \{1, 2, \ldots, m\}$ and m controllers C_i, $i \in P = \{1, 2, \ldots, m\}$. Let the subsystems interact with each other through their states. If subsystem $\mathcal{S}_i$ is affected by $\mathcal{S}_j$, for any $i \in \mathcal{P}$ and $j \in \mathcal{P}$, subsystem $\mathcal{S}_i$ is said to be a downstream subsystem of subsystem $\mathcal{S}_j$, and subsystem $\mathcal{S}_j$ is an upstream system of $\mathcal{S}_i$. Let $\mathcal{P}_{+i}$ denote the set of the subscripts of the upstream systems of $\mathcal{S}_i$ and $\mathcal{P}_{-i}$ is the set of the subscripts of the downstream systems of $\mathcal{S}_i$. Then, using the state-interacted model introduced in Chapter 4, each subsystem $\mathcal{S}_i$ can be expressed as

$$\begin{cases} \mathbf{x}_i(k+1) = \mathbf{A}_{ii}\mathbf{x}_i(k) + \mathbf{B}_{ii}\mathbf{u}_i(k) + \sum\limits_{j\in\mathcal{P}_{+i}} \mathbf{A}_{ij}\mathbf{x}_j(k) \\ \mathbf{y}_i(k) = \mathbf{C}_{ii}\mathbf{x}_i(k) \end{cases} \tag{8.1}$$

where $\mathbf{x}_i \in \mathbb{R}^{n_{xi}}$, $\mathbf{u}_i \in \mathcal{U}_i \subset \mathbb{R}^{n_{ui}}$ and $\mathbf{y}_i \in \mathbb{R}^{n_{yi}}$ are, respectively, the local state, input and output vectors, and $\mathcal{U}_i$ is the feasible set of the input $\mathbf{u}_i$, which is used to bound the input according to the physical constraints on the actuators, the control requirements, or the characteristics of the plant. A nonzero matrix $\mathbf{A}_{ij}$, that is, $j \in \mathcal{P}_{+i}$, indicates that $\mathcal{S}_i$ is affected by $\mathcal{S}_j$. In the concatenated vector form, the system dynamics can be written as

$$\begin{cases} \mathbf{x}(k+1) = \mathbf{A}\mathbf{x}(k) + \mathbf{B}\mathbf{u}(k) \\ \mathbf{y}(k) = \mathbf{C}\mathbf{x}(k) \end{cases} \tag{8.2}$$

where

$$\mathbf{x}(k) = \begin{bmatrix} \mathbf{x}_1^{\mathrm{T}}(k) & \mathbf{x}_2^{\mathrm{T}}(k) & \cdots & \mathbf{x}_m^{\mathrm{T}}(k) \end{bmatrix}^{\mathrm{T}} \in \mathrm{R}^{n_x}$$

$$\mathbf{u}(k) = \begin{bmatrix} \mathbf{u}_1^{\mathrm{T}}(k) & \mathbf{u}_2^{\mathrm{T}}(k) & \cdots & \mathbf{u}_m^{\mathrm{T}}(k) \end{bmatrix}^{\mathrm{T}} \in \mathrm{R} \subset \mathrm{R}^{n_u}$$

$$\mathbf{y}(k) = \begin{bmatrix} \mathbf{y}_1^{\mathrm{T}}(k) & \mathbf{y}_2^{\mathrm{T}}(k) & \cdots & \mathbf{y}_m^{\mathrm{T}}(k) \end{bmatrix}^{\mathrm{T}} \in \mathrm{R}^{n_y}$$

are, respectively, the concatenated state, control input, and output vectors of the overall system $\mathcal{S}$. Also, $\mathbf{u}(k) \in U = U_1 \times U_2 \times \cdots \times U_m$. $\mathbf{A}$, $\mathbf{B}$, and $\mathbf{C}$ are the constant matrices of appropriate dimensions and are defined as follows:

$$\mathbf{A} = \begin{bmatrix} \mathbf{A}_{11} & \mathbf{A}_{12} & \cdots & \mathbf{A}_{1m} \\ \mathbf{A}_{21} & \mathbf{A}_{22} & \cdots & \mathbf{A}_{2m} \\ \vdots & \vdots & \ddots & \vdots \\ \mathbf{A}_{m1} & \mathbf{A}_{m2} & \cdots & \mathbf{A}_{mm} \end{bmatrix}^{\mathrm{T}}$$

$$\mathbf{B} = \begin{bmatrix} \mathbf{B}_{11} & \mathbf{B}_{12} & \cdots & \mathbf{B}_{1m} \\ \mathbf{B}_{21} & \mathbf{B}_{22} & \cdots & \mathbf{B}_{2m} \\ \vdots & \vdots & \ddots & \vdots \\ \mathbf{B}_{m1} & \mathbf{B}_{m2} & \cdots & \mathbf{B}_{mm} \end{bmatrix}^{\mathrm{T}}$$

$$\mathbf{C} = \begin{bmatrix} \mathbf{C}_{11} & \mathbf{C}_{12} & \cdots & \mathbf{C}_{1m} \\ \mathbf{C}_{21} & \mathbf{C}_{22} & \cdots & \mathbf{C}_{2m} \\ \vdots & \vdots & \ddots & \vdots \\ \mathbf{C}_{m1} & \mathbf{C}_{m2} & \cdots & \mathbf{C}_{mm} \end{bmatrix}^{\mathrm{T}}$$

The control objective is to stabilize the overall system $\mathcal{S}$ in a distributed predictive control framework.

8.3 Stabilizing Dual Mode Noncooperative DMPC with Input Constraints

8.3.1 Formulation

In this section, m separate optimal control problems, one for each subsystem, and the LCO-based DMPC algorithm with communicating once a control period are defined. In every distributed optimal control problem, the same constant prediction horizon N, $N \geq 1$, is used. And every distributed MPC law is updated globally synchronously. At each update, every subsystem-based MPC optimizes only for its own open-loop control sequence, given the current states and the estimated inputs of the whole system.

To proceed, we need the following assumption, and we also define the necessary notation in Table 8.1.

Assumption 8.1 For every subsystem $\mathcal{S}_i$, $\forall i \in \mathcal{P}$ there exists a state feedback $\mathbf{u}_{i,k} = \mathbf{K}_i \mathbf{x}_{i,k}$ such that the closed-loop system $\mathbf{x}(k+1) = \mathbf{A}_c \mathbf{x}(k)$ is asymptotically stable, where $\mathbf{A}_c = \mathbf{A} + \mathbf{B}\mathbf{K}$ and $\mathbf{K} = \text{block} - \text{diag}\{\mathbf{K}_1, \mathbf{K}_2, \dots, \mathbf{K}_m\}$

Remark 8.1 *This assumption is usually used in the design of stabilizing DMPC [50]. It presumes that each subsystem is able to be stabilized by a decentralized control $\mathbf{K}_i\mathbf{x}_i$, $i \in \mathcal{P}$, and the decentralized control gain $\mathbf{K}$ can be obtained by LMI or LQR.*

We also define the necessary notation in Table 8.1.

Each subsystem-based MPC minimizes the cost function of its corresponding subsystem. More specifically, the performance index is defined as

$$J_i(k) = \left\| \mathbf{x}_i^{\mathrm{p}}(k+N|k) \right\|_{\mathbf{P}_i}^2 + \sum_{s=0}^{N-1} \left(\left\| \mathbf{x}_i^{\mathrm{p}}(k+s|k) \right\|_{\mathbf{Q}_i}^2 + \left\| \mathbf{u}_i(k+s|k) \right\|_{\mathbf{R}_i}^2 \right) \tag{8.3}$$

where $\mathbf{Q}_i = \mathbf{Q}_i^{\mathrm{T}} > 0$, $\mathbf{R}_i = \mathbf{R}_i^{\mathrm{T}} > 0$ and $\mathbf{P}_i = \mathbf{P}_i^{\mathrm{T}} > 0$. The matrix $\mathbf{P}_i$ is chosen to satisfy the Lyapunov equation

$$\mathbf{A}_{\mathrm{d}i}^{\mathrm{T}} \mathbf{P}_i \mathbf{A}_{\mathrm{d}i} - \mathbf{P}_i = -\hat{\mathbf{Q}}_i$$

where $\hat{\mathbf{Q}}_i = \mathbf{Q}_i + \mathbf{K}_i^{\mathrm{T}} \mathbf{R}_i \mathbf{K}_i$. Denote

$$\begin{aligned}
\mathbf{P} &= \text{block-diag}\{\mathbf{P}_1, \mathbf{P}_2, \dots, \mathbf{P}_m\} \\
\mathbf{Q} &= \text{block-diag}\{\mathbf{Q}_1, \mathbf{Q}_2, \dots, \mathbf{Q}_m\} \\
\mathbf{R} &= \text{block-diag}\{\mathbf{R}_1, \mathbf{R}_2, \dots, \mathbf{R}_m\} \\
\mathbf{A}_{\mathrm{d}} &= \text{block-diag}\{\mathbf{A}_{\mathrm{d}1}, \mathbf{A}_{\mathrm{d}2}, \dots, \mathbf{A}_{\mathrm{d}m}\}
\end{aligned}$$

Table 8.1 Notations in this chapter

Notation	Explanation
i	The subscript denoting all downstream subsystems of $\mathcal{S}_i$
$+i$	The subscript denoting all upstream subsystems of $\mathcal{S}_i$
$\mathbf{x}_i^{\mathrm{p}}(k+s\|k)$	The predicted state sequence of $\mathcal{S}_i$, calculated by $\mathcal{C}_i$ at time k, $\mathbf{x}_i^{\mathrm{p}}(k+s\|k) = \mathbf{x}_{i,i}^{\mathrm{p}}(k+s\|k)$
$\mathbf{u}_i^{p}(k+s\|k)$	The predicted control sequence of $\mathcal{S}_i$, calculated by $\mathcal{C}_i$ at time k
$\hat{\mathbf{x}}_i(k+s\|k)$	The presumed state sequence of $\mathcal{S}_i$, calculated by $\mathcal{C}_i$ at time k, $\hat{\mathbf{x}}_i(k+s\|k) = \hat{\mathbf{x}}_{i,i}(k+s\|k)$
$\hat{\mathbf{u}}_i(k+s\|k)$	The presumed control sequence of $\mathcal{S}_i$, defined by $\mathcal{C}_i$ at time k
$\mathbf{x}_i^{\mathrm{f}}(k+s\|k)$	The feasible state sequence of $\mathcal{S}_i$, calculated by $\mathcal{C}_i$ at time k, $\mathbf{x}_i^{\mathrm{f}}(k+s\|k) = \mathbf{x}_{i,i}^{\mathrm{f}}(k+s\|k)$
$\mathbf{u}_i^{\mathrm{f}}(k+s\|k)$	The feasible control sequence of $\mathcal{S}_i$, defined by $\mathcal{C}_i$ at time k, $\mathbf{u}_i^{\mathrm{f}}(k+s\|k) = \mathbf{u}_{i,i}^{\mathrm{f}}(k+s\|k)$

Then, it follows that

$$\mathbf{A}_{\mathrm{d}}^{\mathrm{T}}\mathbf{P}\mathbf{A}_{\mathrm{d}} - \mathbf{P} = -\hat{\mathbf{Q}}$$

where $\hat{\mathbf{Q}} = \mathbf{Q} + \mathbf{K}^{\mathrm{T}}\mathbf{R}\mathbf{K} > 0$.

To get the predicted state sequence $\mathbf{x}_i^{\mathrm{p}}(k+s|k)$ of subsystem $\mathcal{S}_i$ under the control decision sequence of $\mathbf{u}_i(k+s|k)$ in (8.3), the system evolution model should be deduced first. Since every subsystem-based controller is updated synchronously, the state and control sequences of other subsystems are unknown to subsystem $\mathcal{S}_i$. Thus, at the time instant k, the presumed state sequence $\{\hat{\mathbf{x}}_j(k|k), \hat{\mathbf{x}}_j(k+1|k), \cdots, \hat{\mathbf{x}}_j(k+N|k)\}$ of $\mathcal{S}_j$ is used in the predictive model of the MPC in $\mathcal{S}_i$, which is given as

$$\mathbf{x}_i^{\mathrm{p}}(k+l|k) = \mathbf{A}_{ii}^{l}\mathbf{x}_i^{\mathrm{p}}(k|k) + \sum_{h=1}^{l}\mathbf{A}_{ii}^{l-h}\mathbf{B}_{ii}\mathbf{u}_i(k+l|k)$$
$$+\sum_{j\in\mathcal{P}_{+i}}\sum_{h=1}^{l}\mathbf{A}_{ii}^{l-h}\mathbf{A}_{ij}\hat{\mathbf{x}}_j(k+h-1|k) \tag{8.4}$$

Given $\mathbf{x}_i^{\mathrm{p}}(k|k) = \mathbf{x}_i(k|k)$, the presumed control sequence for subsystem $\mathcal{S}_i$ for any $i \in \mathcal{P}$ is given by

$$\hat{\mathbf{u}}_i(k+s-1|k) = \begin{cases} \mathbf{u}_i\,(k+s-1|k-1), s = 1, 2, \ldots, N-1 \\ \mathbf{K}_i\mathbf{x}_i^{\mathrm{p}}(k+N-1|k-1), s = N \end{cases} \tag{8.5}$$

Set each presumed state sequence $\hat{\mathbf{x}}_i$ to be the remainder of the sequence predicted at time instant $k-1$, concatenated with the closed-loop response under the state feedback control $\hat{\mathbf{u}}_i(k+s-1|k) = \mathbf{K}_i\mathbf{x}_i^{\mathrm{p}}(k+N-1|k-1)$, that is

$$\begin{cases} \hat{\mathbf{x}}_i\,(k+s-1|k) = \mathbf{x}_i^{\mathrm{p}}(k+s-1|k-1), s = 1, 2, \ldots, N, \\ \hat{\mathbf{x}}_i(k+N+1-1|k) = \mathbf{A}_{\mathrm{d}i}\mathbf{x}_i^{\mathrm{p}}(k+N-1|k-1) \\ \qquad\qquad + \sum_{j\in\mathcal{P}_{+i}}\mathbf{A}_{ij}\mathbf{x}_j^{\mathrm{p}}(k+N-1|k-1) \end{cases} \tag{8.6}$$

It should be noticed that since $\hat{\mathbf{x}}_i(k+N|k)$ is only a media viable here, $\hat{\mathbf{x}}_i(k+N|k)$ is not equal to the solution of substituting $\hat{\mathbf{u}}_i(k+N-1|k)$ into (8.4) based on the presumed state sequence obtained at time $k-1$.

In MPC systems, it is an important proposition to focus on the systems sequence feasibility and the stability. The same things happen in the distributed MPC systems. To enlarge the feasible region, a terminal state constraint is included in each subsystem-based MPC, which guarantees that the terminal controllers are stabilizing inside a terminal set. To define this terminal state set, we need to make an assumption and establish a technical lemma.

Assumption 8.2 The block-diagonal matrix

$$\mathbf{A}_{\mathrm{d}} = \text{block-diag}\{\mathbf{A}_{\mathrm{d}1}, \mathbf{A}_{\mathrm{d}2}, \ldots, \mathbf{A}_{\mathrm{d}m}\}$$

and the off-diagonal matrix $\mathbf{A}_{\mathrm{o}} = \mathbf{A}_{\mathrm{c}} - \mathbf{A}_{\mathrm{d}}$ satisfy the following inequality:

$$\mathbf{A}_{\mathrm{o}}^{\mathrm{T}}\mathbf{P}\mathbf{A}_{\mathrm{o}} + \mathbf{A}_{\mathrm{o}}^{\mathrm{T}}\mathbf{P}\mathbf{A}_{\mathrm{d}} + \mathbf{A}_{\mathrm{d}}^{\mathrm{T}}\mathbf{P}\mathbf{A}_{\mathrm{o}} < \hat{\mathbf{Q}}/2$$

where $\hat{\mathbf{Q}} = \mathbf{Q} + \mathbf{K}^{\mathrm{T}}\mathbf{R}\mathbf{K} > 0$

It, along with Assumption 8.1, helps with the design of the terminal set. This assumption quantifies how strengthening the coupling among subsystems is sufficient so that the overall system can be stabilized by the proposed DMPC here. This assumption is not necessary, and some systems that do not satisfy this assumption may also be stabilized by the proposed DMPC, and the more relaxing condition is still remaining to be designed in the future work.

Lemma 8.1 *Under Assumptions 8.1 and 8.2, for any positive scalar c the set*

$$\Omega(c) = \{\mathbf{x} \in \mathbb{R}^{n_x} : \|\mathbf{x}\|_{\mathbf{P}} \le c\}$$

is a positive invariant region of attraction for the closed-loop system $\mathbf{x}(k+1) = \mathbf{A}_c\mathbf{x}(k)$. Additionally, there exists a small enough positive scalar ε such that $\Omega(\varepsilon)$ is in the feasible input set $\mathcal{U} \in \mathbb{R}^{n_u}$ for all $\mathbf{x} \in \Omega(\varepsilon)$.

Proof. Consider the function $V(k) = \|\mathbf{x}(k)\|^2_{\mathbf{P}}$. The time difference of $V(k)$ along the trajectories of the closed-loop system $\mathbf{x}(k+1) = \mathbf{A}_c\mathbf{x}(k)$ can be evaluated as

$$\begin{aligned}\Delta V(k) &= \mathbf{x}^T(k)\mathbf{A}_c^T\mathbf{P}\mathbf{A}_c\mathbf{x}(k) - \mathbf{x}^T(k)\mathbf{P}\mathbf{x}(k)\\ &= \mathbf{x}^T(k)(\mathbf{A}_d^T\mathbf{P}\mathbf{A}_d - \mathbf{P} + \mathbf{A}_o^T\mathbf{P}\mathbf{A}_o + \mathbf{A}_o^T\mathbf{P}\mathbf{A}_d + \mathbf{A}_d^T\mathbf{P}\mathbf{A}_o)\mathbf{x}(k)\\ &\le -\mathbf{x}^T(k)\hat{\mathbf{Q}}\mathbf{x}(k) + \frac{1}{2}\mathbf{x}^T(k)\hat{\mathbf{Q}}\mathbf{x}(k)\\ &\le 0 \end{aligned} \tag{8.7}$$

which holds for all $\mathbf{x}(k) \in \Omega(\mathrm{c})\backslash\{0\}$. This implies that all trajectories of the closed-loop system that starts inside Ω (c) will remain inside and converge to the origin.

The existence of an $\varepsilon > 0$ such that $\mathbf{Kx} \in \mathcal{U}$ for all $\mathbf{x} \in \Omega(\varepsilon)$ follows from the fact that $\mathbf{P}$ is a positive definite, which implies that the set $\Omega(\varepsilon)$ shrinks to the origin as ε decreases to zero. This completes the proof.

In the optimization problem of each subsystem-based MPC, the terminal state constraint set for each $\mathcal{S}_i$ can then be set to be

$$\Omega_i(\varepsilon) = \{\mathbf{x}_i \in \mathbb{R}^{n_{xi}} : \|\mathbf{x}_i\|_{\mathbf{P}_i} \le \varepsilon/\sqrt{m}\} \tag{8.8}$$

Clearly, if $\mathbf{x} \in \Omega_1(\varepsilon) \times \cdots \times \Omega_m(\varepsilon)$, then the decoupled controllers will stabilize the system at the origin, since

$$\|\mathbf{x}_i\|^2_{P_i} \le \frac{\varepsilon^2}{m}, \forall i \in \mathsf{P}$$

implies that

$$\sum_{i \in \mathsf{P}} \|\mathbf{x}_i\|^2_{\mathbf{P}_i} \le \varepsilon^2$$

which in turn implies that $\mathbf{x} \in \Omega(\varepsilon)$. Suppose that at some time k_0, $\mathbf{x}_i(k_0) \in \Omega_i(\varepsilon)$ for every subsystem. Then, by Lemma 8.1, stabilization can be achieved if every C_i employs its decoupled static feedback controller $\mathbf{K}_i\mathbf{x}_i(k)$ after time instant k_0.

Thus, the objective of each subsystem-based MPC law is to drive the state of each subsystem $\mathcal{S}_i$ to the set $\Omega_i(\varepsilon)$. Once all subsystems have reached these sets, they switch to their decoupled controllers for stabilization. Such switching from an MPC law to a terminal controller once the state reaches a suitable neighborhood of the origin is referred to as the dual mode MPC [69]. For this reason, the DMPC algorithm we propose in this chapter is a dual mode DMPC algorithm. In addition, in the distributed MPC systems, the subsystems' controllers use the estimates made by the time $k-1$ to predict the future states. There are some deviations between these and estimates made by the current time. It is difficult to construct a feasible solution for the time k because of deviations. We need to add consistency constraints to limit these deviations.

In what follows, we formulate the optimization problem for each subsystem-based MPC.

Problem 8.1 Consider subsystem $\mathcal{S}_i$. Let $\varepsilon > 0$ be as specified in Lemma 8.1. Let the update time be $k \geq 1$. Given $\mathbf{x}_i(k)$ and $\hat{\mathbf{x}}_j(k+s|k)$, $s = 1, 2, \ldots, N$, $\forall j \in \mathcal{P}_{+i}$, find the control sequence $\mathbf{u}_i(k+s|k) : \{0, 1, \ldots, N-1\} \rightarrow \mathcal{U}_i$ that minimizes

$$J_i(k) = \left\|\mathbf{x}_i^{\mathrm{p}}(k+N|k)\right\|_{\mathbf{P}_i}^2 + \sum_{s=0}^{N-1}\left(\left\|\mathbf{x}_i^{\mathrm{p}}(k+s|k)\right\|_{\mathbf{Q}_i}^2 + \left\|\mathbf{u}_i(k+s|k)\right\|_{\mathbf{R}_i}^2\right) \tag{8.9}$$

subject to the following constraints:

$$\sum_{l=1}^{s} \alpha_{s-l}\|\mathbf{x}_i^{\mathrm{p}}(k+l|k) - \hat{\mathbf{x}}_i(k+l|k)\|_2 \leq \frac{\xi\kappa\varepsilon}{2\sqrt{mm_1}}, s = 1, 2, \ldots, N-1 \tag{8.10}$$

$$\left\|\mathbf{x}_i^{\mathrm{p}}(k+N|k) - \hat{\mathbf{x}}_i(k+N|k)\right\|_{\mathbf{P}_i} \leq \frac{\kappa\varepsilon}{2\sqrt{m}} \tag{8.11}$$

$$\left\|\mathbf{x}_i^{\mathrm{p}}(k+s|k)\right\|_{P_i} \leq \left\|\mathbf{x}_i^{\mathrm{f}}(k+s|k)\right\|_{\mathbf{P}_i} + \frac{\varepsilon}{\mu N\sqrt{m}}, s = 1, 2, \ldots, N \tag{8.12}$$

$$\mathbf{u}_i^{\mathrm{p}}(k+s|k) \in \mathsf{U}_i,\ s = 0, 1, \ldots, N-1 \tag{8.13}$$

$$\mathbf{x}_i^{\mathrm{p}}(k+N|k) \in \Omega_i(\varepsilon/2) \tag{8.14}$$

In the constraints above,

$$m_2 = \max_{i \in \mathsf{P}}\{\text{number of elements in } \mathsf{P}_{+i}\} \tag{8.15}$$

$$\alpha_l = \max_{i \in \mathcal{P}}\max_{j \in \mathcal{P}_i}\left\{\lambda_{\max}^{\frac{1}{2}}\left(\left(\mathbf{A}_{ii}^l\mathbf{A}_{ij}\right)^{\mathrm{T}}\mathbf{P}_j\mathbf{A}_{ii}^l\mathbf{A}_{ij}\right)\right\}, l = 0, 1, \ldots, N-1 \tag{8.16}$$

The constants $0 < \kappa < 1$ and $0 < \xi \leq 1$are design parameters whose values will be chosen in the sequel.

Equations (8.10) and (8.11) are referred to as the consistency constraints, which require that each predictive sequence and control variables remain close to their presumed values. These constraints are the keys to proving that $\mathbf{x}_{j,i}^{\mathrm{f}}$ is a feasible state sequence at each update.

Equation (8.12) will be utilized to prove that the LCO-DMPC described by Problem 8.1 is stabilizing, where $\mu > 0$ is a design parameter whose value will be specified later to satisfy Lemma 8.1, $\mathbf{x}_i^{\mathrm{f}}(k+s|k)$ is a feasible state sequence, and $\mathbf{x}_i^{\mathrm{f}}(k+s|k)$ equals to the solution of (8.4) under the initial state of $\mathbf{x}_i(k)$, the presumed state of $\hat{\mathbf{x}}_j(k+s|k), j \in \mathcal{P}_{+i}$ and the feasible control sequence $\mathbf{u}_i^{\mathrm{f}}(k+s-1|k)$ is defined by

$$\mathbf{u}_i^{\mathrm{f}}(k+s-1|k) = \begin{cases} \mathbf{u}_i^{\mathrm{p}}(k+s-1|k-1), & s = 1, 2, \ldots, N-1 \\ \mathbf{K}_i \mathbf{x}_i^{\mathrm{f}}(k+N-1|k), & s = N \end{cases} \tag{8.17}$$

It should be noticed that the terminal constraint in each optimal control problem is $\Omega_i(\varepsilon/2)$, although Lemma 8.1 ensures that the larger $\Omega(\varepsilon)$ suffices for the feasibility of the terminal controllers. In the analysis presented in the next section, it will be shown that tightening the terminal set in this way is required to guarantee the feasibility properties.

8.3.2 Algorithm Design for Resolving Each Subsystem-based Predictive Control

Before stating the constraint LCO-DMPC algorithm, we make the following assumption to facilitate the initialization phase.

Assumption 8.3 At initial time k_0, there exists a feasible control $\mathbf{u}_i(k_0+s|k_0) \in \mathcal{U}_i$, $s = 1, 2, \ldots, N-1$, for each $i \in \mathcal{P}$, such that the solution to the full system $\mathbf{x}(s+1+k_0) = \mathbf{A}\mathbf{x}(s+k_0) + \mathbf{B}\mathbf{u}(s+k_0|k_0)$, denoted as $\mathbf{x}_i^{\mathrm{p}}(\cdot|k_0)$, satisfies $\mathbf{x}_i^{\mathrm{p}}(N+k_0) \in \Omega_i(\varepsilon/2)$ and results in a bounded cost $J_i(k_0)$. Moreover, each subsystem has access to $\mathbf{u}_i(\cdot|k_0)$.

Assumption 8.3 bypasses the difficult task of actually constructing an initially feasible solution in a distributed way. In fact, finding an initially feasible solution for many optimization problems is often a primary obstacle, whether or not such problems are used in a control setting. As such, many centralized implementations of MPC also assume that an initially feasible solution is available [69].

Under the Assumption 8.3, we can get the algorithm of the networked cooperative DMPC.

Algorithm 8.1 *Constraint DMPC algorithm* The dual mode DMPC law for any $\mathcal{S}_i$ is constructed as follows:

Step 1: Initialization.

- Initialize $\mathbf{x}(k_0)$, $\mathbf{u}_i(k_0+s|k_0)$, $s = 1, 2, \ldots, N$, to satisfy
- Assumption 8.3
- At time k_0, if $\mathbf{x}(k_0) \in \Omega(\varepsilon)$, then apply the terminal controller $\mathbf{u}_i(k) = \mathbf{K}_i(\mathbf{x}_i(k))$, for all $k \geq k_0$, else
- Compute $\hat{\mathbf{x}}_i(k_0+s+1|k_0+1)$ according to (8.4) and transmit $\hat{\mathbf{x}}_i(k_0+s+1|k_0+1)$ to $\mathcal{S}_j$, $j \in \mathcal{P}_{-i}$.

Step 2: Communicating at time $k+1$

- Measure $\mathbf{x}_i(k)$, transmit $\mathbf{x}_i(k)$, $\hat{\mathbf{x}}_i(k+s+1|k)$, to $\mathcal{S}_j, j \in \mathcal{P}_{-i}$, and receive $\mathbf{x}_j(k)$, $\hat{\mathbf{x}}_j(k+s|k)$ from $\mathcal{S}_j, j \in \mathcal{P}_{+i}$.

Step 3: Update of control law at time $k+1$

- If $\mathbf{x}(k)\in\Omega(\varepsilon)$, then apply the terminal controller $\mathbf{u}_i(k)=\mathbf{K}_i(\mathbf{x}_i(k))$, else
- Solve Problem 8.1for $\mathbf{u}_i(k|k)$ and apply $\mathbf{u}_i(k|k)$
- Compute $\hat{\mathbf{x}}_i(k+s+1|k+1)$ according to (8.4) and transmit $\hat{\mathbf{x}}_i(k+s+1|k+1)$ to $\mathcal{S}_j, j\in\mathcal{P}_{-i}$.

Step 4: Update of control at time $k+1$

Let $k+1\rightarrow k$, repeat Step 2.

Algorithm 8.1 presumes that all local controllers $\mathcal{C}_i$, $i\in\mathcal{P}$ have access to the full state $\mathbf{x}(k)$. This requirement results solely from the use of the dual mode control, in which the switching occurs synchronously only when $\mathbf{x}(k)\in\Omega(\varepsilon)$, with $\Omega(\varepsilon)$ being as defined in Lemma 8.1. In the next section, it will be shown that the constraint LCO-DMPC policy drives the state $\mathbf{x}(k+s)$ to $\Omega(\varepsilon)$ in a finite number of updates. As a result, if $\Omega_i(\varepsilon)$ is chosen sufficiently small, then MPC can be employed for all time without switching to a terminal controller, eliminating the need of the local controllers to access the full state. Of course, in this case, instead of asymptotic stability at the origin, we can only drive the state toward the small set $\Omega(\varepsilon)$.

The analysis in the next section shows that the constraint LCO-DMPC algorithm is feasible at every update and is stabilizing.

8.4 Analysis

8.4.1 Recursive Feasibility of Each Subsystem-based Predictive Control

The main result of this section is that, provided that an initially feasible solution is available and Assumption 8.3 holds true, for any $\mathcal{S}_i$ and at any time $k\geq 1$, $\mathbf{u}_i(\cdot|k)=\mathbf{u}_i^{\mathrm{f}}(\cdot|k)$ is a feasible control solution to Problem 1. This feasibility result refers that, for any $\mathcal{S}_i$ and at any update $k\geq 1$, the control and state pair $(\mathbf{u}_i^{\mathrm{f}}(\cdot|k),\mathbf{x}_i^{\mathrm{f}}(\cdot|k))$, $j\in\mathcal{P}_i$ satisfy the consistency constraints (8.10) and (8.11), the control constraint (8.13), and the terminal state constraint (8.14). Theorem 8.2 identifies sufficient conditions that ensure $\hat{\mathbf{x}}_i(k+N|k)\in\Omega_i(\varepsilon'/2)$, where $\varepsilon'=(1-\kappa)\varepsilon$. Lemma 8.2 identifies sufficient conditions that ensure $\left\|\mathbf{x}_i^{\mathrm{f}}(s+k|k)-\hat{\mathbf{x}}_i(s+k|k)\right\|_{\mathbf{P}_i}\leq\kappa\varepsilon/(2\sqrt{m})$ for all $i\in\mathcal{P}$. Lemma 8.3 establishes that the control constraint is satisfied. Finally, the results in Lemma 8.2– 8.4 are combined to arrive at the conclusion that, for any $i\in\mathcal{P}$, the control and state pair $(\mathbf{u}_i^{\mathrm{f}}(\cdot|k),\mathbf{x}_i^{\mathrm{f}}(\cdot|k))$ are a feasible solution to Problem 8.1 at any update $k\geq 1$.

Lemma 8.2 *Suppose Assumptions 8.1–8.3 hold and $\mathbf{x}(k_0)\in\mathcal{X}$. For any $k\geq 0$, if Problem 8.1 has a solution at time $k-1$, and $\hat{\mathbf{x}}_i(k+N-1|k-1)\in\Omega_i(\varepsilon/2)$ for any $i\in\mathcal{P}$, then*

$$\hat{\mathbf{x}}_i(k+N-1|k)\in\Omega_i(\varepsilon/2)$$

and

$$\hat{\mathbf{x}}_i(k+N|k)\in\Omega_i(\varepsilon'/2)$$

Provided $\boldsymbol{P}_i$ and $\hat{\mathbf{Q}}_i$ satisfy that

$$\max_{i\in\mathcal{P}}(\rho_i)\leq 1-\kappa \tag{8.18}$$

where, $\varepsilon'=(1-\kappa)\varepsilon$, $\rho=\lambda_{\max}\sqrt{(\hat{\mathbf{Q}}_i\mathbf{P}_i^{-1})^{\mathrm{T}}\hat{\mathbf{Q}}_i\mathbf{P}_i^{-1}}$.

Proof. Since Problem 8.1 has a solution at time $k-1$, by construction (8.6), it has

$$\left\|\hat{\mathbf{x}}_i(k+N-1)\,|k\right\|_{\mathbf{P}_j} = \left\|\mathbf{x}_i^{\mathrm{p}}(k+N-1)\,|k-1\right\|_{\mathbf{P}_i} \le \frac{\varepsilon}{2\sqrt{m}}$$

In addition, since

$$\begin{aligned}\hat{\mathbf{x}}_i(k+N|k) &= \mathbf{A}_{\mathbf{d}i}\mathbf{x}_i^{\mathrm{p}}(k+N-1|k-1) + \sum_{j\in\mathcal{P}_{+i}}\mathbf{A}_{ij}\mathbf{x}_j^{\mathrm{p}}(k+N-1|k-1)\\ &= \mathbf{A}_{\mathrm{d}i}\hat{\mathbf{x}}_i(k+N-1|k) + \sum_{j\in\mathcal{P}_{+i}}\mathbf{A}_{ij}\hat{\mathbf{x}}_j(k+N-1|k)\end{aligned}$$

It has

$$\left\|\hat{\mathbf{x}}_i(k+N|k)\right\|_{\mathbf{P}_i} = \left\|\mathbf{A}_{\mathrm{d}i}\hat{\mathbf{x}}_i(k+N-1|k) + \sum_{j\in\mathcal{P}_{+i}}\mathbf{A}_{ij}\hat{\mathbf{x}}_j(k+N-1|k)\right\|_{\mathbf{P}_i}$$

Consider Assumption 8.2, $\mathbf{A}_{\mathrm{o}}^{\mathrm{T}}\mathbf{P}\mathbf{A}_{\mathrm{o}} + \mathbf{A}_{\mathrm{o}}^{\mathrm{T}}\mathbf{P}\mathbf{A}_{\mathrm{d}} + \mathbf{A}_{\mathrm{d}}^{\mathrm{T}}\mathbf{P}\mathbf{A}_{\mathrm{o}} < \hat{\mathbf{Q}}/2$
Thus if, then

$$\begin{aligned}\left\|\hat{\mathbf{x}}_i(k+N|k)\right\|_{\mathbf{P}_i} &\le \left\|\hat{\mathbf{x}}_i(k+N-1|k)\right\|_{\hat{\mathbf{Q}}/2}\\ &\le \lambda_{\max}\sqrt{(\hat{\mathbf{Q}}_i\mathbf{P}_i^{-1})^{\mathrm{T}}\hat{\mathbf{Q}}_i\mathbf{P}_i^{-1}}\left\|\hat{\mathbf{x}}_i(k+N-1|k)\right\|_{\mathbf{P}_i}\\ &\le (1-\kappa)\frac{\varepsilon}{2\sqrt{m}}\end{aligned}$$

This completes the proof.

Lemma 8.3 *Suppose Assumptions 8.1–8.3 hold and* $\mathbf{x}(k_0)\in\mathcal{X}, \forall k\ge 0$, *if Problem 8.1 has a solution at every update time l, l = 1, 2, …, k − 1, then*

$$\left\|\mathbf{x}_i^{\mathrm{f}}(k+s|k) - \hat{\mathbf{x}}_i(k+s|k)\right\|_{\mathbf{P}_i} \le \frac{\kappa\varepsilon}{2\sqrt{m}} \tag{8.19}$$

for all $i\in\mathcal{P}$ *and all s = 1, 2, …, N, provided that (8.18) and the following parametric condition hold:*

$$\frac{\sqrt{m_2}}{\xi\lambda_{\min}(P)}\sum_{l=0}^{N-2}\alpha_l \le 1 \tag{8.20}$$

where α_l *is as defined in (8.16). Furthermore, the feasible control* $\mathbf{u}_i^{\mathrm{f}}(k+s|k)$ *and the feasible state* $\mathbf{x}_i^{\mathrm{f}}(k+s|k)$ *satisfy constraints (8.10) and (8.11).*

Proof. We will prove (8.19) first. Since a solution exists at update time $1, 2, \ldots, k-1$, according to (8.4), (8.5), and (8.17), for any $s = 1, 2, \ldots, N-1$, the feasible state is given by

$$\begin{aligned}\mathbf{x}_i^{\mathrm{f}}(k+l|k) &= \mathbf{A}_{ii}^{l}\mathbf{x}_i^{\mathrm{f}}(k|k) \\ &\quad + \sum_{h=1}^{l} \mathbf{A}_{ii}^{l-h}\mathbf{B}_{ii}\mathbf{u}_i^{\mathrm{f}}(k+l|k) \\ &\quad + \sum_{j\in\mathcal{P}_{+i}} \sum_{h=1}^{l} \mathbf{A}_{ii}^{l-h}\mathbf{A}_{ij}\hat{\mathbf{x}}_j(k+h-1|k) \\ &= \mathbf{A}_{ii}^{l}\Bigg(\mathbf{A}_{ii}^{l}\mathbf{x}_i(k-1|k-1) \\ &\quad + \mathbf{B}_{ii}\mathbf{u}_i(k-1|k-1) + \sum_{j\in\mathcal{P}_{+i}} \mathbf{A}_{ij}\mathbf{x}_j(k-1|k-1)\Bigg) \\ &\quad + \sum_{h=1}^{l} \mathbf{A}_{ii}^{l-h}\mathbf{B}_{ii}\hat{\mathbf{u}}_i(k+l|k) \\ &\quad + \sum_{j\in\mathcal{P}_{+i}} \sum_{h=1}^{l} \mathbf{A}_{ii}^{l-h}\mathbf{A}_{ij}\mathbf{x}_j^{\mathrm{P}}(k+h-1|k-1)\end{aligned} \tag{8.21}$$

and

$$\begin{aligned}\hat{\mathbf{x}}(k+l|k) &= \mathbf{A}_{ii}^{l}\mathbf{x}_i(k|k-1) \\ &\quad + \sum_{h=1}^{l} \mathbf{A}_{ii}^{l-h}\mathbf{B}_{ii}\mathbf{u}_i(k+l|k-1) \\ &\quad + \sum_{j\in\mathcal{P}_{+i}} \sum_{h=1}^{l} \mathbf{A}_{ii}^{l-h}\mathbf{A}_{ij}\hat{\mathbf{x}}_j(k+h-1|k-1) \\ &= \mathbf{A}_{ii}^{l}\Bigg(\mathbf{A}_{ii}^{l}\mathbf{x}_i(k-1|k-1) \\ &\quad + \mathbf{B}_{ii}\mathbf{u}_i(k-1|k-1) + \sum_{j\in\mathcal{P}_{+i}} \mathbf{A}_{ij}\hat{\mathbf{x}}_j(k-1|k-1)\Bigg) \\ &\quad + \sum_{h=1}^{l} \mathbf{A}_{ii}^{l-h}\mathbf{B}_{ii}\hat{\mathbf{u}}_i(k+l|k) \\ &\quad + \sum_{j\in\mathcal{P}_{+i}} \sum_{h=1}^{l} \mathbf{A}_{ii}^{l-h}\mathbf{A}_{ij}\hat{\mathbf{x}}_j(k+h-1|k-1)\end{aligned} \tag{8.22}$$

Subtracting (8.22) from (8.21), and from the definition of (8.16), we obtain the discrepancy between the feasible state sequence and the presumed state sequence as

$$\begin{aligned}
&\left\|\mathbf{x}^{\mathrm{f}}_{j,i}(k+s|k) - \hat{\mathbf{x}}_{j,i}(k+s|k)\right\|_{\mathbf{P}_j} \\
&\quad = \left\|\sum_{l=1}^{s} \mathbf{A}_{ii}^{s-l}\mathbf{A}_{ij}\left(\mathbf{x}^{\mathrm{p}}_i(k+l-1|k-1) - \hat{\mathbf{x}}_i(k+l-1|k-1)\right)\right\|_{\mathbf{P}_i} \\
&\quad \le \sum_{l=1}^{s} \left\|\mathbf{A}_{ii}^{s-l}\mathbf{A}_{ij}\left(\mathbf{x}^{\mathrm{p}}_i(k+l-1|k-1) - \hat{\mathbf{x}}_i(k+l-1|k-1)\right)\right\|_{\mathbf{P}_i} \\
&\quad \le \sum_{l=1}^{s} \alpha_{s-l}\left\|\mathbf{x}^{\mathrm{p}}_i(k+l-1|k-1) - \hat{\mathbf{x}}_i(k+l-1|k-1)\right\|_2
\end{aligned} \tag{8.23}$$

Let the subsystems, which respectively maximize the following functions, as $\mathcal{S}_g$

$$\sum_{l=1}^{s} \alpha_{s-l}\left\|\mathbf{x}^{\mathrm{p}}_i(k-1+l|k-1) - \hat{\mathbf{x}}_i(k-1+l|k-1)\right\|_2, i \in \mathcal{P}$$

Then, the following equation can be deduced from (8.23):

$$\begin{aligned}
&\left\|\mathbf{x}^{f}_j(k+s|k) - \hat{\mathbf{x}}_j(k+s|k)\right\|_{\mathbf{P}_i} \\
&\quad \le \sqrt{m_1}\sum_{l=1}^{s} \alpha_{s-l}\left\|\mathbf{x}^{\mathrm{p}}_g(k+l-1|k-1) - \hat{\mathbf{x}}_g(k+l-1|k-1)\right\|_2
\end{aligned}$$

Since $\mathbf{x}^{\mathrm{p}}_i(l|k-1)$ satisfy constraints (8.10) for all times $l = 1, 2, \ldots, k-1$, the following equation can be deduced:

$$\begin{aligned}
&\left\|\mathbf{x}^{\mathrm{f}}_i(k+s|k) - \hat{\mathbf{x}}_i(k+s|k)\right\|_{\mathbf{P}_i} \\
&\quad \le \frac{(1-\xi)(1-\kappa)\varepsilon}{2\sqrt{m}} + \frac{\xi(1-\kappa)\varepsilon}{2\sqrt{m}} \\
&\quad = \frac{\kappa\varepsilon}{2\sqrt{m}}
\end{aligned} \tag{8.24}$$

Thus, (8.19) holds for all $s = 1, 2, \ldots, N-1$.

When $l = N$, we can get

$$\mathbf{x}^{\mathrm{f}}_i(k+N|k) = \mathbf{A}_{\mathrm{d},i}\mathbf{x}^{\mathrm{f}}_i(k+N-1|k) + \sum_{j\in\mathcal{P}_{+i}} \mathbf{A}_{ij}\hat{\mathbf{x}}_j(k+N-1|k) \tag{8.25}$$

$$\hat{\mathbf{x}}_i(k+N|k) = \mathbf{A}_{\mathrm{d},i}\hat{\mathbf{x}}_i(k+N-1|k) + \sum_{j\in\mathcal{P}_{+i}} \mathbf{A}_{ij}\hat{\mathbf{x}}_j(k+N-1|k) \tag{8.26}$$

From the subtraction of the two equations, then, the discrepancy between the feasible state $\mathbf{x}^{\mathrm{f}}_{j,i}(k+N|k)$ and the presumed state $\hat{\mathbf{x}}_{j,i}(k+N|k)$ is

$$\mathbf{x}^{\mathrm{f}}_i(k+N|k) - \hat{\mathbf{x}}_i(k+N|k) = \mathbf{A}_{\mathrm{d},i}(\mathbf{x}^{\mathrm{f}}_i(k+N-1|k) - \hat{\mathbf{x}}_i(k+N-1|k)) \tag{8.27}$$

This completes the proof of (8.19).

In what follows we will prove that the feasible state $\mathbf{x}_i^{\mathrm{f}}(k+s|k)$ satisfies the constraints (8.10) and (8.11) when (8.19) holds.

When $l=1,2,\ldots,N-1$, substitute $\mathbf{x}_i^{\mathrm{f}}(k+l|k)$ in the constraint (8.10) with considering (8.20), we can get

$$\begin{aligned}
&\sum_{l=1}^{s}\alpha_{s-l}\left\|\mathbf{x}_i^{\mathrm{f}}(k+l|k)-\hat{\mathbf{x}}_i(k+l|k)\right\|_2\\
&\le\frac{1}{\lambda_{\min}(\mathbf{P}_i)}\sum_{l=1}^{s}\alpha_{s-l}\left\|\mathbf{x}_i^{\mathrm{f}}(k+l|k)-\hat{\mathbf{x}}_i(k+l|k)\right\|_{\mathbf{P}_i}\\
&\le\frac{1}{\lambda_{\min}(\mathbf{P})}\sum_{l=1}^{s}\alpha_{s-l}\frac{\sqrt{m_2}}{\xi}\frac{\xi\kappa\varepsilon}{2\sqrt{mm_2}}
\end{aligned}\tag{8.28}$$

Thus, when

$$\frac{\sqrt{m_2}}{\xi\lambda_{\min}(P)}\sum_{l=1}^{s}\alpha_{s-l}\le 1$$

the state $\mathbf{x}_i^{\mathrm{f}}(k+s|k)$, $s=1,2,\ldots,N-1$, satisfies the constraint (8.10).

Finally, when $l=N$, $\mathbf{x}_i^{\mathrm{f}}(k+N|k)$ satisfies the constraint (8.11).

$$\left\|\mathbf{x}_i^{\mathrm{f}}(k+N|k)-\hat{\mathbf{x}}_i(k+N|k)\right\|_{\mathbf{P}_i}\le\frac{\kappa\varepsilon}{2\sqrt{m}}\tag{8.29}$$

which shows that the constraint (8.11) is satisfied. This concludes the proof.

In what follows we establish that, at time k, if conditions (8.18) and (8.20) are satisfied, then $\mathbf{x}_{j,i}^{\mathrm{f}}(k+s|k)$ and $\mathbf{u}_i^{\mathrm{f}}(k+s|k)$, $s=1,2,\ldots,N$, are a feasible solution of Problem 8.1.

Lemma 8.4 *Suppose Assumptions 8.1–8.3 hold, $\mathbf{x}(k_0)\in\mathbb{R}^{n_x}$, and conditions (8.18) and (8.20) are satisfied. For any $k\ge 0$, if Problem 8.1 has a solution at every update time l, $l=1,2,\ldots,k-1$, then $\mathbf{u}_i^{\mathrm{f}}(k+s|k)\in\mathcal{U}$, for all $s=1,2,\ldots,N-1$.*

Proof. Since Problem 8.1 has a feasible solution at $l=1,2,\ldots,k-1$, and $\mathbf{u}_i^{\mathrm{f}}(k+s-1|k)=\mathbf{u}_i^{\mathrm{p}}(k+s-1|k-1)$ for all $s=1,2,\ldots,N-1$, we only need to show that $\mathbf{u}_i^{\mathrm{f}}(k+N-1|k)\in\mathcal{U}$.

Since ε has been chosen to satisfy the conditions of Lemma 8.1, $\mathbf{K}_i\mathbf{x}_i\in\mathcal{U}$ for all $i\in\mathcal{P}$ when $\mathbf{x}\in\Omega(\varepsilon)$. Consequently, a sufficient condition for $\mathbf{u}_i^{\mathrm{f}}(k+N-1|k)\in\mathcal{U}$ is that $\mathbf{x}_i^{\mathrm{f}}(k+N-1|k)\in\Omega(\varepsilon)$.

In view of Lemmas 8.2 and 8.3, using the triangle inequality, we have

$$\begin{aligned}
\|\mathbf{x}_i^{\mathrm{f}}(k+N-1|k)\|_{\mathbf{P_i}}&\le\left\|\mathbf{x}_i^{\mathrm{f}}(k+N-1|k)-\hat{\mathbf{x}}_i(k+N-1|k)\right\|_{\mathbf{P}_i}\\
&\quad+\left\|\hat{\mathbf{x}}_i(k+N-1|k)\right\|_{\mathbf{P}_i}\\
&\le\frac{\varepsilon}{2(q+1)\sqrt{m}}+\frac{\varepsilon}{2\sqrt{m}}\\
&\le\frac{\varepsilon}{\sqrt{m}}
\end{aligned}\tag{8.30}$$

that is, $\mathbf{x}_i^{\mathrm{f}}(k+N|k)\in\Omega_i(\varepsilon)$. This completes the proof.

Lemma 8.5 *Suppose Assumptions 8.1–8.3 hold,* $\mathbf{x}(k_0) \in \mathbb{R}^{n_x}$*, and conditions (8.18) and (8.20) are satisfied. For any* $k \geq 0$*, if Problem 8.1 has a solution at every update time* l*,* $l = 1, 2, \ldots, k-1$*, then* $\mathbf{x}_i^{\mathrm{f}}(k+N|k) \in \Omega(\varepsilon/2)$*,* $\forall i \in \mathcal{P}$*.*

Proof. In view of Lemmas 8.2 and 8.3, using the triangle inequality, we have

$$\begin{aligned}\left\|\mathbf{x}_i^{\mathrm{f}}(k+N|k)\right\|_{\mathbf{P}_i} &\leq \left\|\mathbf{x}_i^{\mathrm{f}}(k+N|k) - \hat{\mathbf{x}}_i(k+N|k)\right\|_{\mathbf{P}_i} \\ &\quad + \left\|\hat{\mathbf{x}}_i(k+N|k)\right\|_{\mathbf{P}_i} \\ &\leq \frac{\kappa\varepsilon}{2\sqrt{m}} + \frac{(1-\kappa)\varepsilon}{2\sqrt{m}} = \frac{\varepsilon}{2\sqrt{m}}\end{aligned} \tag{8.31}$$

for each $j \in \mathcal{P}_i$, $i \in \mathcal{P}$. This shows that the terminal state constraint is satisfied. This completes the proof.

Theorem 8.1 *Suppose Assumptions 8.1–8.3 hold,* $\mathbf{x}(k_0) \in \mathbb{R}^{n_x}$ *and constraints (8.10), (8.11), and (8.13) are satisfied at* k_0*. Then, for every* $i \in \mathcal{P}$*, the control* $\mathbf{u}_i^{\mathrm{f}}(\cdot|k)$ *and state* $\mathbf{x}_{j,i}^{\mathrm{f}}(\cdot|k)$*, respectively, defined by (8.17) and (8.4), are a feasible solution of Problem 8.1 at every update k.*

Proof. We will prove the theorem by induction.

First, consider the case of $k = 1$. The state sequence $\mathbf{x}_{j,i}^{\mathrm{p}}(\cdot|1) = \mathbf{x}_{j,i}^{\mathrm{f}}(\cdot|1)$ trivially satisfies the dynamic equation (8.4), the stability constraint (8.12), and the consistency constraints (8.10) and (8.11).

Observe that

$$\hat{\mathbf{x}}_i(1|1) = \mathbf{x}_i^{\mathrm{p}}(1|0) = \mathbf{x}_i^{\mathrm{f}}(1|1) = \mathbf{x}_i(1), i \in \mathcal{P},$$

and that

$$\mathbf{x}_i^{\mathrm{f}}(1+s|1) = \mathbf{x}_i^{\mathrm{p}}(1+s|0)$$

$$s = 1, 2, \ldots, N-1$$

Thus, $\mathbf{x}_i^{\mathrm{f}}(N|1) \in \Omega_i(\varepsilon/2)$. By the invariance of $\Omega(\varepsilon)$ under the terminal controller and the conditions in Lemma 8.1, it follows that the terminal state and control constraints are also satisfied. This completes the proof of the case of $k = 1$.

Now suppose $\mathbf{u}_i^{\mathrm{p}}(\cdot|l) = \mathbf{u}_i^{\mathrm{f}}(\cdot|l)$ is a feasible solution for $l = 1, 2, \ldots, k-1$. We will show that $\mathbf{u}_i^{\mathrm{f}}(\cdot|k)$ is a feasible solution at update k.

As before, the consistency constraint (8.10) is trivially satisfied, and $\mathbf{x}_i^{\mathrm{f}}(\cdot|k)$ is the corresponding state sequence that satisfies the dynamic equation. Since there is a solution for Problem 8.1 at updates $l = 1, 2, \ldots, k-1$, Lemmas 8.2–8.5 can be invoked. Lemma 8.4 guarantees control constraint feasibility. Lemma 8.5 shows that the terminal state constraint is satisfied and the proof of Theorem 8.1 is completed.

8.4.2 Stability Analysis of Entire Closed-loop System

The stability of the closed-loop system is analyzed in this subsection.

Theorem 8.2 *Suppose Assumptions 8.1–8.3 hold,* $\mathbf{x}(k_0) \in \mathbb{R}^{n_x}$*, constraints (8.10), (8.11), and (8.13) are satisfied, and the following parametric condition holds*

$$\frac{(N-1)\kappa}{2} - \frac{1}{2} + \frac{1}{\mu} < 0 \tag{8.32}$$

Then, by application of Algorithm 8.1, the closed-loop system of (8.2) is asymptotically stable at the origin.

Proof. By Algorithm 8.1 and Lemma 8.1, when $\mathbf{x}(k)$ enters $\Omega(\varepsilon)$, the terminal controllers take over to keep it in there and stabilize the system at the origin. Therefore, it remains to show that if $\mathbf{x}(0) \in \mathcal{X}\backslash\Omega(\varepsilon)$, then by the application of Algorithm 8.1, the state of system (8.2) is driven to the set $\Omega(\varepsilon)$ in finite time.

Define the nonnegative function for $\mathcal{S}$

$$V(k) = \sum_{s=1}^{N} \|\mathbf{x}^{\mathrm{p}}(k+s|k)\|_{\mathbf{P}} \tag{8.33}$$

In what follows, we will show that, for any $k \geq 0$, if $\mathbf{x}(k) \in \mathcal{X}\backslash\Omega(\varepsilon)$, then there exists a constant $\eta \in (0, \infty)$ such that $V(k) \leq V(k-1) - \eta$. Constraint (8.12) implies that

$$\|\mathbf{x}^{\mathrm{p}}(k+s|k)\|_{\mathbf{P}} \leq \left\|\mathbf{x}^{\mathrm{f}}(k+s|k)\right\|_{\mathbf{P}} + \frac{\varepsilon}{\mu N} \tag{8.34}$$

Therefore,

$$V(k) \leq \sum_{s=1}^{N} \left\|\mathbf{x}^{\mathrm{f}}(k+s|k)\right\|_{\mathbf{P}} + \frac{\varepsilon}{\mu}$$

Subtracting $V(k-1)$ from $V(k)$ and using $\mathbf{x}^{\mathrm{p}}(k+s|k-1) = \hat{\mathbf{x}}(k+s|k)$, $s = 1, 2, \ldots, N-1$, gives

$$\begin{aligned} V(k) - V(k-1) \leq & -\|\mathbf{x}^{\mathrm{p}}(k|k-1)\|_{P} + \frac{\varepsilon}{\mu} + \left\|\mathbf{x}^{\mathrm{f}}(k+N|k)\right\|_{\mathbf{P}} \\ & + \sum_{s=1}^{N-1} \left(\left\|\mathbf{x}^{\mathrm{f}}(k+s|k)\right\|_{\mathbf{P}} - \|\hat{\mathbf{x}}(k+s|k)\|_{\mathbf{P}}\right) \end{aligned} \tag{8.35}$$

Assuming $\mathbf{x}(k) \in \mathcal{X}\backslash\Omega(\varepsilon)$ yields

$$\|\mathbf{x}^{\mathrm{p}}(k|k-1)\|_{\mathbf{P}} > \varepsilon \tag{8.36}$$

Also, by Theorem 8.1 we have

$$\left\|\mathbf{x}^{\mathrm{f}}(k+N|k)\right\|_{\mathbf{P}} \leq \varepsilon/2 \tag{8.37}$$

and by Lemma 8.3, we have

$$\sum_{s=1}^{N-1}\left(\left\|\mathbf{x}^{\mathrm{f}}(k+s|k)\right\|_{\mathbf{P}}-\|\hat{\mathbf{x}}(k+s|k)\|_{\mathbf{P}}\right)\leq\frac{(N-1)\kappa\varepsilon}{2} \tag{8.38}$$

Using (8.36)–(8.38) in (8.35) then yields

$$V(k)-V(k-1)<\varepsilon\left(-1+\frac{(N-1)\kappa}{2}+\frac{1}{2}+\frac{1}{\mu}\right) \tag{8.39}$$

which, in view of (8.34), implies that $V(k)-V(k-1)<0$. Thus, for any $k\geq 0$, if $\mathbf{x}(k)\in\mathcal{X}\backslash\Omega(\varepsilon)$, there is a constant $\eta\in(0,\infty)$ such that $V(k)\leq V(k-1)-\eta$. It then follows that there exists a finite time k' such that $\mathbf{x}(k')\in\Omega(\varepsilon)$. This concludes the proof.

We have now established the feasibility the DMPC and the stability of the resulting closed-loop system. That is, if an initially feasible solution could be found, subsequent feasibility of the algorithm is guaranteed at every update, and the resulting closed-loop system is asymptotically stable at the origin.

8.5 Example

8.5.1 The System

A distributed system consisting of four interacted subsystems is used to demonstrate the effectiveness of the proposed method. The relationship among these four subsystems is shown in Figure 8.1, where $\mathcal{S}_1$ is impacted by $\mathcal{S}_2$, $\mathcal{S}_3$ is impacted by $\mathcal{S}_1$ and $\mathcal{S}_2$, and $\mathcal{S}_4$ is impacted by $\mathcal{S}_3$. Let $\Delta\mathcal{U}_i$ be defined to reflect both the constraint on the input $u_i\in\left[u_i^{\min}\ u_i^{\max}\right]$ and the constraint on the increment of the input $\Delta u_i\in\left[\Delta u_i^{\min}\ \Delta u_i^{\max}\right]$.

The models of these four subsystems are respectively given by

$$\begin{aligned}
&\mathrm{S}_1: x_1(k+1)=0.62x_1(k)+0.34u_1(k)-0.12x_2(k)\\
&\mathrm{S}_2: x_2(k+1)=0.58x_2(k)+0.33u_2(k)\\
&\mathrm{S}_3: x_3(k+1)=0.60x_3(k)+0.34u_3(k)+0.11x_1(k)-0.07x_2(k)\\
&\mathrm{S}_4: x_4(k+1)=0.65x_4(k)+0.35u_4(k)+0.13x_3(k)
\end{aligned} \tag{8.40}$$

For the purpose of comparison, both the centralized MPC and the LCO-DMPC are applied to this system.

Here, the simulation program is developed with MATLAB. And the optimizing tool, FMINCON, is used to solve each subsystem-based MPC in every control period. The tool of

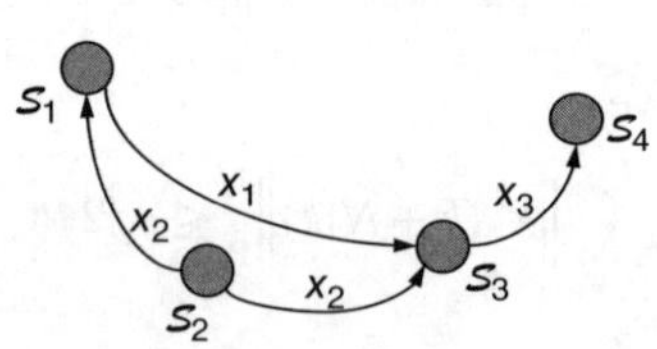

Figure 8.1 The interaction relationship among subsystems

Table 8.2 Parameters of the LCO-DMPC

Subsystem	$\mathbf{K}_i$	$\mathbf{P}_\mathrm{i}$	$\mathbf{Q}_i$	$\mathbf{R}_i$	$\Delta u_i^{\max}$ $\Delta u_i^{\min}$	$u_i^{\max}$ $u_i^{\min}$
$\mathcal{S}_1$	−0.35	5.36	4	0.2	± 1	± 2
$\mathcal{S}_2$	−0.25	5.35	4	0.2	± 1	± 2
$\mathcal{S}_3$	−0.28	5.36	4	0.2	± 1	± 2
$\mathcal{S}_4$	−0.43	5.38	4	0.2	± 1	± 2

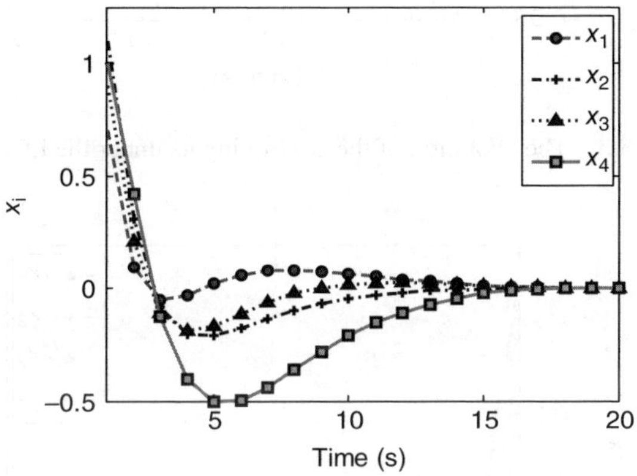

Figure 8.2 The evolution of the states under the LCO-DMPC

FMINCON has already been provided in MATLAB and it is able to solve multivariable cost function with nonlinear constraints.

Some parameters of the controllers are shown in Table 8.2. Among these parameters, $\mathcal{P}_i$ is obtained by solving the Lyapunov function. The eigenvalue of each closed-loop system under the feedback control shown in the Table 8.2 is 0.5. The eigenvalues of $\mathbf{A}_\mathrm{o}^\mathrm{T}\mathbf{P}\mathbf{A}_\mathrm{o} + \mathbf{A}_\mathrm{o}^\mathrm{T}\mathbf{P}\mathbf{A}_\mathrm{d} + \mathbf{A}_\mathrm{d}^\mathrm{T}\mathbf{P}\mathbf{A}_\mathrm{o} - \mathbf{Q}/2$ are $\{-2.42, -2.26, -1.80, -1.29\}$, all of which are negative. Thus Assumption 10.2 is satisfied. Set $\varepsilon = 0.2$. Consequently, if $\|x_i\|_{P_i} \leq \varepsilon/\sqrt{N} \leq 0.1$, then $\|K_i x\|_{i2}$ would be less than 0.1, and the constraints on the inputs and the increments of inputs, as shown in Table 8.2, are satisfied. Set the control horizon of all the controllers to be $N = 10$. Set the initial presumed inputs and states, at time $k_0 = 0$, be the solution calculated by a centralized MPC and the corresponding predictive states.

The state responses and the inputs of the closed-loop system are shown in Figures 8.2 and 8.3, respectively.

The states of all four subsystems converge close to zeroes in about 14 sec. The state of $\mathcal{S}_4$ undershoots by 0.05 before converging to zero.

8.5.2 Performance Comparison with the Centralized MPC

To further demonstrate the performance of the proposed DMPC, a dual model centralized MPC are applied to the system described by (8.40). In what follows, we discuss the performance comparison with the centralized MPC.

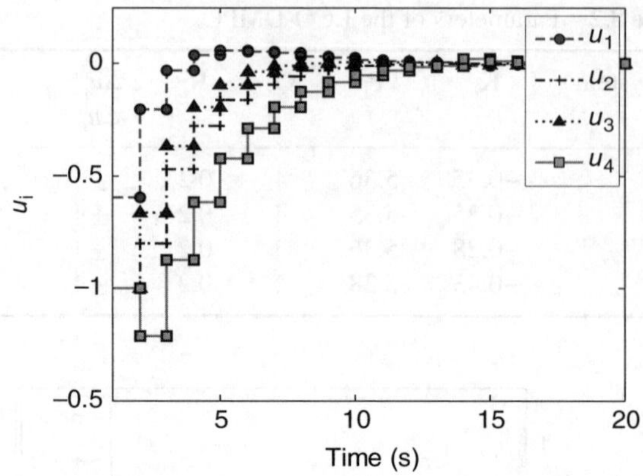

Figure 8.3 The evolution of the control inputs under the LCO-DMPC

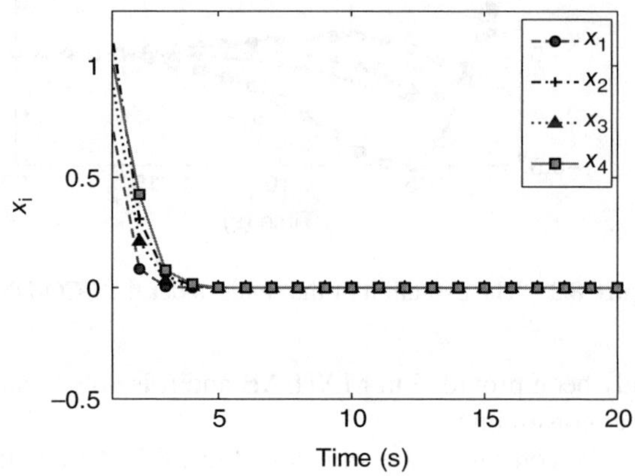

Figure 8.4 The evolution of the states under the centralized MPC

In the centralized MPC, the dual mode strategy is adopted, and the control horizon is set to be $N = 10$. The terminal constraints of the state of all subsystems are $\|x_i\,(k+10|k)\|_{P_i} \leq \varepsilon/2 = 0.1$. MPC switch to the feedback control laws given in Table 8.2 when all states enter the attractive region $\Omega(\varepsilon)$. The bounds of the input and the input increment of each of the four subsystems are $[-2, 2]$ and $[-1, 1]$, respectively.

Figures 8.4 and 8.5 show the state responses and the control inputs of the closed-loop system under the control of the centralized MPC. The shapes of the state response curves under the centralized MPC are similar to those under the LCO-DMPC. Under centralized MPC, all subsystems converge to near zero in 8 s. Under the LCO-DMPC control design, all subsystems converge to near zero in about 14 s. There is no significant undershooting.

Table 8.3 shows the state square errors of the closed-loop system under the control of the centralized MPC and the LCO-based DMPC, respectively. The total errors resulting from the LCO-based DMPC is 6.55 (40.5%) larger than that results from the centralized MPC.

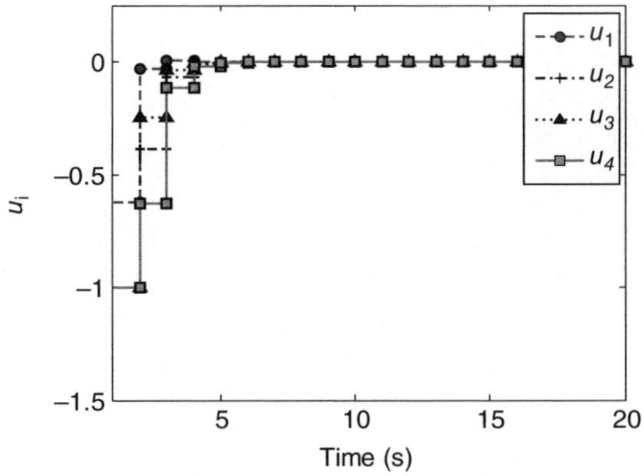

Figure 8.5 The evolution of the control inputs under the centralized MPC

Table 8.3 State square errors of the closed-loop system under the control of the centralized MPC (CMPC) and the LCO-DMPC

Subsystems	CMPC	LCO-DMPC
$\mathcal{S}_1$	2.07	2.22
$\mathcal{S}_2$	5.47	6.26
$\mathcal{S}_3$	3.63	4.12
$\mathcal{S}_4$	5.00	10.12
Total	16.17	22.72

From these simulation results, it can be seen that the proposed algorithm is able to steer the system states to the origin if there is a feasible solution at the initial states.

8.6 Conclusion

In this chapter, a stabilizing distributed implementation of MPC is developed for dynamically coupled distributed systems subject to decoupled input constraints. Each local controller minimizes its own subsystem's cost and uses the state prediction of the previous time instant to approximate the state sequence at the current time instant in computing the optimal solution. Under this coordination strategy, the central element to the feasibility and stability is that the actual and presumed states and control variables are not too far away from one another. In particular, if an initially feasible solution could be found, the subsequent feasibility of the algorithm is guaranteed at every update, and the resulting closed-loop system is asymptotically stable.

9

Cooperative Distributed Predictive Control with Constraints

9.1 Introduction

Consider that the performance of a distributed model predictive control (DMPC) is, in most cases, not as good as that of a centralized MPC. How to improve the performance of the entire system with limited local computation is a problem remained to be solved.

To improve the global performance of the entire system, Chapter 6 provides a commonly used coordination strategy, called as cooperative DMPC (C-DMPC), where each subsystem-based MPC optimizes the cost over the entire system to improve the global performance of closed-loop system. This coordination strategy can dramatically improve the global optimization performance of closed-loop system, when the global information is available for each subsystem-based MPC. Thus it is worthy to develop a stabilized C-DMPC that only communicates once a control period.

As mentioned in Chapter 8, control design of the stabilizing DMPC is an important and challenging problem [27, 50, 51]. In noniterative DMPC, the future state sequences of upstream neighbors, which are calculated based on the solution in the previous time instant, may not be equal to the predictive states calculated by the corresponding subsystem at the current time instant, and the errors between them are hard to estimate. The remaining part of the optimal control sequence calculated at the previous time instant may not be a feasible solution at the current time instant. These factors make it difficult to design DMPC. Farina and Scattolini [50] and Dunbar [51] gave two methods on designing stabilizing LCO-DMPC. However, comparing to LCO-DMPC, both the predictive model and the optimization problem of C-DMPC are different from that of LCO-DMPC; to design a constraint C-DMPC that guarantees the stability of closed-loop system is still an interesting problem remained to be studied.

In this chapter, a stabilizing C-DMPC where each subsystem-based MPC only communicates with the other subsystems once a sampling time is proposed. The consistency constraints, which limit the error between the optimal inputs sequence calculated at the previous time instant, referred to as the presumed inputs, and the optimal inputs sequence calculated at the current time instant to within a prescribed bound, are designed and included in the optimization problem of each subsystem-based MPC. Moreover, a dual mode predictive control

Distributed Model Predictive Control for Plant-Wide Systems, First Edition. Shaoyuan Li and Yi Zheng.

[69, 78] strategy is adopted. These consistency constraints and the dual mode strategy guarantee that the remaining part of the solution at the previous time instant is a feasible solution if there is a feasible solution at initial time instant. They also guarantee the asymptotical stability of the closed-loop system.

The remainder of this chapter is organized as follows: Section 9.2 describes the problem to be solved in this chapter. Section 9.3 presents the design of the stabilizing C-DMPC with communicating once a sampling time. The feasibility of the proposed C-DMPC and the stability of the resulting closed-loop system are analyzed in Section 9.4. Section 9.5 presents the simulation results to demonstrate the effectiveness of the proposed C-DMPC algorithm. Finally, a brief conclusion to the chapter is drawn in Section 9.6.

9.2 System Description

Consider a spatially distributed system, with each subsystem-based controller which in turn is able to exchange information with all other subsystem-based controllers.

Without losing generality, suppose the whole system is composed of m discrete-time linear subsystems $\mathcal{S}_i$, $i \in \mathcal{P}$, $\mathcal{P} = \{1, \ldots, m\}$. Let the subsystems interact with each other through their states. Then, subsystem $\mathcal{S}_i$ can be expressed as

$$\begin{cases} \mathbf{x_i}(k+1) = \mathbf{A}_{ii}x_i(k) + \mathbf{B}_{ii}\mathbf{u}_i(k) + \sum\limits_{j \in P_{+i}} \mathbf{A}_{ij}\mathbf{x}_j(k) \\ \mathbf{y}_i(k) = \mathbf{C}_{ii}\mathbf{x}_i(k) \end{cases}$$

where $\mathbf{x}_i \in \mathbb{R}^{n_{xi}}$, $\mathbf{u}_i \in \mathcal{U}_i \subset \mathbb{R}^{n_{ui}}$, and $\mathbf{y}_i \in \mathbb{R}^{n_{yi}}$ are, respectively, the local state, input and output vectors, and $\mathcal{U}_i$ is the feasible set of the input $\mathbf{u}_i$, which is used to bound the input according to the physical constraints on the actuators, the control requirements, or the characteristics of the plant. A nonzero matrix $\mathbf{A}_{ij}$ indicates that $\mathcal{S}_i$ is affected by $\mathcal{S}_j$, $j \in \mathcal{P}$, and subsystem $\mathcal{S}_j$ is said to be an upstream system of $\mathcal{S}_j$. Let $\mathcal{P}_{+i}$ denote the set of the subscripts of the upstream systems of $\mathcal{S}_i$, that is, $j \in \mathcal{P}_{+i}$, and set $\mathcal{P}_{-i}$ be the set of the subscripts of the downstream systems of $\mathcal{S}_i$. In addition, set $P_i = \{j | j \in \mathcal{P}$, and, $j \neq i\}$. In the concatenated vector form, the system dynamics can be written as

$$\begin{cases} \mathbf{x}(k+1) = \mathbf{A}\mathbf{x}(k) + \mathbf{B}\mathbf{u}(k) \\ \mathbf{y}(k) = \mathbf{C}\mathbf{x}(k) \end{cases}$$

where $\mathbf{x} = [\mathbf{x}_1^{\mathrm{T}}, \mathbf{x}_2^{\mathrm{T}}, \ldots, \mathbf{x}_m^{\mathrm{T}}]^{\mathrm{T}} \in \mathbb{R}^{n_x}$, $\mathbf{u} = [\mathbf{u}_1^{\mathrm{T}}, \mathbf{u}_2^{\mathrm{T}}, \ldots, \mathbf{u}_m^{\mathrm{T}}]^{\mathrm{T}} \in \mathbb{R}^{n_u}$, and $\mathbf{y} = [\mathbf{y}_1^{\mathrm{T}}, \mathbf{y}_2^{\mathrm{T}}, \ldots, \mathbf{y}_m^{\mathrm{T}}]^{\mathrm{T}}$ $\in \mathbb{R}^{n_y}$ are, respectively, the concatenated state, control input, and output vectors of the overall system $\mathcal{S}$, and $\mathbf{A}$, $\mathbf{B}$, and $\mathbf{C}$ are constant matrices of appropriate dimensions. Also, $\mathbf{u} \in \mathcal{U} = \mathcal{U}_1 \times \mathcal{U}_1 \times \cdots \times \mathcal{U}_m$ and $\mathcal{U}$ contain a neighborhood of the origin.

The control objective is to stabilize the overall system $\mathcal{S}$ in a DMPC framework with the limited communication resources. Meanwhile, the achieved performance index of the overall system should be as close as possible to the performance index achievable under a centralized MPC.

When the global information is available for each subsystem-based MPC, the coordination strategy where each subsystem-based MPC optimizes the cost of entire system is very suitable for the control of this class of system, since it is able to achieve a very good global performance.

Considering the limitation on the time consumption of communication, a stabilizing C-DMPC design method, which communicates once a control period, is proposed in the next section.

9.3 Stabilizing Cooperative DMPC with Input Constraints

9.3.1 Formulation

In this section, m separate optimal control problems, one for each subsystem and the C-DMPC algorithm with communicating once a control period, is defined. In every distributed optimal control problem, the same constant prediction horizon N, $N > 1$, is used. And every distributed MPC law is updated globally synchronously. At each update, every subsystem-based MPC optimizes only for its own open-loop control sequence, given the current states and the estimated inputs of the whole system.

To proceed, we need the following assumption, and we also define the necessary notations in Table 9.1.

Assumption 9.1 For every subsystem $i \in \mathcal{P}$ there exists a state feedback $\mathbf{u}_i = \mathbf{K}_i\mathbf{x}$ such that the closed-loop system $\mathbf{x}(k+1) = \mathbf{A}_c\mathbf{x}(k)$ is asymptotically stable, where $\mathbf{A}_c = \mathbf{A} + \mathbf{BK}$ and $\mathbf{K} = \text{block-diag}(\mathbf{K}_1, \mathbf{K}_2, \ldots, \mathbf{K}_m)$.

The state evolution of subsystem $\mathcal{S}_j$, $j \in \mathcal{P}_{-i}$, is affected by the optimal control decision of $\mathcal{S}_i$, and the affection on the control performance of subsystem $\mathcal{S}_j$ may be negative sometime. Thus, the idea of global cost optimization [114] is adopted here, that is, each subsystem-based MPC takes the cost function of all subsystems into account; more specifically, the performance index is defined as

$$\overline{J}_i = \|\hat{\mathbf{x}}(k+N|k,i)\|_{\mathbf{P}} + \sum_{l=0}^{N-1}\left(\|\hat{\mathbf{x}}(k+l|k,i)\|_{\mathbf{Q}} + \left\|\mathbf{u}_i(k+l|k)\right\|_{\mathbf{R}_i}\right)$$

where $\mathbf{Q} = \mathbf{Q}^{\mathrm{T}} > 0$, $\mathbf{R}_j = \mathbf{R}_j^{\mathrm{T}} > 0$, $\mathbf{P} = \mathbf{P}^{\mathrm{T}} > 0$. And the matrix $\mathbf{P}$ is chosen to satisfy the Lyapunov equation

$$\mathbf{A}_c^{\mathrm{T}}\mathbf{P}\mathbf{A}_c - \mathbf{P} = -\hat{\mathbf{Q}}$$

Table 9.1 Notations in this chapter

Notation	Explanation
$\mathcal{P}$	The set of the subscripts of all subsystems
$\mathcal{P}_i$	The set of the subscripts of all subsystems excluding $\mathcal{S}$ itself
$\mathbf{u}_i(k+l-1\,\vert\,k)$	The optimal control sequence of $\mathcal{S}_i$, calculated by $\mathcal{C}_i$ at time k
$\hat{\mathbf{x}}_j(k+l\vert k,i)$	The predicted state sequence of $\mathcal{S}_j$, calculated by $\mathcal{C}_i$ at time k
$\hat{\mathbf{x}}(k+l\vert k,i)$	The predicted state sequence of all, subsystems calculated by v at time k
$\mathbf{u}_i^{\mathrm{f}}(k+l-1\vert k)$	The feasible control at time k+l-1 of $\mathcal{S}_i$, defined by $\mathcal{C}_i$ at time k
$\mathbf{x}_j^{\mathrm{f}}(k+l\vert k,i)$	The predictive feasible state sequence of $\mathcal{S}_j$, defined by $\mathcal{C}_i$ at time k
$\mathbf{x}^{\mathrm{f}}(k+l\,\vert\,k,i)$	The predictive feasible state sequence of all subsystems calculated by $\mathcal{C}_i$ at time k
$\mathbf{x}^{\mathrm{f}}(k+l\,\vert\,k)$	The predictive feasible state sequence of all subsystems, and $\mathbf{x}^{\mathrm{f}}(k+l\vert k) = [\mathbf{x}_1^{\mathrm{f}}(k+l\vert k), \mathbf{x}_2^{\mathrm{f}}(k+l\vert k), \ldots, \mathbf{x}_m^{\mathrm{f}}(k+l\vert k)]^{\mathrm{T}}$
$\Vert\cdot\Vert_{\mathbf{P}}$	Refer to the $\mathbf{P}$ norm, $\mathbf{P}$ is any positive matrix, and $\vert\vert\mathbf{z}\vert\vert_P = \sqrt{\mathbf{x}^{\mathrm{T}}(k)\mathbf{P}\mathbf{x}(k)}$

where $\hat{\mathbf{Q}} = \mathbf{Q} + \mathbf{K}^{\mathrm{T}}\mathbf{R}\mathbf{K}$, and

$$\mathbf{R} = \text{block-diag}\{\mathbf{R}_1, \mathbf{R}_2, \ldots, \mathbf{R}_m\}$$

Since every subsystem-based controller updates synchronously, the control sequences $\mathcal{S}_j$, $j \in \mathcal{P}_i$ are unknown to subsystem $\mathcal{S}_i$. Thus, at the time instant k, presume the control sequence of $\mathcal{S}_j, j \in \mathcal{P}_i$ be the optimal control sequence calculated by $\mathcal{C}_j$ at time $k-1$ concatenated with the feedback control law, that is

$$\begin{aligned}&[\mathbf{u}_j(k|k-1), \mathbf{u}_j(k+1|k-1),\\ &\quad \ldots, \mathbf{u}_j(k+N-2|k-1), K_j\hat{\mathbf{x}}(k+N-1|k-1, j)]\end{aligned}$$

Then, the predictive model in the MPC for $\mathcal{S}_i$ can be expressed as

$$\begin{aligned}\hat{\mathbf{x}}(k+l|k,i) &= \mathbf{A}^l\mathbf{x}(k) + \sum_{h=1}^{l} \mathbf{A}^{l-h}\overline{\mathbf{B}}_i\mathbf{u}_i(k+h-1|k)\\ &\quad + \sum_{j\in \mathrm{P}_i}\sum_{h=1}^{l} \mathbf{A}^{l-h}\overline{\mathbf{B}}_j\mathbf{u}_j(k+h-1|k-1)\end{aligned}$$

where, for $\forall\, i$ and $j \in \mathcal{P}_i$,

$$\overline{\mathbf{B}}_i = \left[\mathbf{0}^{n_{ui}\times\Sigma_{j<i}n_{xj}} \ \ \mathbf{B}_i \ \ \mathbf{0}^{n_{ui}\times\Sigma_{j>i}n_{xj}}\right]^{\mathrm{T}}$$

In addition, to enlarge the feasible region, a terminal-state constraint is included in each subsystem-based MPC. The terminal-state constraint set should guarantee that the terminal controllers are stabilizing inside it.

Lemma 9.1 *Under Assumption 9.1, for any positive scalar c the set* $\Omega(c) = \{x \in \mathbb{R}^{n_x} : \|x\|_P \le c\}$ *is a positive invariant region of attraction for the closed-loop system* $\mathbf{x}(k+1) = \mathbf{A}_c\mathbf{x}(k)$. *Additionally, there exists a small enough positive scalar* ε *such that* $\mathbf{Kx}$ *is in the feasible input set* $\mathcal{U} \subset \mathbb{R}^{n_u}$, *for all* $\mathbf{x} \in \Omega(\varepsilon)$.

Proof. From Assumption 9.1, for all $\mathbf{x}(k) \in \Omega(c)\backslash\{0\}$, the closed-loop system $\mathbf{x}(k+1) = \mathbf{A}_c\mathbf{x}(k)$ is asymptotically stable. This implies that all trajectories of the closed-loop system that start inside $\Omega(c)$will remain inside and converge to the origin with $\mathbf{P}$ satisfying the Lyapunov equation.

The existence of an $\Omega(c)\varepsilon > 0$ such that $\mathbf{Kx} \in \mathcal{U}$ for all $\mathbf{x} \in \Omega(\varepsilon)$ follows from the fact that P is positive definite, which implies that the set $\Omega(\varepsilon)$ shrinks to the origin as ε decreases to zero. This completes the proof.

In the optimization problem of each subsystem-based MPC, the terminal-state constraint set for $\mathcal{S}$ can then be set to be

$$\Omega(\varepsilon) = \{\mathbf{x} \in \mathbb{R}^{n_x} | \|\mathbf{x}\|_{\mathbf{P}} \le \varepsilon\}$$

Suppose at some time k_0, $\mathbf{x}(k_0) \in \Omega(\varepsilon)$ for every subsystem. Then, by Lemma 9.1, stabilization can be achieved if every $\mathcal{C}_i, i \in \mathcal{P}$ employs its static feedback controller $\mathbf{K}_i\mathbf{x}_i(k)$ for all time $k \ge k_0$.

Thus, the objective of the MPC law is to drive state of all subsystem to the set $\Omega(\varepsilon)$. Once all subsystems have reached this set, they switch to their decoupled controllers for stabilization.

Switching from MPC to a terminal controller once the state reaches a suitable neighborhood of the origin is referred to as dual mode MPC [71, 78]. For this reason, the implementation here is considered a dual mode distributed MPC algorithm.

In what follows, we formulate the optimization problem for each subsystem-based MPC.

Problem 9.1 Consider subsystem $\mathcal{S}_i$. Let $\varepsilon > 0$ be as specified in Lemma 9.1. Let the update time be $k \geq 1$. Given $\mathbf{x}(k)$, and $\mathbf{u}(k+l\,|\,k-1)$, $l = 1, 2, \ldots, N-1$, find the control sequence $\mathbf{u}_i(k+l\,|\,k) : \{0, 1, \ldots, N-1\} \rightarrow U_i$ that minimizes

$$\overline{J}_i = \left\|\hat{\mathbf{x}}_j(N|k,i)\right\|_{\mathrm{P}} + \sum_{l=0}^{N-1}\left(\|\hat{\mathbf{x}}(k+l|k,i)\|_{\mathrm{Q}} + \left\|\mathbf{u}_i(k+l|k)\right\|_{R_i}\right)$$

Subject to the constraints in (9.6)

$$\sum_{h=0}^{l}\beta_{l-h}\left\|\mathbf{u}_i(k+h|k) - \mathbf{u}_i(k+h|k-1)\right\|_2$$

$$\leq \frac{\gamma\kappa\alpha e}{m-1}, l = 1, 2, \ldots, N-1;$$

$$\mathbf{u}_i(k+l-1|k) \in \mathcal{U}_i,\ l = 0, 1, \ldots, N-1;$$

$$\hat{\mathbf{x}}(k+N|k,i) \in \Omega(\alpha\varepsilon)$$

In the constraints above,

$$\beta_l = \max_{i \in P}\left(\lambda_{\max}\left(\left(\mathbf{A}^l\overline{\mathbf{B}}_i\right)^T\mathbf{P}\mathbf{A}^l\overline{\mathbf{B}}_i\right)^{\frac{1}{2}}\right),$$

$$l = 0, 1, \ldots, N-1$$

$$\lambda_{\max}\left(\sqrt{\mathbf{A}_c^{\mathrm{T}}\mathbf{A}_c}\right) \leq 1 - \kappa$$

$$0 < 1 - \kappa < 1$$

The constant $0 < \kappa < 1$, $0 < \alpha < 0.5$, and $\gamma > 0$ are design parameters whose value will be chosen in the sequel.

Equation (9.8) is referred to as the consistency constraints, which requires that each predictive manipulated variables remain close to the presumed sequence. It is a key equation in proving that $\mathbf{x}^{\mathrm{f}}(\cdot\,|k,i)$ is a feasible state sequence at each update.

Note that the terminal constraint in each optimal control problem is $\Omega(\alpha\varepsilon)$, $0 \leq \alpha < 0.5$, although Lemma 9.1 ensures that the larger terminal set $\Omega(\varepsilon)$ suffices as a collective region of attraction for the terminal controllers. In the analysis presented in the next section, it is shown that tightening the terminal set in this way is required to guarantee the feasibility properties.

9.3.2 Constraint C-DMPC Algorithm

Before stating the C-DMPC algorithm, an assumption is made to facilitate the initialization phase.

Assumption 9.2 At initial time k_0, there exists a feasible control $\mathbf{u}_i(k_0+l) \in \mathcal{U}_i$, $l \in \{1, \ldots, N\}$, for each $\mathcal{S}_i$, such that the solution to the full system $\mathbf{x}(l+1+k_0) = \mathbf{A}\mathbf{x}(l+k_0) + \mathbf{B}\mathbf{u}(l+k_0)$, denoted $\hat{\mathbf{x}}(\cdot|k_0, i)$, satisfies $\hat{\mathbf{x}}(N+k_0|k_0, i) \in \Omega(\alpha\varepsilon)$ and results in a bounded cost $\overline{J}_i(k_0)$. Moreover, each subsystem has access to $\mathbf{u}_i(\cdot|k_0)$.

Assumption 9.2 bypasses the task of actually constructing an initially feasible solution in a distributed way. In fact, finding an initially feasible solution for many optimization problems is often a primary obstacle, whether or not such problems are used in a control setting [69, 115]. One possible way to obtain an initially feasible solution can be to solve the corresponding centralized MPC solution at the initial time instant.

The dual mode C-DMPC law for any $\mathcal{S}_i$, which communicates once every update, is as follows.

Algorithm 9.1

Step 1: Initialization at time k_0

- Initialize $\mathbf{x}(k_0), \mathbf{u}(k_0+l-1 \,|\, k_0)$, where $l = 1, 2, \ldots, N$, to satisfy Assumption 9.1
- Transmit $\mathbf{u}_i(k_0+l|k_0)$ and $\mathbf{x}_j(k_0)$ to all other subsystems, receive $\mathbf{u}_j(k_0+l-1 \,|\, k_0)$ and $\mathbf{x}_j(k_0), j \in \mathcal{P}_i$ from all $\mathcal{S}_j$
- At time k_0, if $\mathbf{x}(k_0) \in \Omega(\varepsilon)$, then apply the terminal controller $\mathbf{u}_i(k_0) = \mathbf{K}_i\mathbf{x}(k_0)$, for all $k \geq k_0$, else
- Solve Algorithm 9.1 for $\mathbf{u}_i(k_0+l-1 \,|\, k_0)$ and apply $\mathbf{u}_i(k_0|k_0)$ to $\mathcal{S}_i$

Step 2: Update control law at time k

- Measure $\mathbf{x}_i(k)$, transmit $\mathbf{x}_i(k)$ and $\mathbf{u}_i(k+l|k)$ to all other subsystems, receive $\mathbf{x}_j(k)$ and $\mathbf{u}_j(k+l-1 \,|\, k-1)$, $j \in \mathcal{P}_i$ from all $\mathcal{S}_j$
- If $\mathbf{x}(k) \in \Omega(\varepsilon)$, then apply the terminal controller $\mathbf{u}_i(k) = \mathbf{K}_i\mathbf{x}(k)$, else
- Solve Algorithm 9.1 for $\mathbf{u}_i(k+l-1 \,|\, k)$ and apply $\mathbf{u}_i(k|k)$ to $\mathcal{S}_i$

Step 3: Update control at time $k+1$

- Let $k+1 \rightarrow k$ repeat Step 2

Algorithm 9.1 presumes that $\mathcal{C}_i, \forall i \in \mathcal{P}$ have access to the full state $\mathbf{x}(k)$. In the next section, it is shown that the C-DMPC policy drives the state $\mathbf{x}(k+l)$ to $\Omega(\varepsilon)$ in a finite number of updates and the state remains in $\Omega(\varepsilon)$ for all future time. And the analysis of the feasibility and stability of C-DMPC algorithm is preceded as follows.

9.4 Analysis

In this section, feasibility is analyzed in the first subsection, followed by stability in the second subsection.

9.4.1 *Feasibility*

The main result of this section is that, provided an initial feasible solution is available and Assumption 9.2 holds true, for any $\mathcal{S}_i$ and at any time $k \geq 1$, $u_i(\cdot|k) = u_i^f(\cdot|k)$ is a feasible control solution to Problem 9.1. Here, $u_i^f(\cdot|k)$ is the reminder of previous control concatenating with a feedback control, that is,

$$\mathbf{u}_i^{\mathrm{f}}(k+l-1|k) = \begin{cases} \mathbf{u}_i\,(k+l-1|k-1) & l=1,\ldots,N-1 \\ \mathbf{K}_i\mathbf{x}^{\mathrm{f}}(k+N-1|k,i) & l=N \end{cases}$$

$\mathbf{x}^{\mathrm{f}}(k+l|k,i)$, $l=1,2,\ \ldots,N$, equals to the solution of (9.6) under the initial state of $x(k)$ and the control sequence $\mathbf{u}_i^{\mathrm{f}}(k+l-1|k)$ and $\mathbf{u}_j(k+l-1\,|\,k-1), j\in\mathcal{P}_i$, and can be expressed as

$$\mathbf{x}^{\mathrm{f}}(k+l|k,i) = \mathbf{A}^l\mathbf{x}(k) + \sum_{h=1}^{l}\mathbf{A}^{l-h}\overline{\mathbf{B}}_i\mathbf{u}_i^{\mathrm{f}}(k+h-1|k)$$

$$+\sum_{j\in \mathrm{P}_i}\sum_{h=1}^{l}\mathbf{A}^{l-h}\overline{\mathbf{B}}_j\mathbf{u}_j(k+h-1|k-1)$$

Substituting (9.13) into (9.14), we have

$$\mathbf{x}^{\mathrm{f}}(k+l|k,i) = \mathbf{x}^{\mathrm{f}}(k+l|k,j) = \mathbf{x}^{\mathrm{f}}(k+l|k),$$
$$\forall i,j\in \mathrm{P}, l=1,2,\ldots,N$$

and

$$x^{\mathrm{f}}(k+N|k) = \mathbf{A}_{\mathrm{c}}\mathbf{x}^{\mathrm{f}}(k+N-1|k)$$

The control $\mathbf{u}_i^{\mathrm{f}}(\cdot|k)$ is a feasible solution to Problem 9.1 for any $\mathcal{S}_i$ and at any update $k\geq 1$ refers that the control $\mathbf{u}_i^{\mathrm{f}}(\cdot|k)$ satisfies Equation (9.8) and the control constraints (9.9), and the corresponding state $\mathbf{x}^{\mathrm{f}}(k+N|k)$ satisfies the terminal-state constraint (9.10). To establish this feasibility result, define that the state $\hat{\mathbf{x}}(k+N|k-1,i)$ be the closed-loop response of

$$\hat{\mathbf{x}}(k+N|k-1,i) = \mathbf{A}_c\hat{\mathbf{x}}(k+N-1|k-1,i)$$

Here, the state $\hat{\mathbf{x}}(k+N|k-1,i)$ is not equal to the result of substituting $\mathbf{u}_i(k+N-1\,|\,k-1)$, defined in (9.5), into system Equation (9.6). It is because that $\hat{\mathbf{x}}(k+N|k-1,i)$ is only a middle variable used in the proof of feasibility, and has no impact on the optimization problem and stability. Thus we can assume it as (9.15).

Figure 9.1 shows the basic idea of how to guarantee that $\mathbf{x}^{\mathrm{f}}(k+N|k)$ satisfies the terminal constraint (9.10). If the difference between the input sequences of two neighboring update time is bounded, then the discrepancy between the presumed state sequence $\{\hat{\mathbf{x}}(k+1|k,i),\ \hat{\mathbf{x}}(k+2|k,i),\ldots\}$ and the state sequence $\{\mathbf{x}^{\mathrm{f}}(k+1\,|\,k,i), \mathbf{x}^{\mathrm{f}}(k+2|k,i),\ \ldots\ \}$ will be limited. When an appropriate bound is selected, $\hat{\mathbf{x}}(k+N|k,i)$ and $\mathbf{x}^{\mathrm{f}}(k+s|k)$ can be sufficiently close to each other such that $\mathbf{x}^{\mathrm{f}}(k+N|k)$ is within the indicated ellipsoid $\Omega(\alpha\varepsilon)$.

In this section, Lemma 9.2 identifies sufficient conditions to ensure that $\left\|\mathbf{x}^{\mathrm{f}}\,(k+l|k)-\hat{\mathbf{x}}(k+l|k,i)\right\|_P\leq\kappa\alpha\varepsilon$, for every $i\in\mathcal{P}$. Lemma 9.3 establishes the control constraint feasibility. Lemma 9.4 presents the terminal constraint feasibility. Finally, Theorem 9.1 combines the results in Lemmas 9.2–9.4 to arrive at the conclusion that, for any $i\in\mathcal{P}$, the control $u_i^f(\cdot|k)$ is a feasible solution to Problem 9.1 at any update $k\geq 1$.

Consider that $\Omega(\varepsilon)$ is the ε-level set of the Lyapunov function of the closed-loop dynamics $\mathbf{x}(k+1)=\mathbf{A}_{\mathrm{c}}x(k)$. Therefore under the condition (9.12), it is trivial to show $\mathbf{x}(k+N)\in\Omega(\alpha\varepsilon')$, provided that $\hat{\mathbf{x}}(k+N-1|k-1,i)\in\Omega(\alpha\varepsilon)$ and $\varepsilon'=(1-\kappa)\,\varepsilon$.

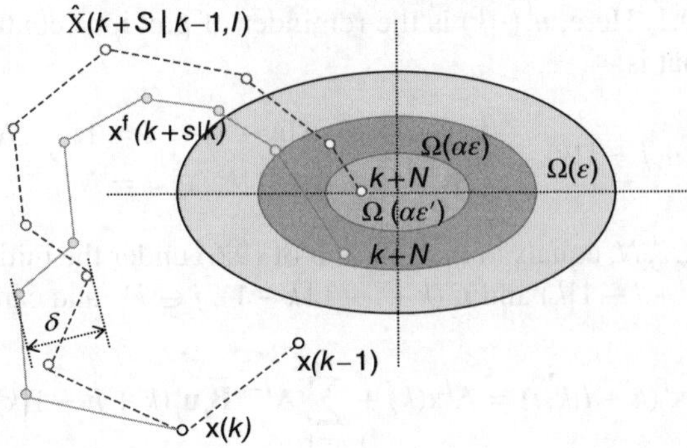

Figure 9.1 Schematic of the discrepancy among feasible state sequence and presumed state sequence

Lemma 9.2 *Suppose Assumptions 9.1 and 9.2 hold and* $\mathbf{x}(k_0) \in \mathcal{X}$, *for any* $k \geq 0$, *if Problem 9.1 has a solution at every update time 0, …, k − 1, then*

$$\|\mathbf{x}^{\mathrm{f}}(k+l|k) - \hat{\mathbf{x}}(k+l|k,i)\|_{\mathbf{P}} \leq \gamma\kappa\alpha\varepsilon$$

where $0 < \gamma < 1$ *is a design parameter, for* $\forall i \in \mathcal{P}, j \in \mathcal{P}_i$, *and all* $l \in \{1, \ldots, N\}$. *In addition,* $\mathbf{u}_i^f(k+l-1|k)$, $l = 1, 2, \ldots, N-1$ *satisfies the constraint (9.8).*

Proof. First we will prove (9.18), provided there is a solution at update time $0, 1, 2, \ldots, k-1$. Substitute (9.13) into (9.14) and consider that

$$\mathbf{x}(k) = \mathbf{A}\mathbf{x}(k-1) + \sum_{i\in\mathcal{P}} \overline{\mathbf{B}}_i \mathbf{u}_i(k-1|k-1)$$

for any $l = 1, 2, \ldots, N-1$, the feasible state is given by

$$\begin{aligned}
&\mathbf{x}^{\mathrm{f}}(k+l|k) \\
&= \mathbf{A}^{l+1}\mathbf{x}(k-1) + \mathbf{A}^{l}\overline{\mathbf{B}}_i u(k-1|k-1) \\
&\quad + \sum_{h=1}^{l} \mathbf{A}^{l-h}\overline{\mathbf{B}}_i \mathbf{u}_i^{\mathrm{f}}(k+h-1|k) \\
&\quad + \sum_{j\in \mathrm{P}_i}\sum_{h=0}^{l} \mathbf{A}^{l-h}\overline{\mathbf{B}}_j \mathbf{u}_j(k+h-1|k-1) \\
&= \mathbf{A}^{l+1}\mathbf{x}(k-1) + \sum_{h=0}^{l} \mathbf{A}^{l-h}\overline{\mathbf{B}}_i \mathbf{u}_i(k+h-1|k-1) \\
&\quad + \sum_{j\in \mathrm{P}_i}\sum_{h=0}^{l} \mathbf{A}^{l-h}\overline{\mathbf{B}}_j \mathbf{u}_j(k+h-1|k-1)
\end{aligned}$$

The state predicted at time $k-1$ is given by

$$\begin{aligned}
&\hat{\mathbf{x}}(k+l|k-1,i) \\
&= \mathbf{A}^{l+1}\mathbf{x}(k-1) + \sum_{h=0}^{l}\mathbf{A}^{l-h}\overline{\mathbf{B}}_i\mathbf{u}_i(k+h-1|k-1) \\
&+ \sum_{j\in P_i}\sum_{h=0}^{l}\mathbf{A}^{l-h}\overline{\mathbf{B}}_j\mathbf{u}_j(k+h-1|k-2)
\end{aligned}$$

Subtracting (9.20) from (9.19), the discrepancy between the feasible state sequence and the state sequence predicted at time $k-1$ is obtained as

$$\begin{aligned}
&\left\|\mathbf{x}^f(k+l|k) - \hat{\mathbf{x}}(k+l|k-1,i)\right\|_{\mathsf{P}} \\
&= \left\|\sum_{h=0}^{l}\sum_{j\in \mathsf{P}_i}\mathbf{A}^{l-h}\overline{\mathbf{B}}_j\left(\mathbf{u}_j(k+h-1|k-1) - \mathbf{u}_j(k+h-1|k-2)\right)\right\|_{\mathsf{P}}
\end{aligned}$$

Let $\mathcal{S}_r$ be the subsystem that maximizes

$$\sum_{h=0}^{l}\beta_{l-h}\left\|\mathbf{u}_i(k+h-1|k-1) - \mathbf{u}_i(k+h-1|k-2)\right\|_2, i\in \mathsf{P}$$

Then, the following equation can be deduced from (9.21):

$$\begin{aligned}
&\left\|\mathbf{x}^f(k+l|k) - \hat{\mathbf{x}}(k+l|k-1,i)\right\|_{\mathsf{P}} \\
&\le \sum_{h=0}^{l}\beta_{l-h}\left\|\mathbf{u}_r(k+h-1|k-1) - \mathbf{u}_r(k+h-1|k-2)\right\|_2
\end{aligned}$$

Since there is a solution at update time $0, 1, 2, \ldots, k-1, \forall i \in \mathcal{P}$ satisfied the constraint (9.8), for all $l=1, 2, \ldots, N-1$, it has

$$\sum_{h=0}^{l}\beta_{l-h}\left\|\mathbf{u}_r(k+h-1|k-1) - \mathbf{u}_r(k+h-1|k-2)\right\|_2 \le \gamma\kappa\alpha\varepsilon/(m-1)$$

Then, for $l=1, 2, \ldots, N-1$, following equation can be deduced:

$$\left\|\mathbf{x}^{\mathrm{f}}(k+l|k) - \hat{\mathbf{x}}(k+l|k-1,i)\right\|_{\mathsf{P}} \le \gamma\kappa\alpha\varepsilon$$

Thus, (9.18) holds for all $l=1, 2, \ldots, N-1$.

When $l=N$, from (9.16) and (9.17), it has

$$\begin{aligned}
&\left\|\mathbf{x}^{\mathrm{f}}(k+N|k) - \hat{\mathbf{x}}(k+N|k-1,i)\right\|_{\mathbf{P}} \\
&\le \lambda_{\max}(\mathbf{A}_c^{\mathrm{T}}\mathbf{A}_c)\left\|\mathbf{x}^f(k+N-1|k) - \hat{\mathbf{x}}(k+N-1|k-1,i)\right\|_{\mathbf{P}} \\
&\le (1-\kappa)\gamma\kappa\alpha\varepsilon
\end{aligned}$$

Consequently, (9.18) holds for all $l=1, 2, \ldots, N$.

In addition, from definition (9.13), it has $\mathbf{u}_i^{\mathrm{f}}(k+l-1|k)-\mathbf{u}_i(k+l-1|k-1)=0$. Thus $\mathbf{u}_i^{\mathrm{f}}(k+l-1|k)$ satisfied the constraint (9.8) when $l=1,2,\ldots,N-1$, concluding the proof.

In what follows we establish that, at time k, if the condition (9.18) is satisfied, then $\mathbf{u}_i^{\mathrm{f}}(k+l-1|k)$, $l=1,2,\ldots,N$, is a feasible solution of Problem 9.1.

Lemma 9.3 *Suppose Assumptions 9.1 and 9.2 hold, $\mathbf{x}(k_0)\in\mathcal{X}$, and the conditions (9.12) and (9.18) are satisfied. For any $k\geq 0$, if Problem 9.1 has a solution at every update time t, $t=0,\ldots,k-1$, then $\mathbf{u}_i^{\mathrm{f}}(k+l-1|k)\in\mathcal{U}$ for all $l=1,2,\ldots,N$, and for any $i\in\mathcal{P}$.*

Proof. Since Problem 9.1 has a feasible solution at $k-1$, and $\mathbf{u}_i^{\mathrm{f}}(k+l-1|k)=\mathbf{u}_i(k+l-1|k-1)$ for all $l\in\{1,\ldots,N-1\}$, it only needs to be shown that $\mathbf{u}_i^{\mathrm{f}}(k+N-1|k)$ is in $\mathcal{U}$.

Since ε is chosen to satisfy the conditions of Lemma 9.1, $\mathbf{K}_i\mathbf{x}\in\mathcal{U}$ for all $i\in\mathcal{P}$, when $\mathbf{x}\in\Omega(\varepsilon)$. Consequently, a sufficient condition for $\mathbf{u}_i^{\mathrm{f}}(k+N-1|k)$ is that $\mathbf{x}^{\mathrm{f}}(k+N-1\,|\,k)\in\Omega(\varepsilon)$.

In view of Lemma 9.2 and $\alpha\leq 0.5$, using the triangle inequality, it has

$$\begin{aligned}
&\left\|\mathbf{x}^{\mathrm{f}}(k+N-1|k)\right\|_P\\
&\leq\left\|\mathbf{x}^{\mathrm{f}}(k+N-1|k)-\hat{\mathbf{x}}(k+N-1|k-1)\right\|_{\mathbf{P}}\\
&\quad+\|\hat{\mathbf{x}}(k+N-1|k-1)\|_{\mathbf{P}}\\
&\leq\gamma\kappa\alpha\varepsilon+\alpha\varepsilon\\
&\leq\varepsilon
\end{aligned}$$

that is, $\mathbf{x}^{\mathrm{f}}(k+N-1\,|\,k)\in\Omega(\varepsilon)$ for any $i\in\mathcal{P}$. This concludes the proof.

Lemma 9.4 *Suppose Assumptions 9.1 and 9.2 hold, $\mathbf{x}(k_0)\in\mathcal{X}$, and the conditions (9.12) and (9.18) are satisfied. For any $k\geq 0$, if Problem 9.1 has a solution at every update time, $t=0,\ldots,k-1$, then the terminal-state constraint is satisfied, for any $i\in\mathcal{P}$.*

Proof. Since there is a solution for Problem 9.1 at updates $t=1,\ldots,k-1$, Lemmas 9.2–9.4 can be invoked. According to Lemma 9.2, conditions (9.12) and (9.18), and using the triangle inequality, it has

$$\begin{aligned}
&\left\|\mathbf{x}^{\mathrm{f}}(k+N|k)\right\|_{\mathbf{P}}\\
&\leq\left\|\mathbf{x}^{\mathrm{f}}(k+N|k)-\hat{\mathbf{x}}(k+N|k-1,i)\right\|_{\mathbf{P}}\\
&\quad+\|\hat{\mathbf{x}}(k+N|k-1,i)\|_{\mathbf{P}}\\
&\leq(1-\kappa)\gamma\kappa\alpha\varepsilon+(1-\kappa)\alpha\varepsilon\\
&\leq\alpha\varepsilon
\end{aligned}$$

for each $i\in\mathcal{P}$. This shows that the terminal-state constraint is satisfied.

Theorem 9.1 *Suppose Assumptions 9.1 and 9.2 hold,* $\mathbf{x}(k_0) \in \mathcal{X}$ *and Equations (9.8), (9.9), and (9.10) are satisfied at* k_0*. Then, for every* $i \in \mathcal{P}$*, the control* $\mathbf{u}_i^{\mathrm{f}}(\cdot|k)$ *and state* $\mathbf{x}^{\mathrm{f}}(\cdot \mid k)$*, defined by (9.13), (9.14) and (9.16), are feasible solutions to Problem 9.1 at every update* $k \geq 1$*.*

Proof. First consider the case of $k = 1$. The state sequence $\hat{\mathbf{x}}(\cdot|1, i) = \mathbf{x}^{\mathrm{f}}(\cdot|1, i)$ trivially satisfies the dynamic equation (9.14) and the consistency constraint (9.8).

Now, suppose is a feasible solution for $t = 1, \ldots, k-1$. Lemmas 9.2–9.4 can be invoked. The consistency constraints (9.8) are trivially satisfied, the feasibility of control constraint and the terminal state constraint is guaranteed, and the proof of Theorem 9.1 is completed.

9.4.2 *Stability*

The stability of the closed-loop system is now analyzed in this subsection.

Theorem 9.2 *Suppose that Assumptions 9.1 and 9.2 hold,* $\mathbf{x}(k_0) \in \mathcal{X}$*, and Equations (9.8), (9.9), and (9.10) are satisfied at* k_0*, and the following parametric conditions hold*

$$\rho - \alpha(0.42 + ((N-1)\rho' + 1)\gamma\kappa) > 0$$

where

$$\rho = \lambda_{\min}\left(P^{-\frac{1}{2}} Q P^{-\frac{1}{2}}\right)^{\frac{1}{2}}$$

$$\rho' = \lambda_{\max}\left(P^{-\frac{1}{2}} Q P^{-\frac{1}{2}}\right)^{\frac{1}{2}}$$

Then, by application of Algorithm 9.1, the closed-loop system (9.2) is asymptotically stabilized to the origin.

Proof. By Algorithm 9.1 and Lemma 9.1, if $\mathbf{x}(k) \in \Omega(\varepsilon)$ for any $k \geq 0$, the terminal controllers take over and stabilize the system to the origin. Therefore, it remains to show that if $\mathbf{x}(k_0) \in \mathcal{X} \backslash \Omega(\varepsilon)$, then by application of Algorithm 9.1, the closed-loop system (9.2) is driven to the set in finite time.

Define the nonnegative function V_k for all system $\mathcal{S}$, $V_k = \sum_{i=1}^{m} V_{k,i}$, and

$$\begin{aligned} V_{k,i} &= \|\hat{\mathbf{x}}\,(k+N|k,i)\|_{\mathbf{P}} \\ &\quad + \sum_{l=0}^{N-1} \left(\|\hat{\mathbf{x}}\,(k+l|k,i)\|_{\mathbf{Q}} + \left\|\mathbf{u}_i\,(k+l|k)\right\|_{\mathbf{R}_i} \right) \end{aligned}$$

In what follows, we will show that for any $k \geq 0$, if $\mathbf{x}(k) \in \mathcal{X} \backslash \Omega(\varepsilon)$, then there exists a constant $\eta \in (0, \infty)$ such that $V_k \leq V_{k-1} - \eta$.

Since the performance index of $\mathcal{S}_i$, $\forall i \in \mathcal{P}$, with the optimal solution of $\mathbf{u}_i(\cdot|k)$ must be not more than the performance index of $\mathcal{S}_i$ with the feasible solution of $\mathbf{u}_i^{\mathrm{f}}(\cdot|k)$. Thus it has

$$\begin{aligned}
&V_{k,i} - V_{k-1,i}\\
&\le -\|\hat{\mathbf{x}}(k-1|k-1,i)\|_{\mathbf{Q}} - \|\mathbf{u}_i(k-1|k-1)\|_{\mathbf{R}_i}\\
&+\sum_{l=0}^{N-2}\left(\left\|\mathbf{x}^{\mathrm{f}}(k+l|k)\right\|_{\mathbf{Q}} + \left\|\mathbf{u}_i^{\mathrm{f}}(k+l|k)\right\|_{\mathbf{R}_i}\right)\\
&+\left(\left\|\mathbf{x}^{\mathrm{f}}(k+N-1|k)\right\|_{\mathbf{Q}} + \left\|\mathbf{u}_i^{\mathrm{f}}(k+N-1|k)\right\|_{\mathbf{R}_i}\right)\\
&+\left\|\mathbf{x}^{\mathrm{f}}(k+N|k)\right\|_{\mathbf{P}}\\
&-\sum_{l=0}^{N-2}\left(\|\hat{\mathbf{x}}(k+l|k-1,i)\|_{\mathbf{Q}} + \|\hat{\mathbf{u}}_i(k+l|k-1)\|_{\mathbf{R}_i}\right)\\
&-\left(\|\hat{\mathbf{x}}(k+N-1|k-1,i)\|_{\mathbf{P}}\right)
\end{aligned}$$

assuming $\mathbf{x}(k) \in \mathcal{X}\backslash\Omega(\varepsilon)$, that is $\|\hat{\mathbf{x}}(k-1|k-1,i)\|_{\mathbf{Q}} \ge \rho\varepsilon$. Considering that $\|\mathbf{u}_i(k-1\,|\,k-1)\|_{\mathbf{R}} > 0$ and substituting Equation (9.18) into Equation (9.31) yields

$$\begin{aligned}
V_{k,i} - V_{k-1,i} \le\ & -\rho e + \rho'(N-1)\gamma\kappa\alpha e\\
&+\left\|\mathbf{x}^{\mathrm{f}}(k+N-1|k)\right\|_{\mathbf{Q}} + \left\|\mathbf{u}_i^{\mathrm{f}}(k+N-1|k)\right\|_{\mathbf{R}}\\
&+\left\|\mathbf{x}^{\mathrm{f}}(k+N|k)\right\|_{\mathbf{P}} - \|\hat{\mathbf{x}}(k+N-1|k-1,i)\|_{\mathbf{P}}
\end{aligned}$$

In the above equation, consider the terms from the third term to the fifth term, it has

$$\begin{aligned}
&\frac{1}{2}\left(\left\|\mathbf{x}^{\mathrm{f}}(k+N-1|k)\right\|_{\mathbf{Q}} + \left\|\mathbf{u}_i^{\mathrm{f}}(k+N-1|k)\right\|_{\mathbf{R}_i} + \left\|\mathbf{x}^{\mathrm{f}}(k+N|k)\right\|_{\mathbf{P}}^2\right)^2\\
&\le \left\|\mathbf{x}^{\mathrm{f}}(k+N-1|k)\right\|_{\mathbf{Q}}^2 + \left\|\mathbf{u}_i^{\mathrm{f}}(k+N-1|k)\right\|_{\mathbf{R}_i}^2 + \left\|\mathbf{x}^{\mathrm{f}}(k+N|k)\right\|_{\mathbf{P}}^2\\
&\le \left\|\mathbf{x}^{\mathrm{f}}(k+N-1|k)\right\|_{\mathbf{Q}}^2 + \left\|\mathbf{u}^{\mathrm{f}}(k+N-1|k)\right\|_{\mathbf{R}}^2 + \left\|\mathbf{x}^{\mathrm{f}}(k+N|k)\right\|_{\mathbf{P}}^2
\end{aligned}$$

Notice that $\|\mathbf{x}^{\mathrm{f}}x^{f}(k+N|k)\|_{\mathbf{P}}^2 = \|\mathbf{A}_c\mathbf{x}^{\mathrm{f}}(k+N-1|k)\|_{\mathbf{P}}^2$ and $\hat{\mathbf{Q}} = \mathbf{Q} + \mathbf{K}^{\mathrm{T}}\mathbf{R}\mathbf{K}$, $\mathbf{A}_c^{\mathrm{T}}\mathbf{P}\mathbf{A}_c - \mathbf{P} = -\hat{\mathbf{Q}}$, it has

$$\begin{aligned}
&\left\|\mathbf{x}^{\mathrm{f}}(k+N-1|k)\right\|_{\mathbf{Q}}^2 + \left\|\mathbf{u}^{\mathrm{f}}(k+N-1|k)\right\|_{\mathbf{R}}^2 + \left\|\mathbf{x}^{\mathrm{f}}(k+N|k)\right\|_{\mathbf{P}}^2\\
&\le \left\|\mathbf{x}^{\mathrm{f}}(k+N-1|k)\right\|_{\hat{\mathbf{Q}}}^2 + \left\|\mathbf{A}_c\mathbf{x}^{\mathrm{f}}(k+N-1|k)\right\|_{\mathbf{P}}^2\\
&= \left\|\mathbf{x}^{\mathrm{f}}(k+N-1|k)\right\|_{\mathbf{P}}^2
\end{aligned}$$

Consider

$$\sqrt{2}\left\|\mathbf{x}^{\mathrm{f}}(k+N-1|k)\right\|_{\mathbf{P}} - \|\hat{\mathbf{x}}(k+N-1|k-1,i)\|_{\mathbf{P}}$$
$$\leq 0.42\left\|\mathbf{x}^{\mathrm{f}}(k+N-1|k)\right\|_{\mathbf{P}}$$
$$+\left\|\mathbf{x}^{\mathrm{f}}(k+N-1|k) - \hat{\mathbf{x}}(k+N-1|k-1,i)\right\|_{\mathbf{P}}$$
$$\leq 0.42\alpha e + \gamma\kappa\alpha e$$

and substitute Equations (9.33)–(9.35) into Equation (9.32), it yields

$$V_{k,i} - V_{k-1,i}$$
$$\leq -\rho e + (N-1)\rho'\gamma\kappa\alpha e + 0.42\alpha e + \gamma\kappa\alpha e$$
$$= -e(\rho - \alpha(0.42 + ((N-1)\rho' + 1)\gamma\kappa))$$

which, in view of (9.28), implies that

$$V_{k,i} - V_{k-1,i} < 0$$

Thus, for any $k \geq 0$, if $\mathbf{x}(k) \in \mathcal{X}\backslash\Omega(\varepsilon)$, then there is a constant $\eta_i \in (0, \infty)$ such that $V_{k,i} \leq V_{k-1,i} - \eta_i$. Furthermore, we have the inequality of $V_k \leq V_{k-1} - \eta$, where $\eta = \sum_{i=1}^{m} \eta_i$ since m is limited. From this inequality, it follows by contradiction that there exists a finite time k' such that $\mathbf{x}(k') \in \Omega(\varepsilon)$. If this is not the case, the inequality implies $V_k \to -\infty$ as $k \to \infty$. However, $V_k \geq 0$, therefore, there exists a finite time k' such that $\mathbf{x}(k') \in \Omega(\varepsilon)$. This concludes the proof.

So far, the feasibility and stability of proposed C-DMPC are analyzed. Provided an initially feasible solution could be found, subsequent feasibility of the algorithm is guaranteed at every update, and the resulting closed-loop system is asymptotically stable at the origin.

9.5 Simulation

The multizone building temperature regulation systems are a class of typical spatially distributed systems, as shown in Figure 9.2, which are composed of many physically interacted subsystems (rooms or zones) labeled with $\mathcal{S}_1, \mathcal{S}_2, \ldots$, respectively. The thermal influences between rooms of the same building occur through internal walls (the internal walls isolation is weak) and/or door openings. A thermal-meter and a heater (or air condition) are installed in each zone, which is used to measure and adjust the temperature of the multizone building.

For simplicity, the seven-zones building is taken as the example. The relationship among these seven zones is also shown in Figure 9.2, where zone $\mathcal{S}_1$ is impacted by zone $\mathcal{S}_2$ and zone $\mathcal{S}_7$; zone $\mathcal{S}_2$ is impacted by zone $\mathcal{S}_1$, $\mathcal{S}_3$, and zone $\mathcal{S}_7$; zone $\mathcal{S}_3$ is impacted by zone $\mathcal{S}_2$, $\mathcal{S}_4$, and zone $\mathcal{S}_7$; zone $\mathcal{S}_4$ is impacted by zone $\mathcal{S}_3$, $\mathcal{S}_5$, and zone $\mathcal{S}_7$; zone $\mathcal{S}_5$ is impacted by zone $\mathcal{S}_4$, $\mathcal{S}_6$, and zone $\mathcal{S}_7$; zone $\mathcal{S}_6$ is impacted by zone $\mathcal{S}_5$ and zone $\mathcal{S}_7$; zone $\mathcal{S}_4$ is impacted by all the other zones. Let $\mathcal{U}_i$ be defined to reflect both the constraint on the input $u_i \in [u_{i,\mathrm{L}}, u_{i,\mathrm{U}}]$ and the constraint on the increment of the input $\Delta u_i \in [\Delta u_{i,\mathrm{L}}, \Delta u_{i,\mathrm{U}}]$.

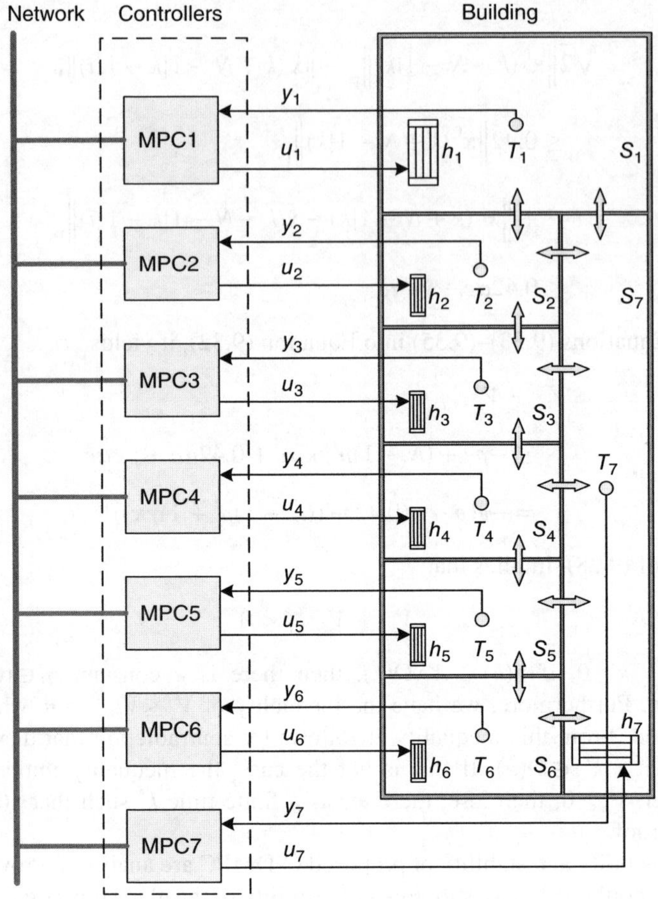

Figure 9.2 The multizone building temperature regulation system

The models of these seven subsystems are, respectively, given by

$$\begin{aligned}
\mathcal{S}_1 : x_1(k+1) &= 0.574x_1(k) + 0.384u_1(k) \\
&\quad + 0.029x_2(k) + 0.057x_7(k) \\
\mathcal{S}_2 : x_2(k+1) &= 0.535x_2(k) + 0.372u_2(k) \\
&\quad + 0.054x_1(k) + 0.054x_3(k) + 0.054x_7(k) \\
\mathcal{S}_3 : x_3(k+1) &= 0.547x_3(k) + 0.376u_3(k) \\
&\quad + 0.055x_2(k) + 0.055x_4(k) + 0.055x_7(k) \\
\mathcal{S}_4 : x_4(k+1) &= 0.606x_4(k) + 0.394u_4(k) \\
&\quad + 0.061x_3(k) + 0.061x_5(k) + 0.061x_7(k)
\end{aligned}$$

$$
\begin{aligned}
\mathcal{S}_5: x_5(k+1) &= 0.681x_5(k) + 0.415u_5(k) \\
&\quad + 0.068x_4(k) + 0.068x_6(k) + 0.068x_7(k) \\
\mathcal{S}_6: x_6(k+1) &= 0.548x_6(k) + 0.376u_6(k) \\
&\quad + 0.055x_5(k) + 0.055x_7(k) \\
\mathcal{S}_7: x_7(k+1) &= 0.716x_7(k) + 0.425u_7(k) \\
&\quad + 0.018x_1(k) + 0.018x_2(k) + 0.018x_3(k) \\
&\quad + 0.018x_4(k) + 0.018x_5(k) + 0.018x_6(k)
\end{aligned}
$$

For the purpose of comparison, the centralized MPC, local cost optimization based MPC, and the C-DMPC are all applied to this system.

Some parameters of the controllers in proposed C-DMPC are shown in Table 9.2. Among these parameters, P_i is obtained by solving the Lyapunov function. The eigenvalue of each closed-loop system under the feedback control shown in Table 9.2 is 0.5. Set $\varepsilon = 0.15$, and set the control horizon of all the controllers to be $N = 10$. Set the initial presumed inputs and states, at time $k_0 = 0$, be zeros.

In both the centralized MPC and the subsystem-based MPCs of the local cost optimization based DMPC, the dual mode strategy is adopted, and set the parameters, the initial states, and the initial presumed inputs to be the same as those used in C-DMPC.

The state responses and the inputs of the closed-loop system under the control of the centralized MPC, C-DMPC, and local optimization based DMPC are shown in Figures 9.3 and 9.4, respectively. The shape of the state response curves under the control C-DMPC is similar to those under the centralized MPC. Under the C-DMPC control design, when set-point changed, there is no significant overshooting, but some fluctuations exist in the trajectories of states of the interacting subsystems. Under the LCO-DMPC control design, the states of all subsystems could converge to set-point, but there exists much larger overshooting comparing to those under the control of C-DMPC and centralized MPC, and there are larger amplitude in the fluctuating state than those under the control of C-DMPC.

Figure 9.5 shows the difference of the absolute value of the state and inputs of each subsystem between under the control of LCO-DMPC and under the control of centralized MPC, and between under the control of C-DMPC and under the control of centralized DMPC. Figure 9.6

Table 9.2 Parameters of C-DMPC

Subsystem	$\mathbf{K}_i$	$\mathbf{P}_i$	$\mathbf{Q}_i$	$\mathbf{R}_i$	$\Delta\mathbf{u}_{i,\mathrm{U}}$ $\Delta\mathbf{u}_{i,\mathrm{L}}$
$\mathcal{S}_1$	−0.44	5.38	4	0.2	±1
$\mathcal{S}_2$	−0.34	5.36	4	0.2	±1
$\mathcal{S}_3$	−0.37	5.37	4	0.2	±1
$\mathcal{S}_4$	−0.52	5.40	4	0.2	±1
$\mathcal{S}_5$	−0.68	5.46	4	0.2	±1
$\mathcal{S}_6$	−0.37	5.37	4	0.2	±1
$\mathcal{S}_7$	−0.76	5.49	4	0.2	±1

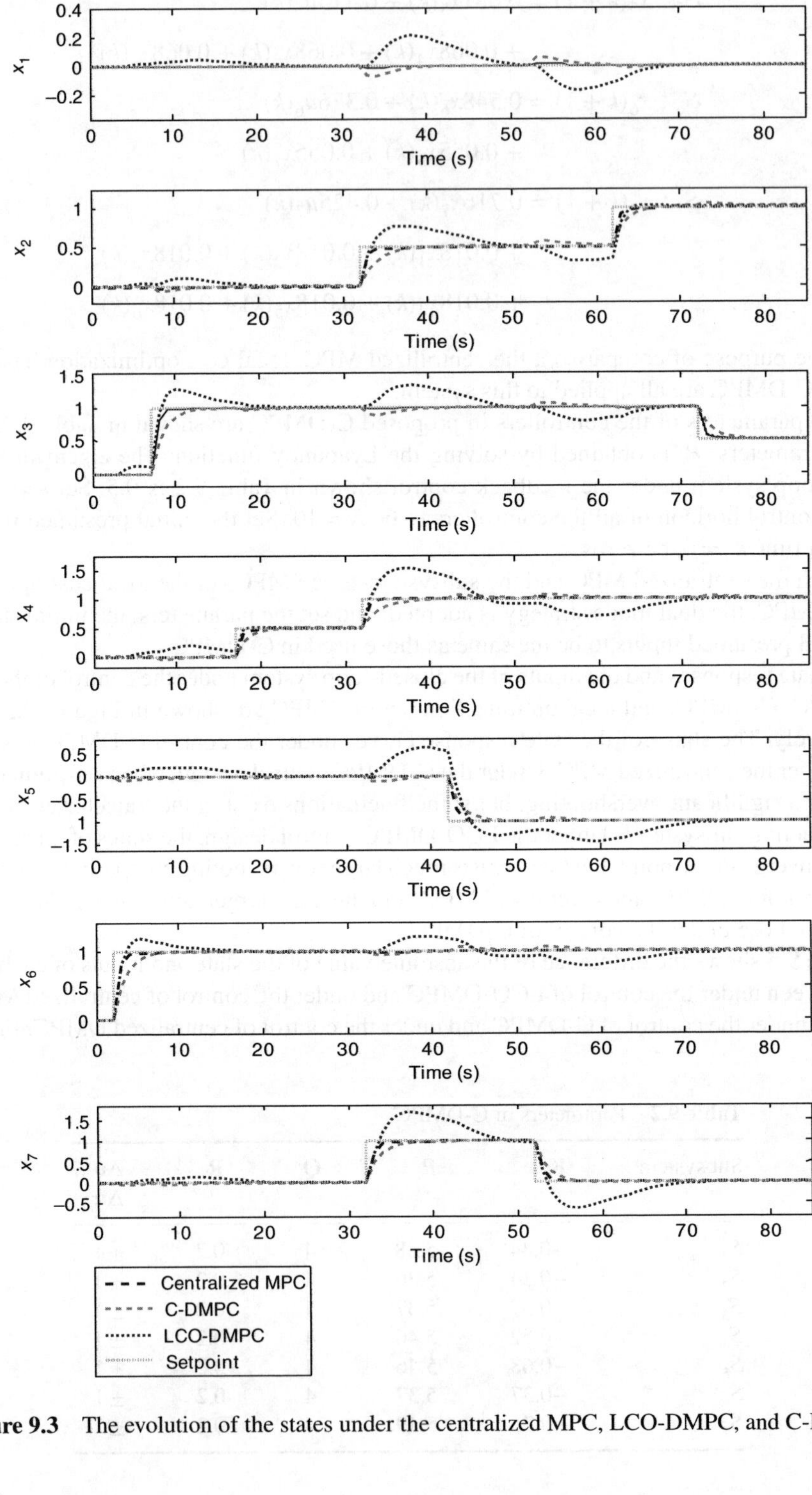

Figure 9.3 The evolution of the states under the centralized MPC, LCO-DMPC, and C-DMPC

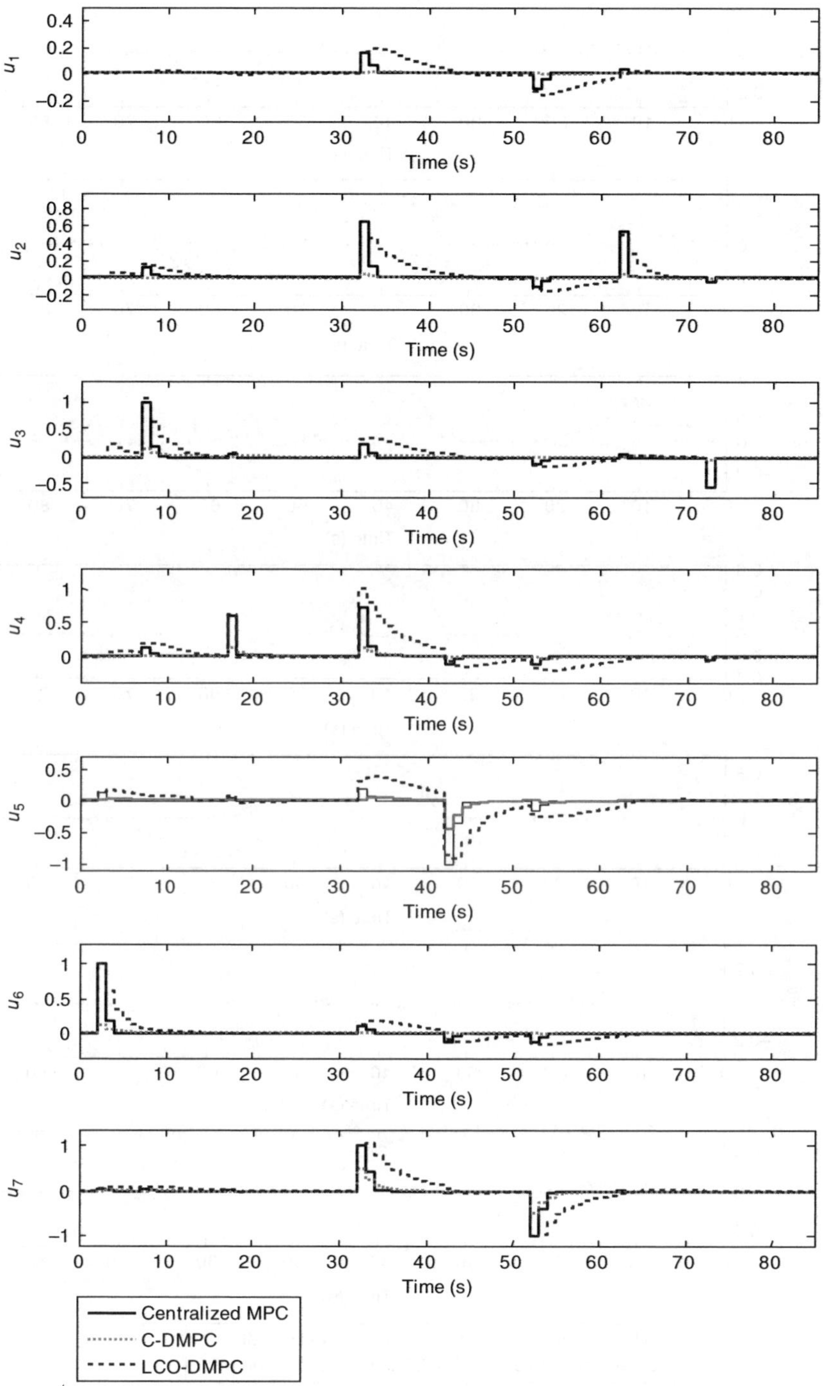

Figure 9.4 The evolution of the inputs under the centralized MPC, LCO-DMPC, and C-DMPC

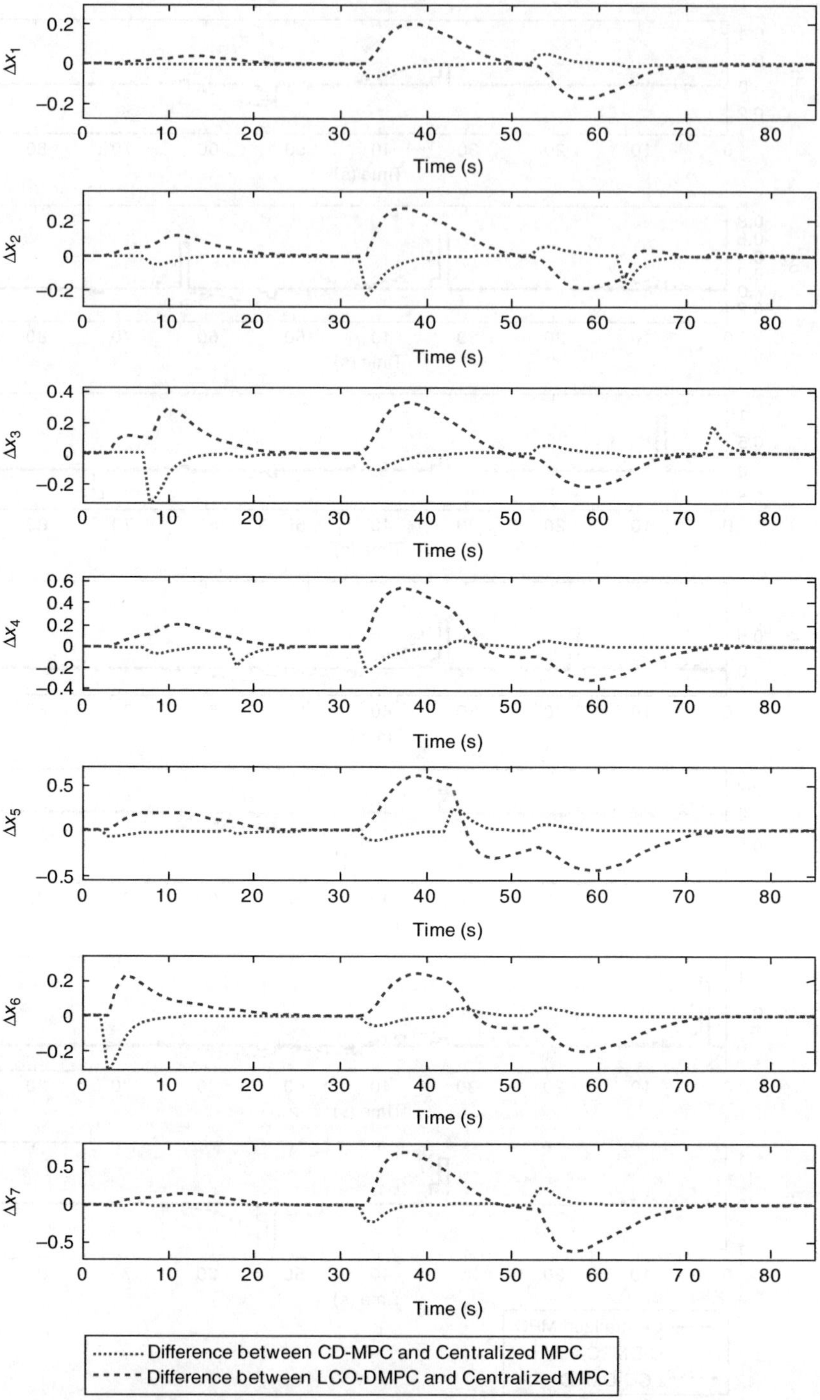

Figure 9.5 Differences of the absolute value of the state and inputs of each subsystem between under the control of LCO-DMPC and under the control of centralized MPC, and between under the control of C-DMPC and under the control centralized DMPC

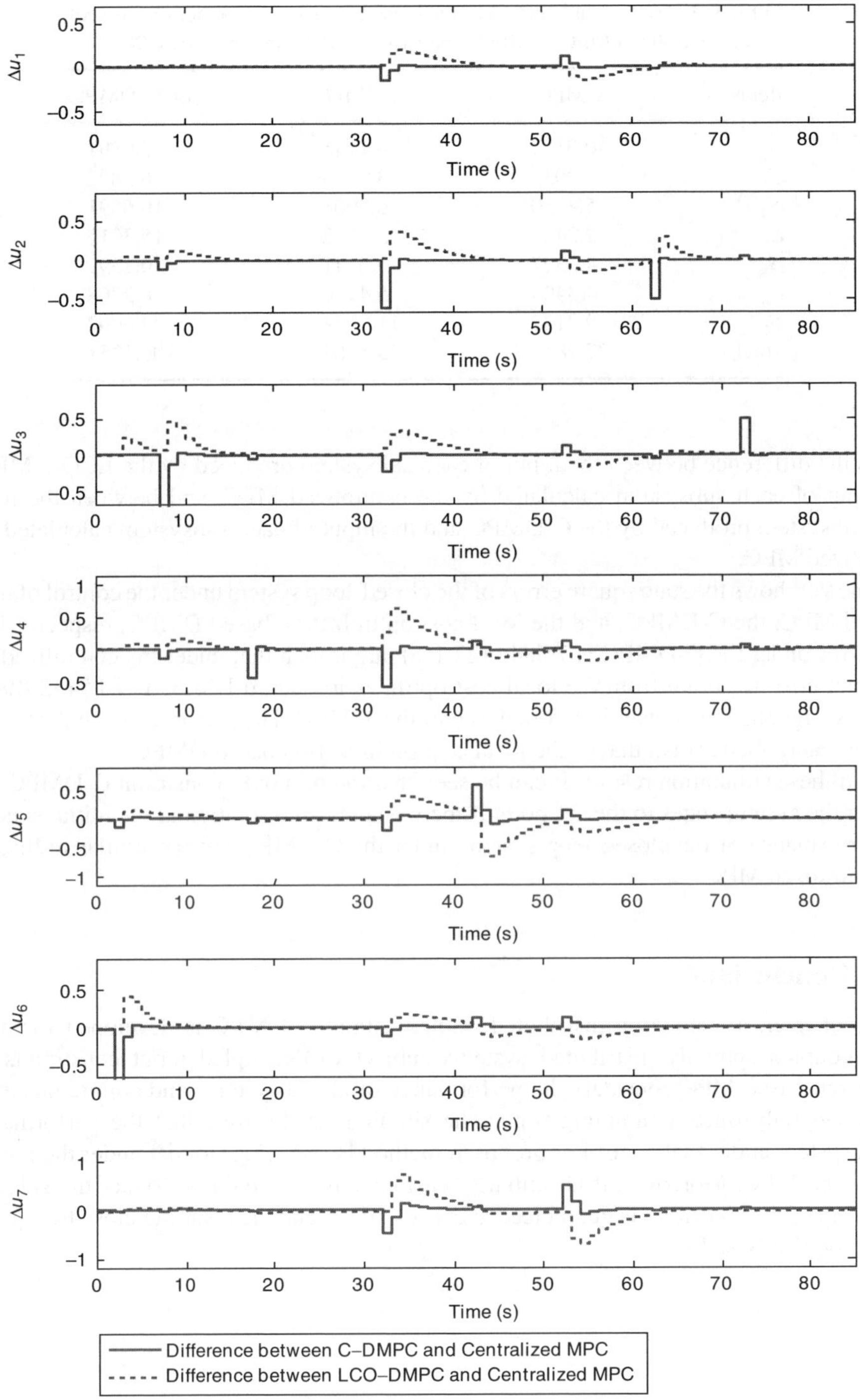

Figure 9.6 The difference between the input of each subsystem produced by the LCO-DMPC and the input of each subsystem calculated by the centralized MPC, and between the input of each subsystem produced by the C-DMPC and the input of each subsystem calculated by the centralized MPC

Table 9.3 State square errors of the closed-loop system under the control of the centralized MPC (CMPC), the LCO-DMPC, and the C-DMPC

Items	CMPC	C-DMPC	LCO-DMPC
S_1	0.0109	0.1146	2.0891
S_2	2.2038	3.0245	6.2892
S_3	5.4350	6.9908	10.6391
S_4	2.2480	3.2122	15.3015
S_5	4.5307	5.6741	30.2392
S_6	4.3403	5.4926	8.2768
S_7	9.2132	11.0574	33.6902
Total	27.9819	35.5663	106.5251

shows the difference between the input of each subsystem produced by the LCO-DMPC and the input of each subsystem calculated by the centralized MPC, and between the input of each subsystem produced by the C-DMPC and the input of each subsystem calculated by the centralized MPC.

Table 9.3 shows the state square errors of the closed-loop system under the control of the centralized MPC, the C-DMPC, and the local cost optimization based DMPC, respectively. The total error under the C-DMPC is 7.5844 (27.1%) larger than that under the centralized MPC. The total error resulting from the local cost optimization based DMPC is 78.5432 (280.7%) which is larger than that resulting from the centralized MPC. The performance of the C-DMPC is significantly better than that of the local cost optimization based DMPC.

From these simulation results, it can be seen that the proposed constraint C-DMPC is able to steer the system states to the set-point if there is a feasible solution at the initial states, and the performance of the closed-loop system under the C-DMPC is very similar to that under the centralized MPC.

9.6 Conclusion

In this chapter, a stabilizing distributed implementation of MPC is developed for dynamically coupled spatially distributed systems subject to decoupled input constraints. Each subsystem-based MPC considers the performance of all subsystems and communicates with each other only once a sampling time. The simulations illustrate that the performance of global system under that control of proposed method is very close to that under the control of centralized MPC. Moreover, if an initially feasible solution could be found, the subsequent feasibility of the algorithm is guaranteed at every update, and the resulting closed-loop system is asymptotically stable.

10

Networked Distributed Predictive Control with Inputs and Information Structure Constraints

10.1 Introduction

As mentioned in Chapter 7, the main advantage of distributed model predictive control (DMPC) is that it has the characteristics of good flexibility and error tolerance. These characteristics are based on the fact that the subsystem-based controllers are relevantly independent to each other. It means that if the number of subsystems that each subsystem-based MPC communicates will decrease, the flexibility and the ability of error tolerance of the whole closed-loop control system will increase. In addition, in some fields or processes the global information are unavailable to controllers (e.g., in multi-intelligent vehicle system) for the management or the system scale reasons. Thus to design a DMPC that could significantly improve the global performance of the closed-loop system with limited information structure constraints is valuable.

In an effort to achieve a tradeoff between the global performance of the entire system and the computational burden, an intuitively appealing strategy is proposed in Chapter 7, where each subsystem-based MPC only considers the cost of its own subsystem and those of the subsystems it directly impacts on. Such a design can be referred to as networked distributed MPC (N-DMPC). Chapter 7 applies this design idea to a metallurgy system and explains why this coordination strategy could improve the global performance. Numerical and practical experiments show that this coordination strategy could obtain a performance close to that of a classical centralized MPC. However, the method introduced in Chapter 7 does not take constraints into consideration in the DMPC design.

Under the DMPC framework, Ref. [51] provides a design for nonlinear continuous systems, which uses a constraint to limit the error between the future state sequences (or called

Distributed Model Predictive Control for Plant-Wide Systems, First Edition. Shaoyuan Li and Yi Zheng.

presumed sequences) of upstream neighbors, which are calculated based on the solution in the previous time instant, and the predictive states calculated by the corresponding subsystem in the current time instant. Then the stability is ensured by judiciously integrating designs of the bound of the error between presumed state sequence and predictive state sequence [51], the terminal cost, the constraint set, and the local controllers [69]. Farina and Scattolini [50] gave another design for linear system, which uses a fixed reference trajectory with a moving window to substitute the presumed state/input of upstream neighbors used in Ref. [51]. Both these methods are design for DMPC in which each subsystem-based MPC optimizes the cost of the corresponding subsystem itself. As for the DMPC, which uses the global cost function, some convergence conditions are deduced if using iterative algorithms, then the distributed problems can be reformulated into a centralized problem, and the stabilizing DMPC can be designed with similar method of centralized MPC. For the coordination strategy used here, there is no global model that can be used. And except that there are errors between the presumed state/input sequences and predictive state sequences of upstream neighbors, the predictive state sequences of downstream neighbors calculated by current subsystem may not equal to those calculated by the downstream neighbors themselves; these error are difficult to estimate. In the presence of constraints, the remaining part of the optimal control sequence calculated at the previous time instant may not be a feasible solution at the current time instant. All these make it difficult to design a stabilizing N-DMPC that takes constraints into consideration.

In this chapter, the coordination strategy that optimizes the impacted-region cost in each subsystem-based MPC is adopted to achieve a DMPC performance that is close to that of a centralized MPC. The consistency constraints, which limit the error between the presumed state and the state predicted at the current time instant within a prescribed bound, are designed and included in the optimization problem of each subsystem-based MPC. These constraints can bound the error between the presumed state and the predictive state of upstream neighbors and the error between the predictive state of downstream neighbor calculated by current subsystem-based MPC and that of the downstream neighbors themselves. And these constraints guarantee that the remaining part of the solution at the current time instant is a feasible solution at the next time instant. In the meantime, stabilization constraints and the dual mode predictive control strategy are adopted to result in a stabilizing N-DMPC.

The remainder of this chapter is organized as follows: Section 10.2 describes the problem to be solved in this chapter. Section 10.3 presents the design of the stabilizing N-DMPC. The feasibility of the proposed N-DMPC and the stability of the resulting closed-loop system are analyzed in Section 10.4. Section 10.5 discusses DMPC formulations under other coordination strategies. Section 10.6 presents the simulation results to demonstrate the effectiveness of the proposed DMPC algorithm. Finally, a brief conclusion to the chapter is drawn in Section 10.7.

10.2 Problem Description

A distributed system, as illustrated in Figure 4.1, is considered here. Suppose the distributed system $\mathcal{S}$ is composed of m discrete-time linear subsystems $\mathcal{S}_i$, $i \in \mathcal{P} = \{1, 2, \dots, m\}$ and m controllers $\mathcal{C}_i$, $i \in \mathcal{P} = \{1, 2, \dots, m\}$. Let the subsystems interact with each other through their states. If subsystem $\mathcal{S}_i$ is affected by $\mathcal{S}_j$, for any $i \in \mathcal{P}$ and $j \in \mathcal{P}$ subsystem $\mathcal{S}_i$ is said to be a downstream system of subsystem $\mathcal{S}_j$, and subsystem $\mathcal{S}_j$ is an upstream system of $\mathcal{S}_i$. Let $\mathcal{P}_{+i}$ denote the set of the subscripts of the upstream systems of $\mathcal{S}_i$, $\mathcal{P}_{-i}$ is the set of the subscripts of the downstream systems of $\mathcal{S}_i$. Then, subsystem $\mathcal{S}_i$ can be expressed as

$$\begin{cases} \mathbf{x}_i(k+1) = \mathbf{A}_{ii}\mathbf{x}_i(k) + \mathbf{B}_{ii}\mathbf{u}_i(k) + \sum_{j \in \mathcal{P}_{+i}} \mathbf{A}_{ij}\mathbf{x}_j(k) \\ \mathbf{y}_i(k) = \mathbf{C}_{ii}\mathbf{x}_i(k) \end{cases} \tag{10.1}$$

where $\mathbf{x}_i \in \mathbb{R}^{n_{xi}}$, $\mathbf{u}_i \in \mathcal{U}_i \subset \mathbb{R}^{n_{ui}}$, and $\mathbf{y}_i \in \mathbb{R}^{n_{yi}}$ are, respectively, the local state, input, and output vectors, and $\mathcal{U}_i$ is the feasible set of the input $\mathbf{u}_i$, which is used to bound the input according to the physical constraints on the actuators, the control requirements or the characteristics of the plant. A nonzero matrix $\mathbf{A}_{ij}$, that is, $j \in \mathcal{P}_{+i}$, indicates that $\mathcal{S}_i$ is affected by $\mathcal{S}_j$. In the concatenated vector form, the system dynamics can be written as

$$\begin{cases} \mathbf{x}(k+1) = \mathbf{A}\mathbf{x}(k) + \mathbf{B}\mathbf{u}(k) \\ \mathbf{y}(k) = \mathbf{C}\mathbf{x}(k) \end{cases} \tag{10.2}$$

where

$$\mathbf{x}(k) = \left[\mathbf{x}_1^{\mathrm{T}}(k)\ \mathbf{x}_2^{\mathrm{T}}(k)\ \ldots\ \mathbf{x}_m^{\mathrm{T}}(k)\right]^{\mathrm{T}} \in \mathrm{R}^{n_x}$$

$$\mathbf{u}(k) = \left[\mathbf{u}_1^{\mathrm{T}}(k)\ \mathbf{u}_2^{\mathrm{T}}(k)\ \ldots\ \mathbf{u}_m^{\mathrm{T}}(k)\right]^{\mathrm{T}} \in \mathrm{U} \subset \mathrm{R}^{n_u}$$

$$\mathbf{y}(k) = \left[\mathbf{y}_1^{\mathrm{T}}(k)\ \mathbf{y}_2^{\mathrm{T}}(k)\ \ldots\ \mathbf{y}_m^{\mathrm{T}}(k)\right]^{\mathrm{T}} \in \mathrm{R}^{n_y}$$

are, respectively, the concatenated state, control input, and output vectors of the overall system $\mathcal{S}$. Also, $\mathbf{u}(k) \in \mathcal{U} = \mathcal{U}_1 \times \mathcal{U}_2 \times \ldots \times \mathcal{U}_m$. $\mathbf{A}$, $\mathbf{B}$, and $\mathbf{C}$ are the constant matrices of appropriate dimensions and are defined as follows:

$$\mathbf{A} = \begin{bmatrix} \mathbf{A}_{11} & \mathbf{A}_{12} & \cdots & \mathbf{A}_{1m} \\ \mathbf{A}_{21} & \mathbf{A}_{22} & \cdots & \mathbf{A}_{2m} \\ \vdots & \vdots & \ddots & \vdots \\ \mathbf{A}_{m1} & \mathbf{A}_{m2} & \cdots & \mathbf{A}_{mm} \end{bmatrix}^{\mathrm{T}}$$

$$\mathbf{B} = \begin{bmatrix} \mathbf{B}_{11} & \mathbf{B}_{12} & \cdots & \mathbf{B}_{1m} \\ \mathbf{B}_{21} & \mathbf{B}_{22} & \cdots & \mathbf{B}_{2m} \\ \vdots & \vdots & \ddots & \vdots \\ \mathbf{B}_{m1} & \mathbf{B}_{m2} & \cdots & \mathbf{B}_{mm} \end{bmatrix}^{\mathrm{T}}$$

$$\mathbf{C} = \begin{bmatrix} \mathbf{C}_{11} & \mathbf{C}_{12} & \cdots & \mathbf{C}_{1m} \\ \mathbf{C}_{21} & \mathbf{C}_{22} & \cdots & \mathbf{C}_{2m} \\ \vdots & \vdots & \ddots & \vdots \\ \mathbf{C}_{m1} & \mathbf{C}_{m2} & \cdots & \mathbf{C}_{mm} \end{bmatrix}^{\mathrm{T}}$$

The control objective is to stabilize the overall system $\mathcal{S}$ in a DMPC framework. And the performance of entire closed-loop system should be as close as that under the control of centralized MPC and the communication cost should not be too high.

The coordination strategy as proposed in Chapter 7 is a preferable method to tradeoff between the communication burden and the global performance. However, the DMPC under this coordination strategy does not take the constraints into consideration. A noniterative stabilizing DMPC that takes constraints into consideration remains to be developed either under this coordination strategy or some other coordination strategies reviewed earlier. The objective of this chapter is to develop such a DMPC design.

10.3 Constrained N-DMPC

10.3.1 Formulation

In this section, m separate optimal control problems, one for each subsystem, and the N-DMPC algorithm are defined. In each of these optimal control problems, the same constant prediction horizon N, $N \geq 1$, is used. The resulting m separate subsystem-based MPC laws are updated synchronously. At each update, every subsystem-based MPC optimizes only over its own predicted open-loop control, given its impacted region's current states, and their upstream subsystems' estimated inputs and states.

To proceed, we need the following assumption:

Assumption 10.1 For every subsystem $\mathcal{S}_i$, $\forall i \in \mathcal{P}$, there exists a state feedback $\mathbf{u}_{i,k} = \mathbf{K}_i\mathbf{x}_{i,k}$ such that the closed-loop system $\mathbf{x}(k+1) = \mathbf{A}_c\mathbf{x}(k)$ is asymptotically stable, where $\mathbf{A}_c = \mathbf{A} + \mathbf{BK}$ and $\mathbf{K} = \text{block-diag}\{\mathbf{K}_1, \mathbf{K}_2, \ldots, \mathbf{K}_m\}$.

This assumption is usually used in the design of stabilizing DMPC [26, 34]. It presumes that each subsystem is able to be stabilized by a decentralized control $\mathbf{K}_i\mathbf{x}_i$, $i \in \mathcal{P}$.

We also define the necessary notation in Table 10.1.

As the state evolution of the downstream subsystems of $\mathcal{S}_j$ is affected by the optimal control decision of $\mathcal{S}_i$, the performance of these downstream subsystems may be affected negatively by the control decision of $\mathcal{S}_i$. Thus, in the ICO-DMPC, each subsystem-based MPC takes into account the cost functions of its downstream subsystems. More specifically, the performance index is defined as

$$\begin{aligned}\bar{J}_i(k) = &\sum_{j\in\mathcal{P}_i} \left\|\mathbf{x}^p_{j,i}(k+N\,|\,k)\right\|^2_{\mathbf{P}_j} \\ &+ \sum_{s=0}^{N-1}\left(\sum_{j\in\mathcal{P}_i}\left\|\mathbf{x}^p_{j,i}(k+s\,|\,k)\right\|^2_{\mathbf{Q}_j} + \left\|\mathbf{u}^p_i(k+s\,|\,k)\right\|^2_{\mathbf{R}_i}\right)\end{aligned} \tag{10.3}$$

where $\mathbf{Q}_i = \mathbf{Q}_i^{\mathrm{T}} > 0$, $\mathbf{R}_i = \mathbf{R}_i^{\mathrm{T}} > 0$ and $\mathbf{P}_i = \mathbf{P}_i^{\mathrm{T}} > 0$. The matrix $\mathbf{P}_i$ is chosen to satisfy the Lyapunov equation

$$\mathbf{A}_{di}^{\mathrm{T}}\mathbf{P}_i\mathbf{A}_{di} - \mathbf{P}_i = -\hat{\mathbf{Q}}_i \tag{10.4}$$

Table 10.1 Notations in this chapter

Notation	Explanation
i	The subscript denoting all downstream subsystems of $\mathcal{S}_i$
$+i$	The subscript denoting all upstream subsystems of $\mathcal{S}_i$
$\tilde{i}$	The subscript denoting all upstream subsystems of $\mathcal{S}_{-i}$ and $\mathcal{S}_i$, excluding $\mathcal{S}_{-i}$ and $\mathcal{S}_i$ themselves
$\mathbf{x}^{\mathrm{p}}_{j,i}(k+s\,\vert\,k)$	The predicted state sequence of $\mathcal{S}_j$, calculated by $\mathcal{C}_i$ at time
$\mathbf{x}^{p}_{i}(k+s\,\vert\,k)$	The predicted state sequence of $\mathcal{S}_i$, calculated by $\mathcal{C}_i$ at time k, $\mathbf{x}^{\mathrm{p}}_{i}(k+s\,\vert\,k)=\mathbf{x}^{\mathrm{p}}_{i,i}(k+s\,\vert\,k)$
$\mathbf{u}^{p}_{i}(k+s\,\vert\,k)$	The predicted control sequence of $\mathcal{S}_i$, calculated by $\mathcal{C}_i$ at time k
$\hat{\mathbf{x}}_{j,i}(k+s\,\vert\,k)$	The presumed state sequence of $\mathcal{S}_j$, calculated by $\mathcal{C}_i$ at time k
$\hat{\mathbf{x}}_{i}(k+s\,\vert\,k)$	The presumed state sequence of $\mathcal{S}_i$, calculated by $\mathcal{C}_i$ at time k, $\hat{\mathbf{x}}_{i}(k+s\,\vert\,k)=\hat{\mathbf{x}}_{i,i}(k+s\,\vert\,k)$
$\hat{\mathbf{u}}_{i}(k+s\vert k)$	The presumed control sequence of $\mathcal{S}_i$, defined by $\mathcal{C}_i$ at time k
$\hat{\mathbf{u}}_{j,i}(k+s\vert k)$	The presumed control sequence of $\mathcal{S}_j$, defined in $\mathcal{C}_i$ at time k
$\mathbf{x}^{\mathrm{f}}_{j,i}(k+s\,\vert\,k)$	The feasible state sequence of $\mathcal{S}_j$, calculated by $\mathcal{C}_i$ at time k
$\mathbf{x}^{\mathrm{f}}_{i}(k+s\,\vert\,k)$	The feasible state sequence of $\mathcal{S}_i$, calculated by $\mathcal{C}_i$ at time k, $\mathbf{x}^{\mathrm{f}}_{i}(k+s\,\vert\,k)=\mathbf{x}^{\mathrm{f}}_{i,i}(k+s\,\vert\,k)$
$\mathbf{u}^{\mathrm{f}}_{j,i}(k+s\,\vert\,k)$	The feasible control sequence of $\mathcal{S}_j$, defined in $\mathcal{C}_i$ at time k
$\mathbf{u}^{\mathrm{f}}_{i}(k+s\,\vert\,k)$	The feasible control sequence of $\mathcal{S}_i$, defined by $\mathcal{C}_i$ at time k, $\mathbf{u}^{\mathrm{f}}_{i}(k+s\,\vert\,k)=\mathbf{u}^{\mathrm{f}}_{i,i}(k+s\,\vert\,k)$

where $\hat{\mathbf{Q}}_i=\mathbf{Q}_i+\mathbf{K}_i^{\mathrm{T}}\mathbf{R}_i\mathbf{K}_i$. Denote

$$\begin{aligned}
\mathbf{P}&=\text{block-diag}\{\mathbf{P}_1,\mathbf{P}_2,\dots,\mathbf{P}_m\}\\
\mathbf{Q}&=\text{block-diag}\{\mathbf{Q}_1,\mathbf{Q}_2,\dots,\mathbf{Q}_m\}\\
\mathbf{R}&=\text{block-diag}\{\mathbf{R}_1,\mathbf{R}_2,\dots,\mathbf{R}_m\}\\
\mathbf{A}_{\mathrm{d}}&=\text{block-diag}\{\mathbf{A}_{\mathrm{d}1},\mathbf{A}_{\mathrm{d}2},\dots,\mathbf{A}_{\mathrm{d}m}\}
\end{aligned}$$

Then, it follows that

$$\mathbf{A}_{\mathrm{d}}^{\mathrm{T}}\mathbf{P}\mathbf{A}_{\mathrm{d}}-\mathbf{P}=-\hat{\mathbf{Q}}$$

where $\hat{\mathbf{Q}}=\mathbf{Q}+\mathbf{K}^{\mathrm{T}}\mathbf{R}\mathbf{K}>0$.

To get the predicted state sequence $\mathbf{x}^{\mathrm{p}}_{j,i}(k+s\,|\,k)$ of subsystem $\mathcal{S}_j$ under the control decision sequence of $\mathbf{u}^{\mathrm{p}}_{i}(k+s\,|\,k)$ in (10.3), the system evolution model should be deduced first. Since every subsystem-based controller is updated synchronously, the state and control sequences of other subsystems are unknown to subsystem $\mathcal{S}_i$. Thus, at the time instant k, the presumed state sequence $\{\hat{\mathbf{x}}_{\tilde{i}}(k\,|\,k),\hat{\mathbf{x}}_{\tilde{i}}(k+1\,|\,k),\dots,\hat{\mathbf{x}}_{\tilde{i}}(k+N\,|\,k)\}$ of $\mathcal{S}_{\tilde{i}}$ and the presumed control sequence $\{\hat{\mathbf{u}}_{-i}(k|k),\hat{\mathbf{u}}_{-i}(k+1|k),\dots,\hat{\mathbf{u}}_{-i}(k+N|k)\}$ of $\mathcal{S}_{-i}$ are used in the predictive model of the MPC in $\mathcal{S}_i$, which is given as

$$\begin{bmatrix}\mathbf{x}^{\mathrm{p}}_{i}(k+s\,|\,k)\\ \mathbf{x}^{\mathrm{p}}_{\tilde{i},i}(k+s\,|\,k)\end{bmatrix}=\overline{\mathbf{A}}_i^{s}\begin{bmatrix}\mathbf{x}_i(k\,|\,k)\\ \mathbf{x}_{\tilde{i}}(k\,|\,k)\end{bmatrix}+\sum_{l=1}^{s}\overline{\mathbf{A}}_i^{s-l}\overline{\mathbf{B}}_i\mathbf{u}^{\mathrm{p}}_i(k+l-1\,|\,k)$$

$$+\sum_{l=1}^{s}\overline{\mathbf{A}}_i^{s-l}\overline{\mathbf{B}}_{\bar{i}}\hat{\mathbf{u}}_{\bar{i}}(k+l-1\,|\,k)$$

$$+\sum_{l=1}^{s}\overline{\mathbf{A}}_i^{s-l}\tilde{\mathbf{A}}_i\hat{\mathbf{x}}_{\tilde{i}}(k+l-1\,|\,k) \tag{10.5}$$

where

$$\overline{\mathbf{A}}_i=\begin{bmatrix}\mathbf{A}_{ii} & \mathbf{A}_{i\bar{i}}\\ \mathbf{A}_{\bar{i}i} & \mathbf{A}_{\bar{i}\bar{i}}\end{bmatrix}$$

$$\tilde{\mathbf{A}}_i=\begin{bmatrix}\mathbf{A}_{i\tilde{i}}\\ \mathbf{A}_{\bar{i}\tilde{i}}\end{bmatrix}$$

$$\overline{\mathbf{B}}_i=\begin{bmatrix}\mathbf{B}_{ii}\\ \mathbf{0}^{\Sigma_{j\in\mathcal{P}_{+i}}n_{x_j}\times n_{u_i}}\end{bmatrix}$$

$$\overline{\mathbf{B}}_{\bar{i}}=\begin{bmatrix}\mathbf{0}^{n_{x_i}\times\Sigma_{j\in\mathcal{P}_{+i}}n_{u_j}}\\ \mathbf{B}_{\bar{i}\bar{i}}\end{bmatrix}$$

$$\mathbf{A}_{i\bar{i}}=\begin{bmatrix}\mathbf{A}_{ij_1} & \mathbf{A}_{ij_2} & \cdots & \mathbf{A}_{ij_{m_i}}\end{bmatrix}$$

$$\mathbf{A}_{\bar{i}i}=\begin{bmatrix}\mathbf{A}_{j_1i}\\ \mathbf{A}_{j_2i}\\ \mathrm{M}\\ \mathbf{A}_{j_{m_i}i}\end{bmatrix}$$

$$\mathbf{A}_{\bar{i}\bar{i}}=\begin{bmatrix}\mathbf{A}_{j_1j_1} & \mathbf{A}_{j_1j_2} & \cdots & \mathbf{A}_{j_1j_{m_i}}\\ \mathbf{A}_{j_2j_1} & \mathbf{A}_{j_2j_2} & \cdots & \mathbf{A}_{j_1j_{m_i}}\\ \mathrm{M} & \mathrm{M} & \ddots & \mathrm{M}\\ \mathbf{A}_{j_{m_i}j_1} & \mathbf{A}_{j_aj_1} & \cdots & \mathbf{A}_{j_{m_i}j_{m_i}}\end{bmatrix}$$

$$\mathbf{B}_{\bar{i}\bar{i}}=\text{block-diag}(\mathbf{B}_{j_1},\mathbf{B}_{j_2},\ldots,\mathbf{B}_{j_{m_i}})$$

$$\{j_1,j_2,\ldots,j_{m_i}\}=\mathrm{P}_{\bar{i}}$$

Given $\mathbf{x}_i^{\mathrm{p}}(k\,|\,k)=\mathbf{x}_i(k\,|\,k)$, $\mathbf{x}_{\bar{i},i}^{\mathrm{p}}(k\,|\,k)=\mathbf{x}_{\bar{i}}(k\,|\,k)$, the presumed control sequence for subsystem $\mathcal{S}_i$ is given by

$$\hat{\mathbf{u}}_i(k+s-1\,|\,k)=\hat{\mathbf{u}}_{i,i}(k+s-1\,|\,k) \tag{10.6}$$

and

$$\hat{\mathbf{u}}_{j,i}(k+s-1\,|\,k)=\begin{cases}\mathbf{u}_j^p(k+s-1\,|\,k-1), s=1,2,\ldots,N-1\\ \mathbf{K}_j\mathbf{x}_{j,i}^{\mathrm{p}}(k+N-1\,|\,k-1), s=N\end{cases} \tag{10.7}$$

Set each presumed state sequence $\hat{\mathbf{x}}_{j,i}$ to be the remainder of the sequence predicted at time instant $k-1$, concatenated with the closed-loop response under the state feedback control $\hat{\mathbf{u}}_{j,i}(k+s-1\,|\,k) = \mathbf{K}_j\mathbf{x}^{\mathrm{p}}_{j,i}(k+N-1\,|\,k-1)$, that is

$$\hat{\mathbf{x}}_{j,i}(k+s-1\,|\,k) = \mathbf{x}^{\mathrm{p}}_{j,i}(k+s-1\,|\,k-1), s = 1, 2, \ldots, N \tag{10.8}$$

and $\hat{\mathbf{x}}_{j,i}(k+N\,|\,k) = \mathbf{x}^{\mathrm{p}}_{j,i}(k+N\,|\,k-1)$ is calculated by substituting $\hat{\mathbf{u}}_{j,i}(k+N-1|k)$ into (10.5) based on the presumed state sequence obtained at time $k-1$.

In addition, a decoupled terminal-state constraint will be included in each subsystem-based MPC, which guarantees that the terminal controllers are stabilizing inside a terminal set. To define this terminal state set, we need to make an assumption and establish a technical lemma.

Assumption 10.2 The block-diagonal matrix

$$\mathbf{A}_{\mathrm{d}} = \text{block-diag}\{\mathbf{A}_{\mathrm{d}1}, \mathbf{A}_{\mathrm{d}2}, \ldots, \mathbf{A}_{\mathrm{d}m}\}$$

and the off-diagonal matrix $\mathbf{A}_{\mathrm{o}} = \mathbf{A}_{\mathrm{c}} - \mathbf{A}_{\mathrm{d}}$ satisfy the following inequality:

$$\mathbf{A}_{\mathrm{o}}^{\mathrm{T}}\mathbf{P}\mathbf{A}_{\mathrm{o}} + \mathbf{A}_{o}^{\mathrm{T}}\mathbf{P}\mathbf{A}_{\mathrm{d}} + \mathbf{A}_{d}^{\mathrm{T}}\mathbf{P}\mathbf{A}_{\mathrm{o}} < \hat{\mathbf{Q}}/2 \tag{10.9}$$

where $\hat{\mathbf{Q}} = \mathbf{Q} + \mathbf{K}^{\mathrm{T}}\mathbf{R}\mathbf{K} > 0$.

It, along with Assumption 10.1, helps with the design of the terminal set. This assumption quantifies how much strengthening the coupling among subsystems is sufficient so that the overall system can be stabilized by the proposed DMPC here. This assumption is not necessary, and some systems that do not satisfy this assumption may also be stabilized by the proposed DMPC, and the more relaxing condition is still remaining to be designed in the future work.

Lemma 10.1 *Under Assumptions 10.1 and 10.2, for any positive scalar* c *the set*

$$\Omega(c) = \{\mathbf{x} \in \mathbb{R}^{n_x} : \|\mathbf{x}\|_{\mathbf{P}} \le c\}$$

is a positive invariant region of attraction for the closed-loop system $\mathbf{x}(k+1) = \mathbf{A}_{\mathrm{c}}\mathbf{x}(k)$. *Additionally, there exists a small enough positive scalar* ε *such that* $\Omega(\varepsilon)$ *is in the feasible input set* $\mathcal{U} \in \mathbb{R}^{n_u}$ *for all* $\mathbf{x} \in \Omega(\varepsilon)$.

Proof. Consider the function $V(k) = \|\mathbf{x}(k)\|_{\mathbf{P}}^2$. The time difference of $V(k)$ along the trajectories of the closed-loop system $\mathbf{x}(k+1) = \mathbf{A}_{\mathrm{c}}\mathbf{x}(k)$ can be evaluated as

$$\begin{aligned}\Delta V(k) &= \mathbf{x}^{\mathrm{T}}(k)\mathbf{A}_{\mathrm{c}}^{\mathrm{T}}\mathbf{P}\mathbf{A}_{\mathrm{c}}\mathbf{x}(k) - \mathbf{x}^{\mathrm{T}}(k)\mathbf{P}\mathbf{x}(k)\\ &= \mathbf{x}^{\mathrm{T}}(k)(\mathbf{A}_{\mathrm{d}}^{\mathrm{T}}\mathbf{P}\mathbf{A}_{\mathrm{d}} - \mathbf{P} + \mathbf{A}_{\mathrm{o}}^{\mathrm{T}}\mathbf{P}\mathbf{A}_{\mathrm{o}} + \mathbf{A}_{\mathrm{o}}^{\mathrm{T}}\mathbf{P}\mathbf{A}_{\mathrm{d}} + \mathbf{A}_{\mathrm{d}}^{\mathrm{T}}\mathbf{P}\mathbf{A}_{\mathrm{o}})\mathbf{x}(k)\\ &\le -\mathbf{x}^{\mathrm{T}}(k)\hat{\mathbf{Q}}\mathbf{x}(k) + \frac{1}{2}\mathbf{x}^{\mathrm{T}}(k)\hat{\mathbf{Q}}\mathbf{x}(k)\\ &\le 0\end{aligned} \tag{10.10}$$

which holds for all $\mathbf{x}(k) \in \Omega(c)\backslash\{0\}$. This implies that all trajectories of the closed-loop system that start inside $\Omega(c)$ will remain inside and converge to the origin.

The existence of an $\varepsilon > 0$ such that $\mathbf{Kx} \in \mathcal{U}$ for all $\mathbf{x} \in \Omega(\varepsilon)$ follows from the fact that $\mathbf{P}$ is positive definite, which implies that the set $\Omega(\varepsilon)$ shrinks to the origin as ε decreases to zero. This completes the proof.

In the optimization problem of each subsystem-based MPC, the terminal-state constraint set for each $\mathcal{S}_i$ can then be set to be

$$\Omega_i(\varepsilon) = \{\mathbf{x}_i \in \mathbb{R}^{n_{xi}} : \|\mathbf{x}_i\|_{\mathbf{P}_i} \le \varepsilon/\sqrt{m}\} \tag{10.11}$$

Clearly, if $\mathbf{x} \in \Omega_1(\varepsilon) \times \cdots \times \Omega_m(\varepsilon)$, then the decoupled controllers will stabilize the system at the origin, since

$$\|\mathbf{x}_i\|_{P_i}^2 \le \frac{\varepsilon^2}{m}, \forall i \in \mathcal{P}$$

implies that

$$\sum_{i \in \mathcal{P}} \|\mathbf{x}_i\|_{\mathbf{P}_i}^2 \le \varepsilon^2$$

which in turn implies that $\mathbf{x} \in \Omega(\varepsilon)$. Suppose at some time k_0, $\mathbf{x}_i(k_0) \in \Omega_i(\varepsilon)$ for every subsystem. Then, by Lemma 8.1, stabilization can be achieved if every C_i employs its decoupled static feedback controller $\mathbf{K}_i\mathbf{x}_i(k)$ after time instant k_0.

Thus, the objective of each subsystem-based MPC law is to drive the state of each subsystem $\mathcal{S}_i$ to the set $\Omega_i(\varepsilon)$. Once all subsystems have reached these sets, they switch to their decoupled controllers for stabilization. Such switching from an MPC law to a terminal controller once the state reaches a suitable neighborhood of the origin is referred to as the dual mode MPC. For this reason, the DMPC algorithm we propose in this chapter is a dual mode DMPC algorithm.

In what follows, we formulate the optimization problem for each subsystem-based MPC.

Problem 10.1 Consider subsystem $\mathcal{S}_i$. Let $\varepsilon > 0$ be as specified in Lemma 10.1. Let the update time be $k \ge 1$. Given $\mathbf{x}_i(k)$, $\mathbf{x}_{\bar{i}}(k)$ and $\hat{\mathbf{x}}_{\bar{i}}(k+s\,|\,k)$, $s = 1, 2, \ldots, N$, and $\hat{\mathbf{u}}_{\bar{i}}(k+s|k)$, $s = 0, 2, \ldots, N-1$, find the control sequence $\mathbf{u}_i^{\mathrm{P}}(k+s\,|\,k) : \{0, 1, \ldots, N-1\} \to \mathcal{U}_i$ that minimizes

$$\begin{aligned} \bar{J}_i(k) = &\sum_{j \in \mathcal{P}_i} \left\| \mathbf{x}_{j,i}^{p}(k+N\,|\,k) \right\|_{\mathbf{P}_j}^2 \\ &+ \sum_{s=0}^{N-1} \left(\sum_{j \in \mathcal{P}_i} \left\| \mathbf{x}_{j,i}^{p}(k+s\,|\,k) \right\|_{\mathbf{Q}_j}^2 + \left\| \mathbf{u}_i^{p}(k+s\,|\,k) \right\|_{\mathbf{R}_i}^2 \right) \end{aligned} \tag{10.12}$$

subject to the following constraints:

$$\begin{aligned} &\sum_{l=0}^{s} \beta_{s-l} \left\| \mathbf{u}_i^{\mathrm{p}}(k+l\,|\,k) - \hat{\mathbf{u}}_i(k+l\,|\,k) \right\|_2 \\ &\le \frac{(1-\xi)\kappa\varepsilon}{2\sqrt{mm_1}}, s = 1, 2, \ldots, N-1 \end{aligned} \tag{10.13}$$

$$\begin{aligned} &\sum_{l=1}^{s} \alpha_{s-l} \left\| \mathbf{x}_i^{\mathrm{p}}(k+l\,|\,k) - \hat{\mathbf{x}}_i(k+l\,|\,k) \right\|_2 \\ &\le \frac{\xi\kappa\varepsilon}{2\sqrt{mm_2}}, s = 1, 2, \ldots, N-1 \end{aligned} \tag{10.14}$$

$$\left\| \mathbf{x}_i^{\mathrm{p}}(k+N\,|\,k) - \hat{\mathbf{x}}_i(k+N\,|\,k) \right\|_{\mathbf{P}_i} \leq \frac{\kappa\varepsilon}{2\sqrt{m}} \tag{10.15}$$

$$\begin{aligned} &\left\| \mathbf{x}_i^{\mathrm{p}}(k+s\,|\,k) \right\|_{P_i} \\ &\qquad \leq \left\| \mathbf{x}_i^{\mathrm{f}}(k+s\,|\,k) \right\|_{\mathbf{P}_i} + \frac{\varepsilon}{\mu N\sqrt{m}},\ s = 1, 2, \ldots, N \end{aligned} \tag{10.16}$$

$$\mathbf{u}_i^{\mathrm{p}}(k+s\,|\,k) \in \mathcal{U}_i,\ s = 0, 1, \ldots, N-1 \tag{10.17}$$

$$\mathbf{x}_{j,i}^{\mathrm{p}}(k+N\,|\,k) \in \Omega_j(\varepsilon/2),\ j \in \mathcal{P}_i \tag{10.18}$$

In the constraints above,

$$m_1 = \max_{i\in\mathcal{P}}\{\text{number of elements in } \mathcal{P}_{-i}\}, \tag{10.19}$$

$$m_2 = \max_{i\in\mathcal{P}}\{\text{number of elements in } \mathcal{P}_{\tilde{i}}\}, \tag{10.20}$$

$$\alpha_l = \max_{i\in\mathcal{P}} \max_{j\in\mathcal{P}_i} \left\{ \lambda_{\max}^{\frac{1}{2}} \left(\left(\mathbf{L}_{j,i} \overline{\mathbf{A}}_i^{\,l} \tilde{\mathbf{A}}_i \right)^T \mathbf{P}_j \mathbf{L}_{j,i} \overline{\mathbf{A}}_i^{\,l} \tilde{\mathbf{A}}_i \right) \right\} \tag{10.21}$$

$$\beta_l = \max_{i\in\mathcal{P}} \max_{j\in\mathcal{P}_i} \left\{ \lambda_{\max}^{\frac{1}{2}} \left(\left(\mathbf{L}_{j,i} \overline{\mathbf{A}}_i^{\,l} \overline{\mathbf{B}}_{\bar{i}} \right)^T \mathbf{P}_j \mathbf{L}_{j,i} \overline{\mathbf{A}}_i^{\,l} \overline{\mathbf{B}}_{\bar{i}} \right) \right\} \tag{10.22}$$

$$l = 0, 1, \ldots, N-1,$$

where

$$\mathbf{L}_{i,j} = \begin{bmatrix} \mathbf{0}_{n_{xj}\times \sum\limits_{m_{l,i}<m_{j,i}} n_{xl}} & \mathbf{I}_{n_{xj}} & \mathbf{0}_{n_{xj}\times \sum\limits_{m_{h,i}>m_{j,i}} n_{xh}} \end{bmatrix} \tag{10.23}$$

and where $\mathcal{S}_j$, $\mathcal{S}_l$, and $\mathcal{S}_h$ are, respectively, the $(m_{j,i}-1)$th, $(m_{l,i}-1)$th, and $(m_{h,i}-1)$th subsystem in the downstream region of $\mathcal{S}_i$. Finally, the constants $0<\kappa<1$ and $0<\xi\leq 1$ are design parameters whose values will be chosen in the sequel.

Equations (10.13)–(10.15) are referred to as the consistency constraints, which require that each predictive sequence and control variables remain close to their presumed values. These constraints are keys to proving that $\mathbf{x}_{j,i}^{\mathrm{f}}$ is a feasible state sequence at each update.

Equation (10.16) will be used to prove that the N-DMPC algorithm is stabilizing, where $\mu>0$ is a design parameter whose value will be specified later to satisfy (10.43), $\mathbf{x}_i^{\mathrm{f}}(k+s\,|\,k) = \mathbf{x}_{i,i}^{\mathrm{f}}(k+s\,|\,k)$ is a feasible state sequence, and $\mathbf{x}_{j,i}^{\mathrm{f}}(k+s\,|\,k)$ equals to the solution of (10.5) under the initial state of $\mathbf{x}(k)$ and the feasible control sequence $\mathbf{u}_{j,i}^{\mathrm{f}}(k+s-1\,|\,k)$, $j\in\mathcal{P}_i$, is defined by

$$\mathbf{u}_{j,i}^{\mathrm{f}}(k+s-1\,|\,k) = \begin{cases} \mathbf{u}_j^{\mathrm{p}}(k+s-1\,|\,k-1), & s = 1, 2, \ldots, N-1 \\ \mathbf{K}_j \mathbf{x}_{j,i}^{\mathrm{f}}(k+N-1\,|\,k), & s = N \end{cases} \tag{10.24}$$

It should be noticed that the terminal constraint in each optimal control problem is $\Omega_i(\varepsilon/2)$, although Lemma 10.1 ensures that the larger $\Omega(\varepsilon)$ suffices for the feasibility of the terminal controllers. In the analysis presented in the next section, it will be shown that tightening the terminal set in this way is required to guarantee the feasibility properties.

Comparing with the method proposed in Chapter 8, both the optimization index and the consistent constraints are different. In Problem 10.1, the constraints in (10.13) are necessary since the estimation error cannot be expressed by the states sequence. In addition, the terminal constraint should bound both the final states of corresponding subsystem but also that of the subsystems it directly impacted on.

10.3.2 Algorithm Design for Resolving Each Subsystem-based Predictive Control

Before stating the N-DMPC algorithm, we make the following assumption to facilitate the initialization phase.

Assumption 10.3 At initial time k_0, there exists a feasible control $\mathbf{u}_i^{\mathrm{p}}(k_0+s) \in \mathcal{U}_i$, $s = 1, 2, \ldots, N-1$, for each $i \in \mathcal{P}$, such that the solution to the full system $\mathbf{x}(s+1+k_0) = \mathbf{A}\mathbf{x}(s+k_0) + \mathbf{B}\mathbf{u}^{\mathrm{p}}(s+k_0)$, denoted as $\mathbf{x}_i^{\mathrm{p}}(\cdot\,|\,k_0)$, satisfies $\mathbf{x}_i^{\mathrm{p}}(N+k_0) \in \Omega_i(\varepsilon/2)$ and results in a bounded cost $J_i(k_0)$. Moreover, each subsystem has access to $\mathbf{u}_i^{\mathrm{p}}(\cdot\,|\,k_0)$.

Assumption 10.3 bypasses the difficult task of actually constructing an initially feasible solution in a distributed way. In fact, finding an initially feasible solution for many optimization problems is often a primary obstacle, whether or not such problems are used in a control setting. As such, many centralized implementations of MPC also assume that an initially feasible solution is available.

Algorithm 10.1 (Constraint N-DPC Algorithm) The dual mode N-DPC law for any $\mathcal{S}_i$ is constructed as follows:

Step 1: Initialization

- Initialize $\mathbf{x}(k_0)$, $\mathbf{u}_i^{\mathrm{p}}(k_0+s\,|\,k_0)$, $s = 1, 2, \ldots, N$, to satisfy Assumption 10.3
- At time k_0, if $\mathbf{x}(k_0) \in \Omega(\varepsilon)$, then apply the terminal controller $\mathbf{u}_i(k) = \mathbf{K}_i(\mathbf{x}_i(k))$, for all $k \geq k_0$, else
- Compute $\hat{\mathbf{x}}_{j,i}(k_0+s+1\,|\,k_0+1)$ according to (10.5) and transmit $\hat{\mathbf{x}}_i(k_0+s+1\,|\,k_0+1) = \hat{\mathbf{x}}_{i,i}(k_0+s+1\,|\,k_0+1)$ to $\mathcal{S}_a$, $a \subset \mathcal{P}_{+-i} \cup \mathcal{P}_{-i}$ and $\hat{\mathbf{u}}_i(k_0+s|k_0+1)$ to $\mathcal{S}_j, j \in \mathcal{P}_{+i}$

Step 2: Communicating at time $k+1$

- Communication: Measure $\mathbf{x}_i(k)$, transmit $\mathbf{x}_i(k)$ to $\mathcal{S}_{\bar{i}}$, and receive $\hat{\mathbf{x}}_{\bar{i}}(k)$, from $\mathcal{S}_{\bar{i}}$
- Transmit $\hat{\mathbf{x}}_i(k+s+1\,|\,k) = \hat{\mathbf{x}}_{i,i}(k+s+1\,|\,k)$ to $\mathcal{S}_a$, $a \subset \mathcal{P}_{+-i} \cup \mathcal{P}_{-i}$ and $\hat{\mathbf{u}}_i(k+s|k)$ to $\mathcal{S}_j, j \in \mathcal{P}_{+i}$; receive $\hat{\mathbf{x}}_{\bar{i}}(k+s+1\,|\,k)$ and $\hat{\mathbf{u}}_i(k+s|k)$ from $\mathcal{S}_j, j \in \mathcal{P}_{-i}$

Step 3: Update of control law at time $k+1$

- If $\mathbf{x}(k) \in \Omega(\varepsilon)$, then apply the terminal controller $\mathbf{u}_i(k) = \mathbf{K}_i(\mathbf{x}_i(k))$, else
- Solve Problem 10.1 for $\mathbf{u}_i^{\mathrm{p}}(k\,|\,k)$ and apply $\mathbf{u}_i^{\mathrm{p}}(k\,|\,k)$
- Compute $\hat{\mathbf{x}}_{j,i}(k+s+1\,|\,k+1)$ according to (10.5) and

Step 4: Update of control at time $k+1$

- Let $k+1 \rightarrow k$, repeat Step 2.

Algorithm 10.1 presumes that all local controllers $\mathcal{C}_i$, $i \in \mathcal{P}$, have access to the full state $\mathbf{x}(k)$. This requirement results solely from the use of the dual mode control, in which the switching occurs synchronously only when $\mathbf{x}(k) \in \Omega(\varepsilon)$, with $\Omega(\varepsilon)$ being as defined in Lemma 10.1. In the next section, it will be shown that the N-DPC policy drives the state $\mathbf{x}(k+s)$ to $\Omega(\varepsilon)$ in a finite number of updates. As a result, if $\Omega_i(\varepsilon)$ is chosen sufficiently small, then MPC can be employed for all time without switching to a terminal controller, eliminating the need of the local controllers to access the full state. Of course, in this case, instead of asymptotic stability at the origin, we can only drive the state toward the small set $\Omega(\varepsilon)$.

The analysis in the next section shows that the N-DMPC algorithm is feasible at every update and is stabilizing.

10.4 Analysis

10.4.1 Feasibility

The main result of this section is that, provided that an initially feasible solution is available and Assumption 10.3 holds true, for any $\mathcal{S}_i$ and at any time $k \geq 1$, $\mathbf{u}_i^{\mathrm{p}}(\cdot\,|\,k) = \mathbf{u}_i^{\mathrm{f}}(\cdot\,|\,k)$ is a feasible control solution to Problem 10.1. This feasibility result refers that, for any $\mathcal{S}_i$ and at any update $k \geq 1$, the control and state pair $(\mathbf{u}_i^{\mathrm{f}}(\cdot\,|\,k), \mathbf{x}_{j,i}^{\mathrm{f}}(\cdot\,|\,k))$, $j \in \mathcal{P}_i$ satisfy the consistency constraints (10.13)–(10.15), the control constraint (10.17), and the terminal-state constraint (10.18). Lemma 10.2 identifies sufficient conditions that ensure $\hat{\mathbf{x}}_{j,i}(k+N\,|\,k) \in \Omega_i(\varepsilon'/2)$, where $\varepsilon' = (1-\kappa)\varepsilon$. Lemma 10.2 identifies sufficient conditions that ensure $\left\|\mathbf{x}_{j,i}^{\mathrm{f}}(s+k\,|\,k) - \hat{\mathbf{x}}_{j,i}(s+k\,|\,k)\right\|_{\mathbf{P}_j} \leq \kappa\varepsilon/(2\sqrt{m})$ for all $j \in \mathcal{P}_i$, $i \in \mathcal{P}$. Lemma 10.3 establishes that the control constraint is satisfied. Finally, Theorem 10.1 combines the results in Lemmas 10.2–10.4 to arrive at the conclusion that, for any $j \in \mathcal{P}_i$, $i \in \mathcal{P}$, the control and state pair $(\mathbf{u}_{j,i}^{\mathrm{f}}(\cdot\,|\,k), \mathbf{x}_i^{\mathrm{f}}(\cdot\,|\,k))$ are a feasible solution to Problem 8.1 at any update $k \geq 1$.

Lemma 10.2 *Suppose Assumptions 10.1–10.3 hold and $\mathbf{x}(k_0) \in \mathcal{X}$. For any $k \geq 0$, if Problem 10.1 has a solution at time $k-1$ and $\hat{\mathbf{x}}_{j,i}(k+N-1\,|\,k-1) \in \Omega_j(\varepsilon/2)$ for any $j \in \mathcal{P}_i$, $i \in \mathcal{P}$, then*

$$\hat{\mathbf{x}}_{j,i}(k+N-1\,|\,k) \in \Omega_j(\varepsilon/2)$$

and

$$\hat{\mathbf{x}}_{j,i}(k+N\,|\,k) \in \Omega_j(\varepsilon'/2)$$

provided that $\hat{\mathbf{Q}}_j$ and $\mathbf{P_j}$ satisfy

$$\max_{j\in\mathcal{P}}(\rho_j) \leq (1-\kappa)^2 \tag{10.25}$$

where $\varepsilon' = (1-\kappa)\varepsilon$, and

$$\begin{aligned}\rho_j = {} & \lambda_{\max}(\mathbf{A}_{\mathrm{d}j}^{\mathrm{T}}\mathbf{A}_{\mathrm{d}j}) + 2\lambda_{\max}(\mathbf{A}_{\mathrm{d}j}^{\mathrm{T}}\mathbf{A}_{\mathrm{d}j})^{\frac{1}{2}} \sum_{r\in\mathcal{P}_j} \lambda_{\max}\left(\mathbf{P}_r^{-\frac{1}{2}}\mathbf{A}_{\mathrm{d}j}^{\mathrm{T}}\mathbf{P}_j\mathbf{A}_{jr}\mathbf{P}_r^{-\frac{1}{2}}\right)^{\frac{1}{2}} \\ & + \sum_{r\in\mathcal{P}_{+j}}\sum_{q\in\mathcal{P}_{+j}} \left(\lambda_{\max}\left(\mathbf{P}_r^{-\frac{1}{2}}\mathbf{A}_{jr}^{\mathrm{T}}\mathbf{P}_j\mathbf{A}_{jr}\mathbf{P}_r^{-\frac{1}{2}}\right)\right. \\ & \left. \times \lambda_{\max}\left(\mathbf{P}_q^{-\frac{1}{2}}\mathbf{A}_{jq}^{\mathrm{T}}\mathbf{P}_j\mathbf{A}_{jq}\mathbf{P}_q^{-\frac{1}{2}}\right)\right)^{\frac{1}{2}}\end{aligned}$$

Proof. Since Problem 8.1 has a solution at time $k-1$, by construction, it has

$$\left\|\hat{\mathbf{x}}_{j,i}\left(k+N-1\right)|k\right\|_{\mathbf{P}_j}=\left\|\mathbf{x}^{\mathrm{p}}_{j,i}\left(k+N-1\right)|k-1\right\|_{\mathbf{P}_j}$$

$$\leq\frac{\varepsilon}{2\sqrt{m}}$$

Define the presumed state of $\mathcal{S}_r$, $r\in\mathcal{P}_{+j}$, in controller $\mathcal{C}_i$ as

$$\hat{\mathbf{x}}_{r,i}(k+N-1\,|\,k)=\begin{cases}\hat{\mathbf{x}}_r\left(N+k-1\,|\,k-1\right),r\notin\mathcal{P}_i\\ \hat{\mathbf{x}}_{r,i}(k+N-1\,|\,k),r\in\mathcal{P}_i\end{cases}\tag{10.26}$$

and substitute (10.26) and the definition (10.7) into (10.5) we have

$$\begin{aligned}
&\left\|\hat{\mathbf{x}}_{j,i}\left(k+N\,|\,k\right)\right\|^2_{\mathbf{P}_j}\\
&=\left\|\mathbf{A}_{\mathrm{d}j}\hat{\mathbf{x}}_{j,i}\left(k+N-1\,|\,k\right)+\sum_{r\in\mathcal{P}_{+j}}\mathbf{A}_{jr}\hat{\mathbf{x}}_{r,i}(k+N-1\,|\,k)\right\|^2_{\mathbf{P}_j}\\
&=\hat{\mathbf{x}}^{\mathrm{T}}_{j,i}(k+N-1\,|\,k)\mathbf{A}^{\mathrm{T}}_{\mathrm{d}j}\mathbf{P}_j\mathbf{A}_{\mathrm{d}j}\hat{\mathbf{x}}_{j,i}(k+N-1\,|\,k)\\
&+2\hat{\mathbf{x}}^{\mathrm{T}}_{j,i}(k+N-1\,|\,k)\mathbf{A}^{\mathrm{T}}_{\mathrm{d}j}\mathbf{P}_j\sum_{r\in\mathcal{P}_{+j}}\mathbf{A}_{jr}\hat{\mathbf{x}}_{r,i}(k+N-1\,|\,k)\\
&+\left(\sum_{r\in V_j}\hat{\mathbf{x}}^{\mathrm{T}}_{r,i}\left(k+N-1\,|\,k\right)\mathbf{A}^{\mathrm{T}}_{jr}\right)\mathbf{P}_j\sum_{r\in\mathcal{P}_{+j}}\mathbf{A}_{jr}\hat{\mathbf{x}}_{r,i}(k+N-1\,|\,k)\\
&\leq\frac{\varepsilon^2}{4m}(\lambda_{\max}(\mathbf{A}^{\mathrm{T}}_{\mathrm{d}j}\mathbf{A}_{\mathrm{d}j})\\
&+2\lambda_{\max}(\mathbf{A}^{\mathrm{T}}_{\mathrm{d}j}\mathbf{A}_{\mathrm{d}j})^{\frac{1}{2}}\sum_{r\in\mathcal{P}_{+j}}\lambda_{\max}\left(\mathbf{P}_r^{-\frac{1}{2}}\mathbf{A}^{\mathrm{T}}_{\mathrm{d}j}\mathbf{P}_j\mathbf{A}_{jr}P_r^{-\frac{1}{2}}\right)^{\frac{1}{2}}\\
&+\sum_{r\in\mathcal{P}_{+j}}\sum_{q\in\mathcal{P}_{+j}}\lambda_{\max}\left(\mathbf{P}_r^{-\frac{1}{2}}\mathbf{A}^{\mathrm{T}}_{jr}\mathbf{P}_j\mathbf{A}_{jr}\mathbf{P}_r^{-\frac{1}{2}}\right)\lambda_{\max}\left(\mathbf{P}_q^{-\frac{1}{2}}\mathbf{A}^{\mathrm{T}}_{jq}\mathbf{P}_j\mathbf{A}_{jq}\mathbf{P}_q^{-\frac{1}{2}}\right)\Bigg)^{\frac{1}{2}}\\
&=\frac{\varepsilon^2\rho_j}{4m}
\end{aligned}\tag{10.27}$$

Consequently, to ensure the bound $\left\|\hat{\mathbf{x}}_{j,i}\left(k+N\,|\,k\right)\right\|_{\mathbf{P}_j}\leq(1-\kappa)\varepsilon/(2\sqrt{m})$ holds in all controllers $\mathcal{C}_i$, $i\in\mathcal{P}$, a sufficient condition is that

$$\max_{j\in\mathrm{P}}\rho_j\leq(1-\kappa)^2,j\in\mathrm{P}\tag{10.28}$$

This completes the proof.

Lemma 10.3 *Suppose Assumptions 10.1–10.3 hold and* $\mathbf{x}(k_0) \in \mathcal{X}, \forall k \geq 0$, *if Problem 10.1 has a solution at every update time l, l = 1, 2, …, k − 1, then*

$$\left\| \mathbf{x}^{\mathrm{f}}_{j,i}(k+s\,|\,k) - \hat{\mathbf{x}}_{j,i}(k+s\,|\,k) \right\|_{\mathbf{P}_j} \leq \frac{\kappa\varepsilon}{2\sqrt{m}} \tag{10.29}$$

for all $j \in \mathcal{P}_i, i \in \mathcal{P}$ *and all* s = 1, 2, …, N, *provided that (10.28) and the following parametric condition hold:*

$$\frac{\sqrt{m_2}}{\xi\lambda_{\min}(\mathbf{P})} \sum_{l=0}^{N-2} \alpha_l \leq 1 \tag{10.30}$$

where α_l *is as defined in (10.21). Furthermore, the feasible control* $\mathbf{u}^{\mathrm{f}}_i(k+s\,|\,k)$ *and the feasible state* $\mathbf{x}^{\mathrm{f}}_{j,i}(k+s\,|\,k)$ *satisfy constraints (10.13)–(10.15).*

Proof. We will prove (10.29) first. Since a solution exists at update time $1, 2, \ldots, k-1$, according to (10.5), (10.7), and (10.24), for any $s = 1, 2, \ldots, N-1$, the feasible state is given by

$$\begin{aligned}
\mathbf{x}^{\mathrm{f}}_{j,i}(k+s\,|\,k) = {} & \mathbf{L}_{j,i}\overline{\mathbf{A}}_i^{s} \begin{bmatrix} \mathbf{x}_i(k\,|\,k) \\ \mathbf{x}_{\bar{i}}(k\,|\,k) \end{bmatrix} \\
& + \mathbf{L}_{j,i}\sum_{l=1}^{s} \overline{\mathbf{A}}_i^{s-l}\overline{\mathbf{B}}_i\hat{\mathbf{u}}_i(k+l-1\,|\,k) \\
& + \mathbf{L}_{j,i}\sum_{l=1}^{s} \overline{\mathbf{A}}_i^{s-l}\overline{\mathbf{B}}_{\bar{i}}\hat{\mathbf{u}}_{\bar{i}}(k+l-1\,|\,k) \\
& + \mathbf{L}_{j,i}\sum_{l=1}^{l} \overline{\mathbf{A}}_i^{s-l}\tilde{\mathbf{A}}_i\hat{\mathbf{x}}_{\bar{i}}(k+l-1\,|\,k)
\end{aligned} \tag{10.31}$$

and the presumed state is

$$\begin{aligned}
\hat{\mathbf{x}}_{j,i}(k+s\,|\,k) &= \hat{\mathbf{x}}^{\mathrm{p}}_{j,i}(k+s\,|\,k-1) \\
&= \mathbf{L}_{j,i} \begin{bmatrix} \mathbf{x}^{\mathrm{p}}_i(k\,|\,k-1) \\ \mathbf{x}^{p}_{i,i}(k\,|\,k-1) \end{bmatrix} \\
&\quad + \mathbf{L}_{j,i}\sum_{l=1}^{s} \overline{\mathbf{A}}_i^{s-l}\overline{\mathbf{B}}_i\mathbf{u}^{\mathrm{p}}_i(k+l-1\,|\,k-1) \\
&\quad + \mathbf{L}_{j,i}\sum_{l=1}^{s} \overline{\mathbf{A}}_i^{s-l}\overline{\mathbf{B}}_{\bar{i}}\hat{\mathbf{u}}_{\bar{i}}(k+l-1\,|\,k-1) \\
&\quad + \mathbf{L}_{j,i}\sum_{l=1}^{l} \overline{\mathbf{A}}_i^{s-l}\tilde{\mathbf{A}}_i\hat{\mathbf{x}}_{\bar{i}}(k+l-1\,|\,k-1)
\end{aligned} \tag{10.32}$$

where, according to (10.5),

$$\begin{aligned}
\begin{bmatrix} \mathbf{x}^{\mathrm{p}}_i(k\,|\,k-1) \\ \mathbf{x}^{\mathrm{p}}_{i,i}(k\,|\,k-1) \end{bmatrix} &= \overline{A}_i \begin{bmatrix} \mathbf{x}^{\mathrm{p}}_i(k-1\,|\,k-1) \\ \mathbf{x}^{\mathrm{p}}_{i,i}(k-1\,|\,k-1) \end{bmatrix} + \overline{B}_i\mathbf{u}^{\mathrm{p}}_i(k-1\,|\,k-1) \\
&\quad + \overline{B}_{\bar{i}}\hat{\mathbf{u}}_{\bar{i}}(k-1\,|\,k-1) + \tilde{A}_i\hat{\mathbf{x}}_{\bar{i}}(k-1\,|\,k-1) \\
\hat{\mathbf{x}}_i(k-1\,|\,k-1) &= \mathbf{x}_i(k-1)
\end{aligned} \tag{10.33}$$

Consider that $\mathbf{x}_i^{\mathrm{p}}(k-2\,|\,k-2) = \mathbf{x}_i(k-2)$, $\mathbf{x}_{\bar{i}}^{\mathrm{p}}(k-2\,|\,k-2) = \mathbf{x}_{\bar{i}}(k-2)$, and $\mathbf{u}_i^{\mathrm{p}}(k-2\,|\,k-2) = \mathbf{u}_i(k-2)$, the above equation can be rewritten as

$$\begin{bmatrix} \mathbf{x}_i^{\mathrm{p}}(k\,|\,k-1) \\ \mathbf{x}_{i,i}^{\mathrm{p}}(k\,|\,k+1) \end{bmatrix} = \begin{bmatrix} \mathbf{x}_i(k\,|\,k) \\ \mathbf{x}_{\bar{i}}(k\,|\,k) \end{bmatrix} - \overline{\mathbf{B}}_{\bar{i}}\mathbf{u}_i^{\mathrm{p}}(k-1\,|\,k-1) + \overline{\mathbf{B}}_{\bar{i}}\hat{\mathbf{u}}_{\bar{i}}(k-1\,|\,k-1) \tag{10.34}$$

Subtracting (10.32) from (10.31), and substituting (10.34) into the resulting equation, we obtain the discrepancy between the feasible state sequence and the presumed state sequence as

$$\begin{aligned}
&\left\| \mathbf{x}_{j,i}^{\mathrm{f}}(k+s\,|\,k) - \hat{\mathbf{x}}_{j,i}(k+s\,|\,k) \right\|_{\mathbf{P}_j} \\
&\le \left\| \sum_{l=0}^{s} \mathbf{L}_{j,i}\overline{\mathbf{A}}_i^{s-l}\overline{\mathbf{B}}_{\bar{i}}\left(\mathbf{u}_i^{\mathrm{p}}(k+l-1\,|\,k-1) - \hat{\mathbf{u}}_{\bar{i}}(k+l-1\,|\,k-1)\right) \right\|_{\mathbf{P}_j} \\
&\quad + \left\| \sum_{l=1}^{s} \mathbf{L}_{j,i}\overline{\mathbf{A}}_i^{s-l}\tilde{\mathbf{A}}_i\left(\mathbf{x}_{\bar{i}}^{\mathrm{p}}(k+l-1\,|\,k-1) - \hat{\mathbf{x}}_{\bar{i}}(k+l-1\,|\,k-1)\right) \right\|_{\mathbf{P}_j} \\
&\le \sum_{l=0}^{s} \left\| \mathbf{L}_{j,i}\overline{\mathbf{A}}_i^{s-l}\overline{\mathbf{B}}_{\bar{i}}\left(\mathbf{u}_i^{\mathrm{p}}(k+l-1\,|\,k-1) - \hat{\mathbf{u}}_{\bar{i}}(k+l-1\,|\,k-1)\right) \right\|_{\mathbf{P}_j} \\
&\quad + \sum_{l=1}^{s} \left\| \mathbf{L}_{j,i}\overline{\mathbf{A}}_i^{s-l}\tilde{\mathbf{A}}_i\left(\mathbf{x}_{\bar{i}}^{\mathrm{p}}(k+l-1\,|\,k-1) - \hat{\mathbf{x}}_{\bar{i}}(k+l-1\,|\,k-1)\right) \right\|_{\mathbf{P}_j} \\
&\le \sum_{l=0}^{s} \beta_{s-l}\left\| \mathbf{u}_i^{\mathrm{p}}(k+l-1\,|\,k-1) - \hat{\mathbf{u}}_{\bar{i}}(k+l-1\,|\,k-1) \right\|_2 \\
&\quad + \sum_{l=1}^{s} \alpha_{s-l}\left\| \mathbf{x}_{\bar{i}}^{\mathrm{p}}(k+l-1\,|\,k-1) - \hat{\mathbf{x}}_{\bar{i}}(k+l-1\,|\,k-1) \right\|_2
\end{aligned} \tag{10.35}$$

Let the subsystems, which respectively maximize the following two functions, be $\mathcal{S}_g$ and $\mathcal{S}_h$

$$\sum_{l=0}^{s} \beta_{s-l}\left\| \mathbf{u}_i^{\mathrm{p}}(k-1+l\,|\,k-1) - \hat{\mathbf{u}}_i(k-1+l\,|\,k-1) \right\|_2, i \in \mathrm{P},$$

$$\sum_{l=1}^{s} \alpha_{s-l}\left\| \mathbf{x}_i^{\mathrm{p}}(k-1+l\,|\,k-1) - \hat{\mathbf{x}}_i(k-1+l\,|\,k-1) \right\|_2, i \in \mathrm{P}$$

Then, the following equation can be deduced from (10.35):

$$\begin{aligned}
&\left\| \mathbf{x}_{j,i}^{\mathrm{f}}(k+s\,|\,k) - \hat{\mathbf{x}}_{j,i}(k+s\,|\,k) \right\|_{\mathbf{P}_j} \\
&\le m_1^{\frac{1}{2}} \sum_{l=1}^{s} \beta_{s-l}\left\| \mathbf{u}_g^{\mathrm{p}}(k+l-1\,|\,k-1) - \hat{\mathbf{u}}_g(k+l-1\,|\,k-1) \right\|_2 \\
&\quad + m_2^{\frac{1}{2}} \sum_{l=1}^{s} \alpha_{s-l}\left\| \mathbf{x}_h^{p}(k+l-1\,|\,k-1) - \hat{\mathbf{x}}_h(k+l-1\,|\,k-1) \right\|_2
\end{aligned}$$

Since $\mathbf{x}_i^{\mathrm{p}}(\cdot)$ and $\mathbf{u}_i^{\mathrm{p}}(\cdot)$ satisfy constraints (10.13) and (10.14) for all times $l = 1, 2, \ldots, k-1$, the following equation can be deduced:

$$\begin{aligned}
&\left\|\mathbf{x}^{\mathrm{f}}_{j,i}(k+s\,|\,k) - \hat{\mathbf{x}}_{j,i}(k+s\,|\,k)\right\|_{\mathbf{P}_j} \\
&\le \frac{(1-\xi)(1-\kappa)\varepsilon}{2\sqrt{m}} + \frac{\xi(1-\kappa)\varepsilon}{2\sqrt{m}} \\
&= \frac{\kappa\varepsilon}{2\sqrt{m}}
\end{aligned} \tag{10.36}$$

Thus, (10.29) holds for all $s = 1, 2, \ldots, N-1$.

In what follows, we prove that (10.29) holds for $s = N$. Denote the feasible states of $\mathcal{S}_r, r \in \mathcal{P}_{+j}$, used in controller $\mathcal{S}_i$, as

$$\mathbf{x}^{\mathrm{f}}_{r,i}(k+N-1\,|\,k) = \begin{cases} \mathbf{x}^{\mathrm{f}}_r(N+k-1\,|\,k-1), r \notin \mathcal{P}_{-i} \\ \mathbf{x}^{\mathrm{f}}_{r,i}(k+N-1\,|\,k), r \in \mathcal{P}_{-i} \end{cases}$$

Then, the discrepancy between the feasible state $\mathbf{x}^{\mathrm{f}}_{j,i}(k+N\,|\,k)$ and the presumed state $\hat{\mathbf{x}}_{j,i}(k+N\,|\,k)$ is

$$\begin{aligned}
&\left\|\mathbf{x}^{\mathrm{f}}_{j,i}(k+N\,|\,k) - \hat{\mathbf{x}}_{j,i}(k+N\,|\,k)\right\|_{\mathbf{P}_j} \\
&= \|\mathbf{A}_{\mathrm{d}j}\left(\mathbf{x}^{\mathrm{f}}_{j,i}(k+N-1\,|\,k) - \hat{\mathbf{x}}_{j,i}(k+N-\,|\,k)\right) \\
&\quad + A_{j\check{j}}\left(\mathbf{x}^{f}_{\check{j},i}(k+N-1\,|\,k) - \hat{\mathbf{x}}_{\check{j},i}(k+N-1\,|\,k)\right)\|_{\mathbf{P}_j}
\end{aligned} \tag{10.37}$$

Now consider

$$\left\|\mathbf{x}^{\mathrm{f}}_{j,i}(k+N-1\,|\,k) - \hat{\mathbf{x}}_{j,i}(k+N-1\,|\,k)\right\|_{\mathbf{P}_j} \le \frac{\kappa\varepsilon}{2\sqrt{m}}$$

and the constraint

$$\left\|\mathbf{x}^{\mathrm{p}}_{j}(k+N-1\,|\,k-1) - \hat{\mathbf{x}}_{j}(k+N-1\,|\,k-1)\right\|_{\mathbf{P}_j} \le \frac{\kappa\varepsilon}{2\sqrt{m}}, \ \forall j \in \mathcal{P}$$

Then, in view of (10.25), we have

$$\begin{aligned}
&\left\|\mathbf{x}^{f}_{j,i}(k+N\,|\,k) - \hat{\mathbf{x}}_{j,i}(k+N\,|\,k)\right\|^2_{\mathbf{P}_j} \\
&\le \rho_j \kappa^2 \frac{\varepsilon^2}{4m} \\
&\le \ \kappa^2 \frac{\varepsilon^2}{4m}
\end{aligned} \tag{10.38}$$

This completes the proof of (10.29).

In what follows, we will prove that the feasible control $\mathbf{u}^{\mathrm{f}}_i(k+s\,|\,k)$ and the feasible state $\mathbf{x}^{\mathrm{f}}_{j,i}(k+s\,|\,k)$ satisfy constraints (10.13)–(8.11).

First, for any $s = 1, 2, \ldots, N-1$, $\mathbf{u}_i^{\mathrm{f}}(k+l-1\,|\,k) = \hat{\mathbf{u}}_i(k+l-1\,|\,k)$. Thus, constraint (10.13) is satisfied.

Also,

$$\begin{aligned}
&\sum_{l=1}^{s} \alpha_{s-l} \left\| \mathbf{x}_i^{\mathrm{f}}(k+l\,|\,k) - \hat{\mathbf{x}}_i(k+l\,|\,k) \right\|_2 \\
&\le \frac{1}{\lambda_{\min}(\mathbf{P}_i)} \sum_{l=1}^{s} \alpha_{s-l} \left\| \mathbf{x}_i^{\mathrm{f}}(k+l\,|\,k) - \hat{\mathbf{x}}_i(k+l\,|\,k) \right\|_{\mathbf{P}_i} \\
&\le \frac{1}{\lambda_{\min}(\mathbf{P})} \sum_{l=1}^{s} \alpha_{s-l} \frac{\sqrt{m_2}}{\xi} \frac{\xi \kappa \varepsilon}{2\sqrt{m m_2}}
\end{aligned} \tag{10.39}$$

Thus, when

$$\frac{\sqrt{m_2}}{\xi \lambda_{\min}(P)} \sum_{l=1}^{s} \alpha_{s-l} \le 1$$

state $\mathbf{x}_i^{\mathrm{f}}(k+s\,|\,k)$, $s = 1, 2, \ldots, N-1$, satisfy constraint (10.14).

Finally,

$$\begin{aligned}
&\left\| \mathbf{x}_i^{\mathrm{f}}(k+N\,|\,k) - \hat{\mathbf{x}}_i(k+N\,|\,k) \right\|_{\mathbf{P}_i} \\
&= \left\| \mathbf{x}_{i,i}^{\mathrm{f}}(k+N\,|\,k) - \hat{\mathbf{x}}_{i,i}(k+N\,|\,k) \right\|_{\mathbf{P}_i} \\
&\le \frac{\kappa \varepsilon}{2\sqrt{m}}
\end{aligned} \tag{10.40}$$

which shows that constraint (10.15) is satisfied. This concludes the proof.

In what follows we establish that, at time k, if conditions (8.18) and (8.20) are satisfied, then $\mathbf{x}_{j,i}^{\mathrm{f}}(k+s\,|\,k)$ and $\mathbf{u}_i^{\mathrm{f}}(k+s\,|\,k)$, $s = 1, 2, \ldots, N$, are a feasible solution of Problem 8.1.

Lemma 10.4 *Suppose Assumptions 10.1–10.3 hold, $\mathbf{x}(k_0) \in \mathbb{R}^{n_x}$, and conditions (10.30) and (10.25) are satisfied. For any $k \ge 0$, if Problem 10.1 has a solution at every update time $l, l = 1, 2, \ldots, k-1$, then $\mathbf{u}_i^{\mathrm{f}}(k+s\,|\,k) \in \mathcal{U}$, for all $s = 1, 2, \ldots, N-1$.*

Proof. Since Problem 10.1 has a feasible solution at $l = 1, 2, \ldots, k-1$, and $\mathbf{u}_i^{\mathrm{f}}(k+s-1\,|\,k) = \mathbf{u}_i^{\mathrm{p}}(k+s-1\,|\,k-1)$ for all $s = 1, 2, \ldots, N-1$, we only need to show that $\mathbf{u}_i^{\mathrm{f}}(k+N-1\,|\,k) \in \mathcal{U}$.

Since ε has been chosen to satisfy the conditions of Lemma 10.1, $\mathbf{K}_i\mathbf{x}_i \in \mathcal{U}$ for all $i \in \mathcal{P}$ when $\mathbf{x} \in \Omega(\varepsilon)$. Consequently, a sufficient condition for $\mathbf{u}_i^{\mathrm{f}}(k+N-1\,|\,k) \in \mathcal{U}$ is that $\mathbf{x}_i^{\mathrm{f}}(k+N-1\,|\,k) \in \Omega(\varepsilon)$.

In view of Lemmas 10.2 and 10.3, using the triangle inequality, we have

$$\begin{aligned}
\left\| \mathbf{x}_i^{\mathrm{f}}(k+N-1\,|\,k) \right\|_{\mathbf{P}_i} &\le \left\| \mathbf{x}_i^{\mathrm{f}}(k+N-1\,|\,k) - \hat{\mathbf{x}}_i(k+N-1\,|\,k) \right\|_{\mathbf{P}_i} \\
&\quad + \left\| \hat{\mathbf{x}}_i(k+N-1\,|\,k) \right\|_{\mathbf{P}_i} \\
&\le \frac{\varepsilon}{2(q+1)\sqrt{m}} + \frac{\varepsilon}{2\sqrt{m}}
\end{aligned} \tag{10.41}$$

$$\leq \frac{\varepsilon}{\sqrt{m}}$$

that is, $\mathbf{x}_i^{\mathrm{f}}(k+N\,|\,k) \in \Omega_i(\varepsilon)$. This concludes the proof.

Theorem 10.1 *Suppose Assumptions 10.1–10.3 hold,* $\mathbf{x}(k_0) \in \mathbb{R}^{n_x}$ *and constraints (10.13)–(10.17) and (10.18) are satisfied at* k_0*. Then, for every* $i \in \mathcal{P}$*, the control* $\mathbf{u}_i^{\mathrm{f}}(\cdot\,|\,k)$ *and state* $\mathbf{x}_{j,i}^{\mathrm{f}}(\cdot\,|\,k)$*, respectively, defined by (10.24) and (10.5), are a feasible solution of Problem 10.1 at every update* k*.*

Proof. We will prove the theorem by induction.

First, consider the case of $k=1$. The state sequence $\mathbf{x}_{j,i}^{\mathrm{p}}(\cdot\,|\,1) = \mathbf{x}_{j,i}^{\mathrm{f}}(\cdot\,|\,1)$ trivially satisfies the dynamic equation (10.5), the stability constraint (10.16), and the consistency constraints (10.13)–(10.15).

Observe that

$$\hat{\mathbf{x}}_{j,i}(1\,|\,1) = \mathbf{x}_{j,i}^{\mathrm{p}}(1\,|\,0) = \mathbf{x}_{j,i}^{\mathrm{f}}(1\,|\,1) = \mathbf{x}_{j,i}(1), i \in \mathcal{P}$$

and that

$$\mathbf{x}_{j,i}^{\mathrm{f}}(1+s\,|\,1) = \mathbf{x}_{j,i}^{\mathrm{p}}(1+s\,|\,0)$$
$$s = 1, 2, \ldots, N-1$$

Thus, $\mathbf{x}_{j,i}^{\mathrm{f}}(N\,|\,1) \in \Omega_i(\varepsilon/2)$. By the invariance of $\Omega(\varepsilon)$ under the terminal controller and the conditions in Lemma 10.1, it follows that the terminal state and control constraints are also satisfied. This completes the proof of the case of $k=1$.

Now suppose $\mathbf{u}_i^{\mathrm{p}}(\cdot\,|\,l) = \mathbf{u}_i^{\mathrm{f}}(\cdot\,|\,l)$ is a feasible solution for $l = 1, 2, \ldots, k-1$. We will show that $\mathbf{u}_i^{\mathrm{f}}(\cdot\,|\,k)$ is a feasible solution at update k.

As before, the consistency constraint (10.13) is trivially satisfied, and $\mathbf{u}_{j,i}^{\mathrm{f}}(\cdot\,|\,k)$ is the corresponding state sequence that satisfies the dynamic equation. Since there is a solution for Problem 10.1 at updates $l = 1, 2, \ldots, k-1$, Lemmas 10.2–10.4 can be invoked. Lemma 10.4 guarantees control constraint feasibility. In view of Lemmas 10.2 and 10.3, using the triangle inequality, we have

$$\begin{aligned}\left\|\mathbf{x}_{j,i}^{\mathrm{f}}(k+N\,|\,k)\right\|_{P_i} &\leq \left\|\mathbf{x}_{j,i}^{\mathrm{f}}(k+N\,|\,k) - \hat{\mathbf{x}}_{j,i}(k+N\,|\,k)\right\|_{\mathbf{P}_i} \\ &\quad + \left\|\hat{\mathbf{x}}_{j,i}(k+N\,|\,k)\right\|_{\mathbf{P}_i} \\ &\leq \frac{\kappa\varepsilon}{2\sqrt{m}} + \frac{(1-\kappa)\varepsilon}{2\sqrt{m}} = \frac{\varepsilon}{2\sqrt{m}}\end{aligned} \tag{10.42}$$

for each $j \in \mathcal{P}_i$, $i \in \mathcal{P}$. This shows that the terminal-state constraint is satisfied and the proof of Theorem 10.1 is complete.

10.4.2 Stability

The stability of the closed-loop system is analyzed in this subsection.

Theorem 10.2 *Suppose Assumptions 10.1–10.3 hold,* $\mathbf{x}(k_0) \in \mathbb{R}^{n_x}$*, constraints (10.13)–(10.15) and (10.16) are satisfied, and the following parametric condition holds:*

$$\frac{(N-1)\kappa}{2} - \frac{1}{2} + \frac{1}{\mu} < 0 \tag{10.43}$$

Then, by application of Algorithm 8.1, the closed-loop system of (8.2) is asymptotically stable at the origin.

Proof. By Algorithm 10.1 and Lemma 10.1, when $\mathbf{x}(k)$ enters $\Omega(\varepsilon)$, the terminal controllers take over to keep it in there and stabilize the system at the origin. Therefore, it remains to show that if $\mathbf{x}(0) \in \mathcal{X}\backslash\Omega(\varepsilon)$, then by the application of Algorithm 10.1, the state of system (10.2) is driven to the set $\Omega(\varepsilon)$ in finite time.

Define the nonnegative function for $\mathcal{S}$

$$V(k) = \sum_{s=1}^{N} \|\mathbf{x}^{\mathrm{p}}(k+s\,|\,k)\|_{\mathbf{P}} \tag{10.44}$$

In what follows, we will show that, for any $k \geq 0$, if $\mathbf{x}(k) \in \mathcal{X}\backslash\Omega(\varepsilon)$, then there exists a constant $\eta \in (0,\infty)$ such that $V(k) \leq V(k-1) - \eta$. Constraint (10.16) implies that

$$\|\mathbf{x}^{\mathrm{p}}(k+s\,|\,k)\|_{\mathbf{P}} \leq \left\|\mathbf{x}^{\mathrm{f}}(k+s\,|\,k)\right\|_{\mathbf{P}} + \frac{\varepsilon}{\mu N} \tag{10.45}$$

Therefore,

$$V(k) \leq \sum_{s=1}^{N} \left\|\mathbf{x}^{\mathrm{f}}(k+s\,|\,k)\right\|_{\mathbf{P}} + \frac{\varepsilon}{\mu}$$

Subtracting $V(k-1)$ from $V(k)$ and using $\mathbf{x}^{\mathrm{p}}(k+s\,|\,k-1) = \hat{\mathbf{x}}(k+s\,|\,k)$, $s=1,2,\ldots,N-1$, gives

$$\begin{aligned} V(k) - V(k-1) \leq & -\|\mathbf{x}^{\mathrm{p}}(k\,|\,k-1)\|_{P} + \frac{\varepsilon}{\mu} + \left\|\mathbf{x}^{\mathrm{f}}(k+N\,|\,k)\right\|_{\mathbf{P}} \\ & + \sum_{s=1}^{N-1}\left(\left\|\mathbf{x}^{\mathrm{f}}(k+s\,|\,k)\right\|_{\mathbf{P}} - \|\hat{\mathbf{x}}(k+s\,|\,k)\|_{\mathbf{P}}\right) \end{aligned} \tag{10.46}$$

Assuming $\mathbf{x}(k) \in \mathcal{X}\backslash\Omega(\varepsilon)$ yields

$$\|\mathbf{x}^{\mathrm{p}}(k\,|\,k-1)\|_{\mathbf{P}} > \varepsilon \tag{10.47}$$

Also, by Theorem 10.1 we have

$$\left\|\mathbf{x}^{\mathrm{f}}(k+N\,|\,k)\right\|_{\mathbf{P}} \leq \varepsilon/2 \tag{10.48}$$

and by Lemma 10.3, we have

$$\sum_{s=1}^{N-1}\left(\left\|\mathbf{x}^{\mathrm{f}}(k+s\,|\,k)\right\|_{\mathbf{P}} - \|\hat{\mathbf{x}}(k+s\,|\,k)\|_{\mathbf{P}}\right) \leq \frac{(N-1)\kappa\varepsilon}{2} \tag{10.49}$$

Using (10.47)–(10.49) in (10.46) then yields

$$V(k) - V(k-1) < \varepsilon\left(-1 + \frac{(N-1)\kappa}{2} + \frac{1}{2} + \frac{1}{\mu}\right) \tag{10.50}$$

which, in view of (10.43), implies that $V(k) - V(k-1) < 0$. Thus, for any $k \geq 0$, if $\mathbf{x}(k) \in \mathcal{X}\backslash\Omega(\varepsilon)$, there is a constant $\eta \in (0, \infty)$ such that $V(k) \leq V(k-1) - \eta$. It then follows that there exists a finite time k' such that $\mathbf{x}(k') \in \Omega(\varepsilon)$. This concludes the proof.

We have now established the feasibility of the N-DMPC and the stability of the resulting closed-loop system. That is, if an initially feasible solution could be found, subsequent feasibility of the algorithm is guaranteed at every update, and the resulting closed-loop system is asymptotically stable at the origin.

It should be noticed that, with the increasing number of the downstream neighbors each subsystem-based controller covers, the cost consumed by communication among subsystems will become higher and higher, and the network connectivity of the entire system will become more and more complex. When the time consumed by communicating becomes so large that it cannot be ignored comparing to the control period, the performance of the entire system will be more or less negatively affected. And the increasing network connectivity will inevitably violate the error-tolerance capability of the entire control system. This is undesired in the distributed control framework. Thus the number of the downstream neighbors of each subsystem should not be too large when the network bandwidth is limited or not large enough.

It should also be noticed that a general mathematical formulation is adopted in the N-DMPC algorithm and its analysis. The N-DMPC and the resulting analysis can be used for any coordination policy mentioned in Section 10.2 with a redefinition of $\mathcal{P}_i$. Thus it provides a unified framework for the DMPCs, which adopts the cost function based coordination strategies. This is a very important contribution of the chapter. In the next section, we will present the formulations under other coordination strategies.

10.5 Formulations Under Other Coordination Strategies

10.5.1 Local Cost Optimization Based DMPC

In this coordination strategy, each subsystem-based MPC minimizes its own cost. Redefine $\mathcal{P}_i = \{i\}$. Note that $\mathbf{u}_{j,i}(\cdot)$ and $\mathbf{x}_{j,i}(\cdot)$ are both nonexistent, that is, $\mathcal{P}_{-i}$ is an empty set, and thus $\mathcal{P}_{\bar{i}} = \mathcal{P}_{+i}$. Consequently the optimization problem of the stabilizing DMPC, where each subsystem-based MPC takes the local cost as the performance index, can be derived in the framework of N-DMPC as

$$\begin{aligned}\bar{J}_i(k) = J_i(k) = &\left\|\mathbf{x}_i^{\mathrm{p}}(k+N\,|\,k)\right\|_{\mathbf{P}_i}^2 \\ &+ \sum_{s=1}^{N-1}\left(\left\|\mathbf{x}_i^{\mathrm{p}}(k+s\,|\,k)\right\|_{\mathbf{Q}_i}^2 + \left\|\mathbf{u}_i^{\mathrm{p}}(k+s-1\,|\,k)\right\|_{\mathbf{R}_i}^2\right)\end{aligned} \tag{10.51}$$

subject to the constraints

$$\mathbf{x}_i^{\mathrm{p}}(k+s\,|\,k) = \mathbf{A}_i^s\mathbf{x}_i(k\,|\,k) + \sum_{l=1}^{s}\mathbf{A}_i^{s-l}\mathbf{u}_i^{p}(k+l-1\,|\,k)$$

$$+\sum_{l=1}^{s} \tilde{A}_i^{s-l}\hat{\mathbf{x}}_{\tilde{i}}(k+l-1\,|\,k), s=0,1,2,\dots,N-1$$

$$\sum_{l=1}^{s} \alpha_{s-l}\left\|\mathbf{x}_i^{\mathrm{p}}(k+l\,|\,k)-\hat{\mathbf{x}}_i(k+l\,|\,k)\right\|_2 \le \frac{\xi\kappa\varepsilon}{2\sqrt{mm_2}}, s=1,2,\dots,N-1$$

$$\left\|\mathbf{x}_i^{\mathrm{p}}(k+N\,|\,k)-\hat{\mathbf{x}}_i(k+N\,|\,k)\right\|_{P_i} \le \frac{\kappa\varepsilon}{2\sqrt{m}}$$

$$\left\|\mathbf{x}_i^{\mathrm{p}}(k+s\,|\,k)\right\|_{\mathbf{P}_i} \le \left\|\mathbf{x}_i^{\mathrm{f}}(k+s\,|\,k)\right\|_{\mathbf{P}_i} + \frac{\varepsilon}{\mu N\sqrt{m}}, s=1,2,\dots,N$$

$$\mathbf{u}_i^p(k+s-1\,|\,k) \in \mathrm{U}_i, s=1,2,\dots,N$$

$$\mathbf{x}_i^{\mathrm{p}}(k+N\,|\,k) \in \Omega_i(\varepsilon/2)$$

It should be noted that the consistency constraints in inputs (10.13) do not appear here. This is because there is no input, except for $\mathbf{u}_i(\cdot)$, that appears in the predictive model. This result is consistent with the linear version of what is presented in Ref. [51], which provides a local performance index based DMPC for continuous nonlinear systems.

10.5.2 Cooperative DMPC

In this coordination strategy, each subsystem-based MPC calculates the optimal $\mathbf{u}_i(k)$ to minimize the cost function of the entire system. The predictive model of each subsystem-based MPC includes the state evolutions of all subsystems, and can be deduced as follows according to (10.1):

$$\begin{aligned}\mathbf{x}_{j,i}^{\mathrm{p}}(k+s\,|\,k) &= \mathbf{L}_{j,i}\mathbf{A}^s + \sum_{l=1}^{s}\mathbf{A}^{s-l}\mathbf{u}_i^{\mathrm{p}}(k+l-1\,|\,k)\\ &\quad + \sum_{l=1}^{s}\mathbf{A}^{s-l}\hat{\mathbf{u}}_{j,i}(k+l-1\,|\,k), s=1,2,\dots,N, j\in \mathrm{P}\end{aligned} \tag{10.52}$$

Also, $\mathbf{x}_{\tilde{i}}(\cdot)$ is nonexistent, $m_1 = m-1$, and $\mathcal{P}_i = \mathcal{P}$. Consequently, the optimization problem of the DMPC, where the global cost is minimized at each subsystem-based MPC, can be deduced from the N-DMPC as follows:

$$\bar{J}_i(k) = J_i(k) = \sum_{j\in V}\left\|\mathbf{x}_{j,i}^{p}(k+N\,|\,k)\right\|_{P_j}^2 + \sum_{s=1}^{N-1}\left(\sum_{j\in V}\left\|\mathbf{x}_{j,i}^{p}(k+s\,|\,k)\right\|_{Q_j}^2 + \left\|\mathbf{u}_i^{p}(k+s-1\,|\,k)\right\|_{R_i}^2\right)$$

subject to the constraints (10.52)

$$\sum_{l=0}^{s} \beta_{s-l} \left\| \mathbf{u}_i^{\mathrm{p}}(k+l\,|\,k) - \hat{\mathbf{u}}_i(k+l\,|\,k) \right\|_2 \le \frac{(1-\xi)\kappa\varepsilon}{2\sqrt{m(m-1)}}, s = 1, 2, \ldots, N-1,$$

$$\|\mathbf{x}_i^{\mathrm{p}}(k+s\,|\,k)\|_{P_i} \le \|\mathbf{x}_i^{\mathrm{f}}(k+s\,|\,k)\|_{P_i} + \frac{\varepsilon}{\mu N\sqrt{m}}, s = 1, 2, \ldots, N,$$

$$\mathbf{u}_i^{\mathrm{p}}(k+s-1\,|\,k) \in \mathrm{U}_i, s = 1, 2, \ldots, N,$$

$$\mathbf{x}_{j,i}^{\mathrm{p}}(k+N\,|\,k) \in \Omega_j(\varepsilon/2), j \in \mathrm{L}$$

It can be seen that the optimization problem with C-DMPC is much simpler than Problem 10.1. The constraints (10.13) in Problem 8.1 have not appeared. It is because that the assumed state sequences of other subsystems are not used in the predictive model of each subsystem-based MPC.

10.6 Simulation Results

10.6.1 The System

A distributed system consisting of four interacted subsystems is used to demonstrate the effectiveness of the proposed method. The relationship among these four subsystems is shown in Figure 10.1, where $\mathcal{S}_1$ is impacted by $\mathcal{S}_2$, $\mathcal{S}_3$ is impacted by $\mathcal{S}_1$ and $\mathcal{S}_2$, and $\mathcal{S}_4$ is impacted by $\mathcal{S}_3$. Let $\Delta\mathcal{U}_i$ be defined to reflect both the constraint on the input $u_i \in \left[u_i^{\min}\ u_i^{\max}\right]$ and the constraint on the increment of the input $\Delta u_i \in \left[\Delta u_i^{\min}\ \Delta u_i^{\max}\right]$.

The models of these four subsystems are respectively given by

$$\begin{aligned} \mathcal{S}_1 &: x_1(k+1) = 0.62x_1(k) + 0.34u_1(k) - 0.12x_2(k) \\ \mathcal{S}_2 &: x_2(k+1) = 0.58x_2(k) + 0.33u_2(k) \\ \mathcal{S}_3 &: x_3(k+1) = 0.60x_3(k) + 0.34u_3(k) + 0.11x_1(k) - 0.07x_2(k) \\ \mathcal{S}_4 &: x_4(k+1) = 0.65x_4(k) + 0.35u_4(k) + 0.13x_3(k) \end{aligned} \tag{10.53}$$

For the purpose of comparison, both the centralized MPC and the N-DMPC are applied to this system.

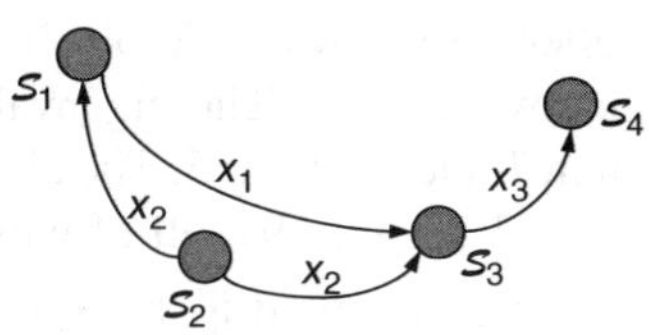

Figure 10.1 The interaction relationship among subsystems

10.6.2 Performance of Closed-loop System under the N-DMPC

According to the interactive relationship of these subsystems and the cost function of (10.3), the corresponding predictive model for each MPC is, respectively, given as follows:

- Predictive model in the subsystem-based MPC for $\mathcal{S}_1$

$$\begin{bmatrix} x_1(k+1) \\ x_3(k+1) \end{bmatrix} = \begin{bmatrix} 0.62 & 0 \\ 0.11 & 0.60 \end{bmatrix} \begin{bmatrix} x_1(k) \\ x_3(k) \end{bmatrix} + \begin{bmatrix} 0.34 \\ 0 \end{bmatrix} u_1(k) + \begin{bmatrix} -0.12 \\ -0.06 \end{bmatrix} \hat{x}_2(k) + \begin{bmatrix} 0 \\ 0.34 \end{bmatrix} \hat{u}_3(k) \tag{10.54}$$

- Predictive model in the subsystem-based MPC for $\mathcal{S}_2$

$$\begin{bmatrix} x_2(k+1) \\ x_1(k+1) \\ x_3(k+1) \end{bmatrix} = \begin{bmatrix} 0.58 & 0 & 0 \\ -0.12 & 0.62 & 0 \\ -0.07 & 0.11 & 0.60 \end{bmatrix} \begin{bmatrix} x_2(k) \\ x_1(k) \\ x_3(k) \end{bmatrix} + \begin{bmatrix} 0.33 \\ 0 \\ 0 \end{bmatrix} u_2(k) + \begin{bmatrix} 0 & 0 \\ 0.34 & 0 \\ 0 & 0.34 \end{bmatrix} \begin{bmatrix} \hat{u}_1(k) \\ \hat{u}_3(k) \end{bmatrix} \tag{10.55}$$

- Predictive model in the subsystem-based MPC for $\mathcal{S}_3$

$$\begin{bmatrix} x_3(k+1) \\ x_4(k+1) \end{bmatrix} = \begin{bmatrix} 0.60 & 0 \\ 0.13 & 0.65 \end{bmatrix} \begin{bmatrix} x_3(k) \\ x_4(k) \end{bmatrix} + \begin{bmatrix} 0.34 \\ 0 \end{bmatrix} u_3(k) + \begin{bmatrix} 0.11 & -0.07 \\ 0 & 0 \end{bmatrix} \begin{bmatrix} \hat{x}_1(k) \\ \hat{x}_2(k) \end{bmatrix} + \begin{bmatrix} 0 \\ 0.35 \end{bmatrix} \hat{u}_4(k) \tag{10.56}$$

- Predictive model in the subsystem-based MPC for $\mathcal{S}_4$

$$x_4(k+1) = 0.65x_4(k) + 0.35u_4(k) + 0.13\hat{x}_3(k) \tag{10.57}$$

Here, the simulation program is developed with MATLAB and the optimizing tool, FMINCON, is used to solve each subsystem-based MPC in every control period. The tool of FMINCON has already been provided in MATLAB and it is able to solve multivariable cost function with nonlinear constraints.

Some parameters of the controllers are shown in Table 10.2. Among these parameters, $\mathcal{P}_i$ is obtained by solving the Lyapunov function. The eigenvalue of each closed-loop system under the feedback control shown in Table 10.2 is 0.5. The eigenvalues of $\mathbf{A}_o^T\mathbf{P}\mathbf{A}_o + \mathbf{A}_o^T\mathbf{P}\mathbf{A}_d + \mathbf{A}_d^T\mathbf{P}\mathbf{A}_o - \mathbf{Q}/2$ are $\{-2.42, -2.26, -1.80, -1.29\}$, all of which are negative. Thus Assumption 10.2 is satisfied. Set $\varepsilon = 0.2$. Consequently, if $\|x_i\|_{P_i} \leq \varepsilon/\sqrt{N} \leq 0.1$, then $\|K_i x_i\|_2$ would be less than 0.1, and the constraints on the inputs and the increments of inputs, as shown in Table 10.2, are satisfied. Set the control horizon of all the controllers to be $N = 10$. Set the

Table 10.2 Parameters of the N-DMPC

Subsystem	$\mathbf{K}_i$	$\mathbf{P}_i$	$\mathbf{Q}_i$	$\mathbf{R}_i$	$\Delta u_i^{\max}$ $\Delta u_i^{\min}$	$u_i^{\max}$ $u_i^{\min}$
$\mathcal{S}_1$	−0.35	5.36	4	0.2	±1	±2
$\mathcal{S}_2$	−0.25	5.35	4	0.2	±1	±2
$\mathcal{S}_3$	−0.28	5.36	4	0.2	±1	±2
$\mathcal{S}_4$	−0.43	5.38	4	0.2	±1	±2

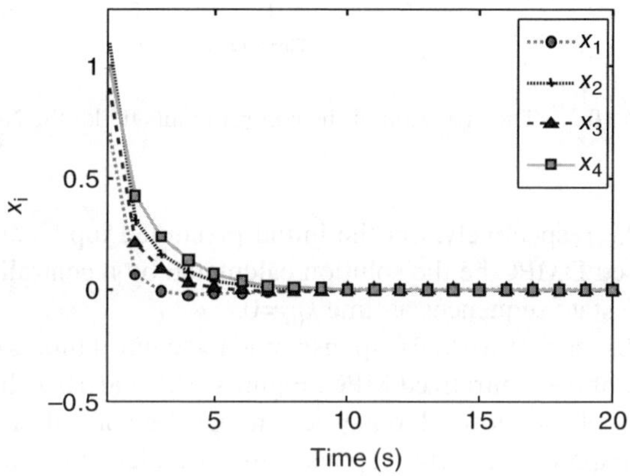

Figure 10.2 The evolution of the states under the N-DMPC

initial presumed inputs and states, at time $k_0 = 0$, be the solution calculated by a centralized MPC and the corresponding predictive states.

The state responses and the inputs of the closed-loop system are shown in Figures 10.2 and 10.3, respectively. The states of all four subsystems converge close to zeros in about 10 s. The state of $\mathcal{S}_3$ undershoots by 0.06 before converging to zero.

10.6.3 Performance Comparison with the Centralized MPC and the Local Cost Optimization based MPC

To further demonstrate the performance of the proposed DMPC, a dual model centralized MPC and a local cost optimization based MPC are applied to the system described by (10.53). In what follows, we discuss the performance comparison with the centralized MPC and the performance comparison with local cost optimization based MPC.

In both the centralized MPC and the subsystem-based MPCs of the local cost optimization based DMPC, the dual mode strategy is adopted, and the control horizon is set to be $N = 10$. The terminal constraints of the state of all subsystems are $\left\|x_i\,(k + 10\,|\,k)\right\|_{P_i} \leq \varepsilon/2 = 0.1$. All MPCs switch to the feedback control laws given in Table 10.2 when all states enter the attractive region $\Omega(\varepsilon)$. The bounds of the input and the input increment of each of the four subsystems

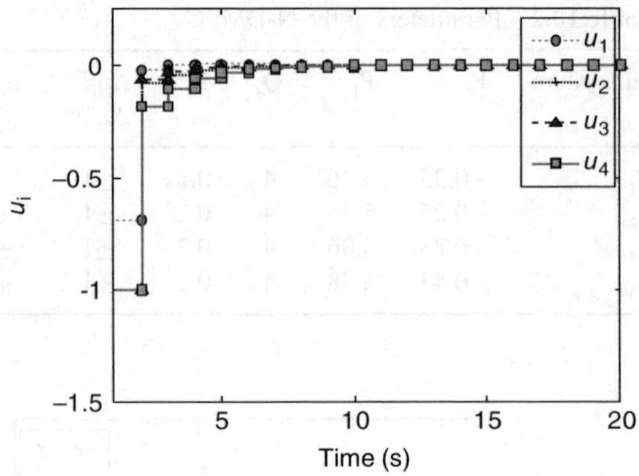

Figure 10.3 The evolution of the control inputs under the N-DMPC

are [−2, 2] and [−1,1], respectively. Set the initial presumed inputs and states of each local cost optimization based DMPC be the solution calculated by a centralized MPC and the corresponding predictive state sequences at time $k_0 = 0$.

Figures 10.4 and 10.5 show the state responses and the control inputs of the closed-loop system under the control of the centralized MPC. Figures 10.6 and 10.7 show the state responses and the control inputs of the closed-loop system under the control of the local optimization based DMPC. The shape of the state response curves under the centralized MPC are similar to those under the N-DMPC, all subsystems converge to near zero in 8 sec. Under the LCO-DMPC control design, all subsystems converge to near zero in about 14 sec. There is no significant undershooting.

Figures 10.8 and 10.9 show the differences between the absolute value of the states and control inputs of the closed-loop system under the centralized MPC and those under the N-DMPC, respectively. The state responses under the N-DMPC are very similar to those

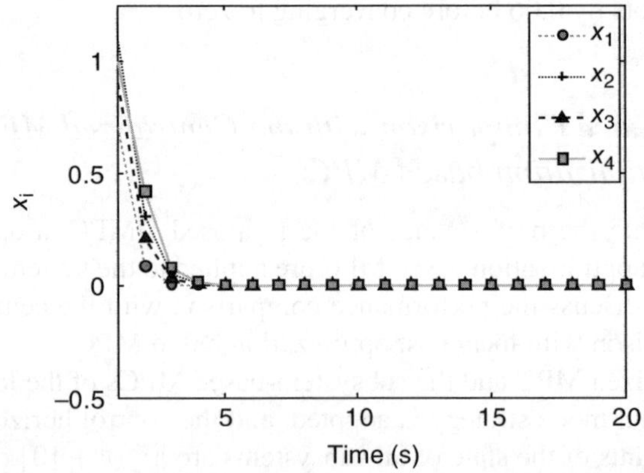

Figure 10.4 The evolution of the states under the centralized MPC

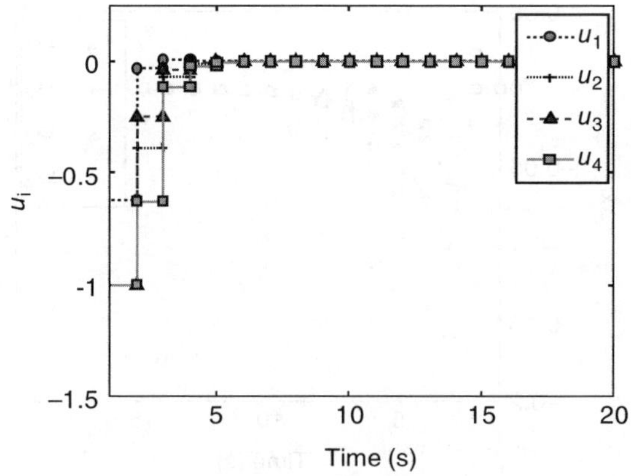

Figure 10.5 The evolution of the control inputs under the centralized MPC

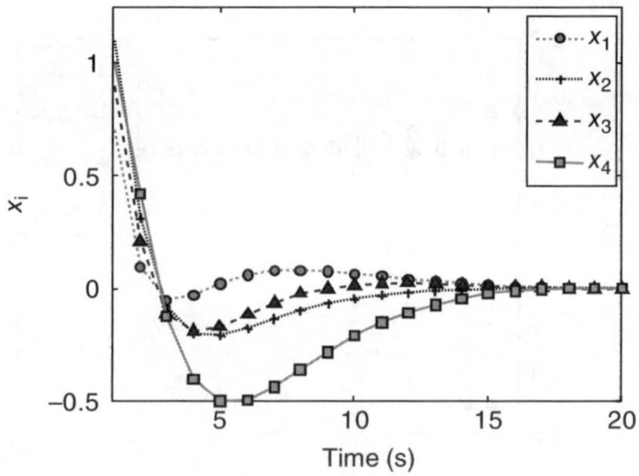

Figure 10.6 The evolution of the states under the LCO-DMPC

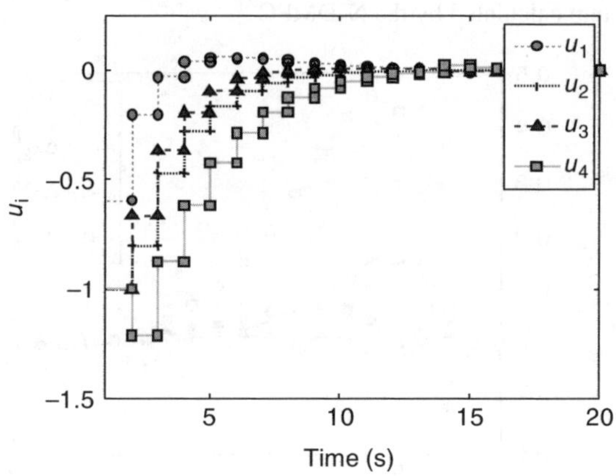

Figure 10.7 The evolution of the control inputs under the LCO-DMPC

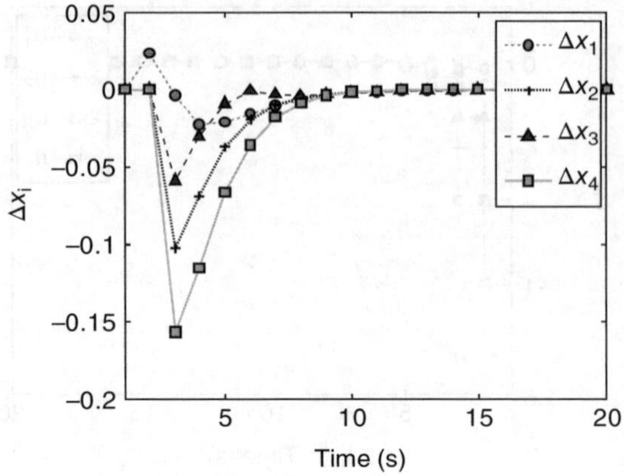

Figure 10.8 The errors between the absolute value of the state of each subsystem under the centralized MPC and the absolute value of the state of each subsystem under the N-DMPC

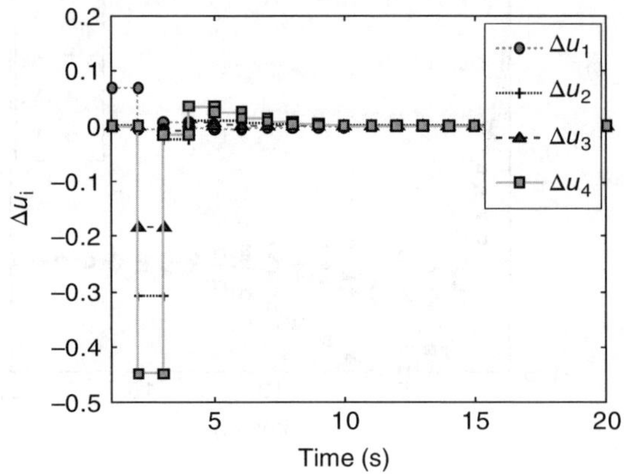

Figure 10.9 The difference between the input of each subsystem produced by the centralized MPC and the input of each subsystem calculated by the N-DMPC

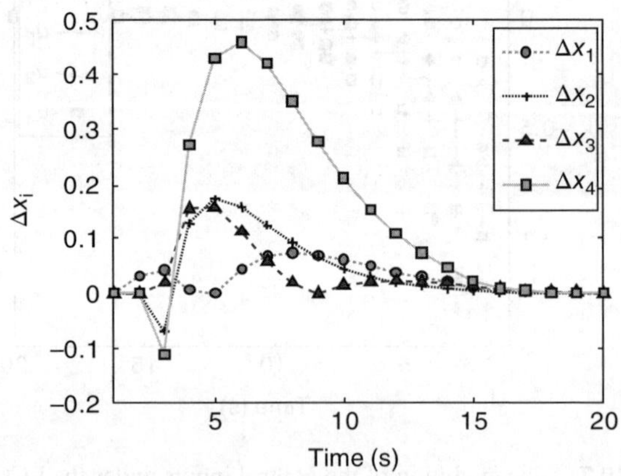

Figure 10.10 The errors between the absolute value of the state of each subsystem under the local cost optimization based DMPC and the absolute value of the state of each subsystem under the N-DMPC

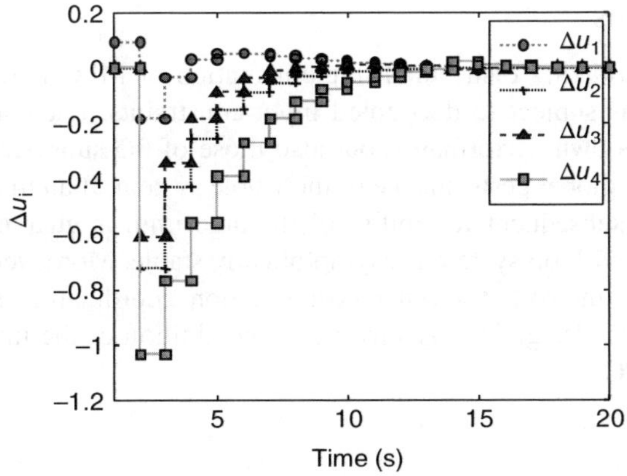

Figure 10.11 The difference between the input of each subsystem produced by the LCO-DMPC and the input of each subsystem calculated by the N-DMPC

under the centralized MPC. The performance of $\mathcal{S}_1$ is slightly better under the N-DMPC than under the centralized MPC, while the performance of $\mathcal{S}_2$, $\mathcal{S}_3$, and $\mathcal{S}_4$ is slightly worse under the N-DMPC than under the centralized MPC. Figures 10.10 and 10.11 show the differences between the absolute value of the states and control inputs of the closed-loop system under the local cost optimization based DMPC and those under the N-DMPC, respectively. The performance of all other subsystems under the N-DMPC is better than that under the local cost optimization based DMPC.

Table 10.3 shows the state square errors of the closed-loop system under the control of the centralized MPC, the N-DMPC, and the local cost optimization based DMPC, respectively. The total errors under the N-DMPC is 0.33 (2.1%) larger than that under the centralized MPC. The total errors resulting from the local cost optimization based DMPC is 6.55 (40.5%) larger than the results from the centralized MPC. The performance of the N-DMPC is significantly better than that of the local cost optimization based DMPC.

From these simulation results, it can be seen that the proposed algorithm is able to steer the system states to the origin if there is a feasible solution at the initial states, and the performance of the closed-loop system under the N-DMPC is very similar to that under the centralized MPC.

Table 10.3 State square errors of the closed-loop system under the control of the centralized MPC(CMPC), the LCO-DMPC, and the N-DMPC

Subsystems	CMPC	LCO-DMPC	N-DMPC
$\mathcal{S}_1$	2.07	2.22	2.08
$\mathcal{S}_2$	5.47	6.26	5.55
$\mathcal{S}_3$	3.63	4.12	3.65
$\mathcal{S}_4$	5.00	10.12	5.22
Total	16.17	22.72	16.50

10.7 Conclusions

In this chapter, a stabilizing distributed implementation of MPC is developed for dynamically coupled systems subject to decoupled input constraints. Each subsystem-based MPC considers not only its own performance but also those of the subsystems in its downstream region to improve the global performance of the whole system. If an initially feasible solution could be found, the subsequent feasibility of the algorithm is guaranteed at every update, and the resulting closed-loop system is asymptotically stable. Moreover, the proposed NDPC provides a unified framework for other cost function coordinating strategies. Under this framework, the DMPC design simply involves a redefinition of the impacted region in each subsystem-based MPC.

Part Four

Application

11

Hot-Rolled Strip Laminar Cooling Process with Distributed Predictive Control

11.1 Introduction

Recently, customers require increasingly better quality for hot-rolled strip products, such as automotive companies expect to gain an advantage from thinner but still very strong types of steel sheeting that make their vehicles more efficient and more environmentally compatible. In addition to the alloying elements, the cooling section is crucial for the quality of products [116]. Hot-rolled strip laminar cooling (HSLC) process is used to cool a strip from an initial temperature of roughly 820–920°C down to a coiling temperature of roughly 400–680°C, according to the steel grade and geometry. The mechanical properties of the corresponding strip are determined by the time–temperature course (or cooling curve) when the strip is cooled down on the runout table [116–117]. The precise and highly flexible control of the cooling curve in the cooling section is therefore extremely important.

Most of the control methods (e.g., Smith predictor control [118], element tracking control [119], self-learning strategy [120], and adaptive control [121]) pursue the precision of coiling temperature and care less about the evolution of strip temperature. In these methods, the control problem is simplified so greatly that only the coiling temperature is controlled by the closed-loop part of the controller. However, it is necessary to regulate the whole evolution procedure of strip temperature if better properties of strip are required. This is a large-scale, MIMO, parameter-distributed complicated system. Therefore, the problem is how to control the whole HSLC process online precisely with the size of HSLC process and the computational efforts required.

Model predictive control (MPC) is widely recognized as a practical control technology with high performance, where a control action sequence is obtained by solving, at each sampling instant, a finite horizon open-loop receding optimization problem and the first control action is applied to the process [26]. An attractive attribute of MPC technology is its ability to systematically account for process constraints. It has been successfully applied to various linear and nonlinear systems in the process industries and is becoming widespread. For large-scale and

Distributed Model Predictive Control for Plant-Wide Systems, First Edition. Shaoyuan Li and Yi Zheng.

relatively fast systems, centralized MPC has been gradually replaced by decentralized or distributed MPC. DMPC accounts for the interactions among subsystems. Each subsystem-based MPC in DMPC, in addition to determining the optimal current response, also generates a prediction of future subsystem behavior. By suitably leveraging this prediction of future subsystem behavior, the various subsystem-based MPCs can be integrated and therefore the overall system performance is improved. Thus the DMPC is a good method to control HSLC.

Consider that the HSLC process is a large-scale system and each subsystem is coupled with its neighbors by states, the DMPC framework should be suitable for the nonlinear system with fast computational speed, appropriate communication burden, and good global performance. Among the DMPC formulations provided in the literatures [27, 29, 33, 38, 41, 47, 122–125], to guarantee performance improvement and the appropriate communication burden among subsystems, the impacted region optimization-based DMPC is adopted here, where the optimization objective of each subsystem-based MPC considers not only the performance of the local subsystem but also those of its neighbors.

In this chapter, each subsystem-based MPC of the DMPC framework proposed is formulated based on the successive online linearization of nonlinear model to overcome the computational obstacle caused by nonlinear model. The prediction model of each MPC is linearized around the current operating point at each time instant. Neighborhood optimization is adopted in each local MPC to improve the global performance of HSLC and lessen the communication burden. Furthermore, since the strip temperature can only be measured at a few positions due to the tough ambient conditions, extended Kalman filter (EKF) is employed to estimate the transient temperature of strip in the water-cooling section.

The contents are organized as follows: Section 11.2 describes the HSLC process and the control problem. Section 11.3 presents the proposed control strategy of HSLC, which includes the modeling of subsystems, the designing of EKF, the functions of predictor, and the development of local MPCs based on the neighborhood optimization for subsystems, as well as the iterative algorithm for solving the proposed DMPC. Both simulation and experiment results are presented in Section 11.4. Finally, a brief conclusion is drawn to summarize the study and potential expansions are explained.

11.2 Laminar Cooling of Hot-rolled Strip

11.2.1 Description

The HSLC process is illustrated in Figure 11.1. Strips enter cooling section at finishing rolling temperature (FT) of 820–920°C, and are coiled by coiler at coiling temperature (CT) of 400–680°C after being cooled in the water-cooling section. The X-ray gauge is used to measure the gauge of strip. Speed tachometers for measuring coiling speed is mounted on the motors of the rollers and the mandrel of the coiler. Two pyrometers are located at the exit of the finishing mill and before the pinch roll, respectively. Strips are 6.30–13.20 mm in thickness and 200–1100 m in length. The runout table has 90 top headers and 90 bottom headers. The top headers are of U-type for laminar cooling and the bottom headers are of straight type for low-pressure spray. These headers are divided into 12 groups. The first nine groups are for the main cooling section and the last three groups are for the fine cooling

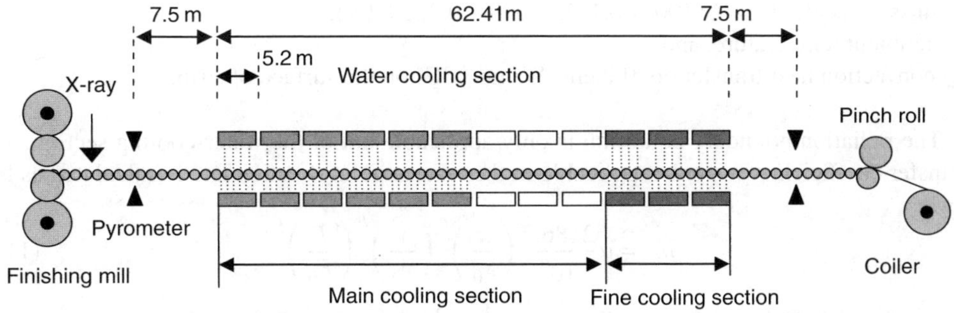

Figure 11.1 Hot-rolled strip laminar cooling process

section. In this HSLC, the number of cooling water header groups and the water flux of each header group are taken as control variables to adjust the temperature distribution of the strip.

11.2.2 Thermodynamic Model

Consider the whole HSLC process from the point of view of geometrically distributed setting system (the limits of which are represented by the geometrical locations of FT and CT, as well as the strip top and bottom sides), a two-dimensional mathematical model for Cartesian coordinates is developed combining academic and industrial research findings [126]. The model assumes that there is no direction dependency for the heat conductivity λ. There is no heat transfer in traverse and rolling direction. The latent heat is considered by using temperature-dependent thermal property developed in [127] and the model is expressed as

$$\dot{x} = \frac{-\lambda}{\rho c_p}\frac{\partial^2 x}{\partial z^2} - \dot{l}\frac{\partial x}{\partial l} \tag{11.1}$$

with the boundary conditions on its top and bottom surfaces

$$\pm\lambda\frac{\partial x}{\partial z} = h(x - x_\infty) \tag{11.2}$$

where the right-hand side of (11.2) is h times $(x - x_\infty)$ and

$$h = h_w\frac{x - x_w}{x - x_\infty} + \sigma_0\varepsilon\frac{x^4 - x_\infty^4}{x - x_\infty} \tag{11.3}$$

and

$x(z, l, t)$ strip temperature at position (z, l),
l,z length coordinate and thickness coordinate, respectively,
ρ density of strip steel,
c_p specific heat capacity;
λ heat conductivity,
σ_0 Stefan–Boltzmann constant (5.67×10^{-8} W/m^2 K^4),

ε emission coefficient $((x/1000)[0.125x/1000-0.38]+1.1)$,
x_∞ ambient temperature, and
h_w convection heat transfer coefficient (W/mm^2 °C) on the surface of strip.

The radiation boundary condition is only applicable out of the water-cooling section. The transfer coefficient h_w is only applicable in the water-cooling section and is calculated as follows:

$$h_w = \alpha \frac{2186.7}{10^6}\left(\frac{x}{x_0}\right)^a \left(\frac{v}{v_0}\right)^b \left(\frac{F}{F_0}\right)^c \tag{11.4}$$

where $x_0 = 1000$°C, $v_0 = 20$ m/s, $F_0 = 350$ m^3/(m^2 min), $a = 1.62$, $b = -0.4$, $c = 1.41$, v is the velocity of strip, and F is the flux of cooling water.

11.2.3 Problem Statement

The technical targets of HSLC refer to CT and the temperature drop curve of strip caused by cooling water. Contemplating the overall system from the point of view of the geometrically distributed setting system, as shown in Figure 11.2, we can transform the desired temperature drop curve of strip into the geometrically location-dependent temperature profile from finishing mill to coiler. Here the temperature on desired cooling curve refers to strip's average temperature in thickness direction. Therefore, the control objective becomes to adjust the average temperature of strip in thickness direction to be consistent with the geometrically location-dependent temperature profile. The manipulated variables of system are the states (opening or closing) and the water fluxes of every header groups.

11.2.3.1 Existing Method

The existing method in industrial manufactory is open-loop and closed-loop method. The open-loop part charges the main cooling section and the closed-loop part charges the fine cooling section. The water fluxes of all opening headers in main cooling section are constant and are same to each other, which are determined by the expert experience according to the cooling rate in the first section of cooling curve. The number of header groups opened in the

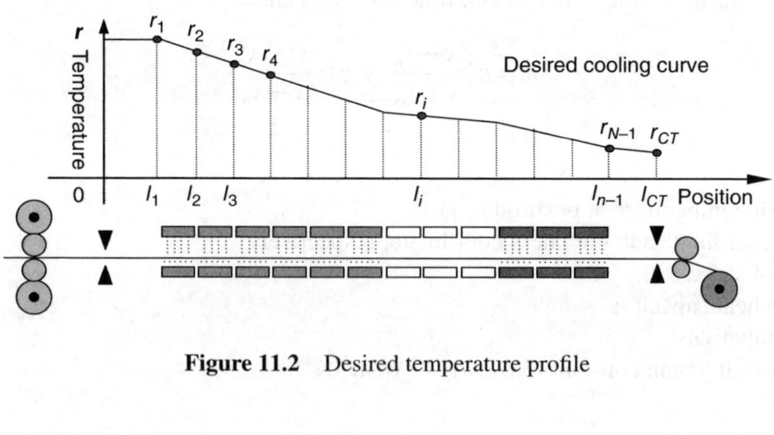

Figure 11.2 Desired temperature profile

main cooling section N_{ho} is used to adjust the strip temperature at the exit of main cooling section, and is calculated online according to the feedforward signal of FT at each control period as follows:

$$N_{ho} = \text{Int}\left(\frac{x_{FT} - x_{cm}}{CR_m}\right)$$

where x_{cm} is the reference temperature at the exit of main cooling section, x_{FT} is the finishing rolling temperature, and CR_m is the reference cooling rate in main cooling section. In the closed-loop part, a PI controller is employed to control the water flux in fine cooling section according to the feedback measurement of CT.

However, in this method, some inaccurate or unreasonable assumptions are applied additionally for the simplification of this large-scale system (e.g., the cooling rate in main cooling section is impossible to be constant if the water flux of every header remains the same), which makes the system less flexible for various cooling curves and be less precise. On the other hand, since the major disturbance (the fluctuation of x_{cm}) does not feed into PI controller in closed-loop part, it is also difficult to achieve an accurate CT. Thus, this control method is difficult to satisfy the increasing quality requirements of steel industry.

11.2.3.2 Why Using DMPC Framework

To obtain productions with better quality, a suitable optimization method is required for making the average temperature of strip in thickness direction consistent with the desired temperature profile at any position of the water-cooling section. To realize it, we should consider the water flux of each header group as an individual manipulated variable rather than consider all of them as one or two manipulated variables simply. That means the control method should be suitable for large-scale, nonlinear, and relatively fast MIMO system. Furthermore, this optimization method should be able to account for the major measurable disturbance of FT online for precision enhancing.

MPC is widely recognized as a practical and well-performed control technology for process control, especially for MIMO system. The measurable disturbance can also be taken into account through the prediction model of MPC. Thus, MPC can be a good selection for this system. As shown in Figure 11.2, the desired geometrically location-dependent temperatures are selected at the locations $l_1, l_2, \dots, l_N$ as the reference temperatures with the notation

$$\mathbf{r} = \begin{bmatrix} r_1 & r_2 & \cdots & r_N \end{bmatrix}^{\mathrm{T}} \tag{11.5}$$

Here, $l_1, l_2, \dots, l_N$ correspond to the locations of the exit of each header group and the locations of the pyrometers used to measure FT and CT. The optimization objective is to minimize

$$J = \sum_{t=1}^{P} \|\mathbf{y}(k+t|k) - \mathbf{r}(k+t)\|_{\mathbf{Q}}^2 + \sum_{h=1}^{M} \|\Delta\mathbf{u}(k+h-1|k)\|_{\mathbf{R}}^2 \tag{11.6}$$

where $\mathbf{y} = [y_1 \ \cdots \ y_N]^{\mathrm{T}}, y_s(s = 1, \dots, N)$ is the average temperature of strip at position l_s, $\mathbf{u}$, the manipulated variable vector, refers to the future sequence of water flux, $\Delta\mathbf{u}$ is the increment of manipulated variable vector $\mathbf{u}$, P is the prediction horizon, and M is the control horizon. The weighting matrices $\mathbf{Q}$ and $\mathbf{R}$ are positive definite and have block-diagonal forms.

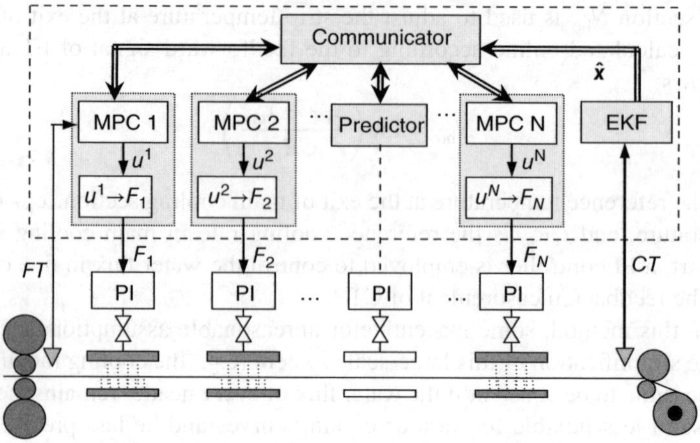

Figure 11.3 The structure of DMPC framework for HSLC

However, for this large-scale, nonlinear, and relatively fast system, the online implementation of centralized MPC is impractical due to the large computation. To decrease the computational burden and guarantee the performance of overall system at the same time, a DMPC framework based on neighborhood optimization and successive linearization is therefore proposed for HSLC.

11.3 Control Strategy of HSLC

Since the major obstacle of accurate online control of HSLC is the large-scale, nonlinear characteristics, the DPMC framework is adopted. The whole system is divided into N subsystems (the sth subsystem ranges from l_{s-1} to $l_s[s = 1, 2, \ldots, N]$ as shown in Figure 11.2), and each subsystem is controlled by a local MPC controller as shown in Figure 11.3. Since the strip temperature can only be measured at a few positions inside the cooling section due to the hard ambient conditions, an EKF is employed to estimate the distribution of strip temperature. Each local MPC calculates the set-point of PI controller based on the current strip temperature estimated by EKF and the future states of its neighbors. Each PI controller regulates water flux to be consistent with the set-point calculated by local MPC. Since there are no manipulated variables in the subsystems with closed header group, a predictor is substituted for local MPC. The predictor estimates the future states of corresponding subsystem and broadcasts the estimations to its neighbors. In this way, the EKF, MPCs and predictors, as well as the PI controllers work together through network information to control the HSLC.

11.3.1 State Space Model of Subsystems

Since it is not easy for MPC to predict the future states using model (11.1), the state space representation of model (11.1) for each subsystem is deduced first in this subsection. Using two-dimensional finite volumes scheme, model (11.1) can be reduced into a finite-dimensional

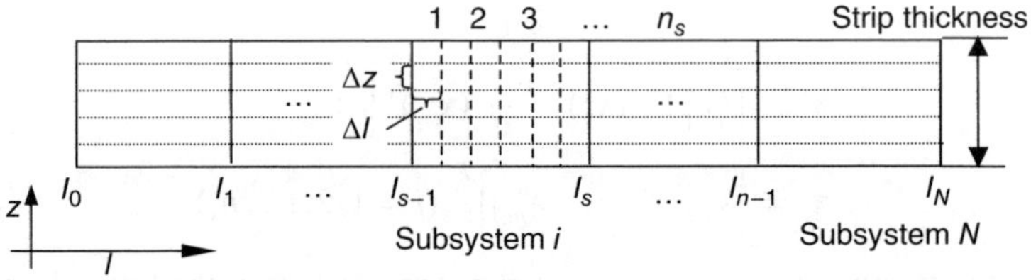

Figure 11.4 The division of each subsystem

problem. In each subsystem s, denote the number of volumes in the l-direction by n_s and the z-direction by m as shown in Figure 11.4. Each volume, denoted by V, equals to $\Delta l \Delta z$. Δl and Δz are the length and thickness of each volume, respectively. Denote the temperature of the ith z-direction and the jth l-direction volume by $x^s_{i,j}$. Let

$$x^0_{i,n_s} = x_{\text{FT}}, i = 1, 2, \ldots, m \tag{11.7}$$

$$x^N_{i,n_s} = x^N_{i,n_s-1}, i = 1, 2, \ldots, m \tag{11.8}$$

The energy balance equation (11.1) being applied to the top surface and bottom surface volumes leads to

$$\dot{x}^s_{1,j} = -\frac{\lambda^s_{1,j}}{\rho^s_{1,j} cp^s_{1,j}} \left(\frac{1}{\Delta z^2} \left(x^s_{2,j} - x^s_{1,j} - \Delta z \frac{h^s_{1,j}}{\lambda^s_{1,j}} \left(x^s_{1,j} - x_\infty \right) \right) \right) - \frac{1}{\Delta l} v(x^s_{1,j} - x^s_{1,j-1}) \tag{11.9}$$

$$\begin{aligned} \dot{x}^s_{m,j} = & -\frac{\lambda^s_{m,j}}{\rho^s_{m,j} cp^s_{m,j}} \left(\frac{1}{\Delta z^2} \left(x^s_{m-1,j} - x^s_{m,j} - \Delta z \frac{h^s_{m,j}}{\lambda^s_{m,j}} \left(x^s_{m,j} - x_\infty \right) \right) \right) \\ & -\frac{1}{\Delta l} v(x^s_{m,j} - x^s_{m,j-1}) \end{aligned} \tag{11.10}$$

For the internal volumes, it is

$$\dot{x}^s_{i,j} = -\frac{1}{\Delta z^2} \frac{\lambda^s_{i,j}}{\rho^s_{i,j} cp^s_{i,j}} (x^s_{i+1,j} - 2x^s_{i,j} + x^s_{i-1,j}) - \frac{1}{\Delta l} v(x^s_{i,j} - x^s_{i,j-1}) \tag{11.11}$$

where $v = \dot{l}$ is the coiling velocity, $x^s_{i,j-1} = x^{s-1}_{i,n_{s-1}}$ when $j = 1$ and $x^s_{i,j+1} = x^{s+1}_{i,1}$ when $j = n_s$.

In industrial application, the measurements are available digitally with a sampling time Δt. Thus the discrete-time version of the subsystem is derived by approximating the derivatives using simple Euler approximation. Since $\rho^s_{i,j}$, $\lambda^s_{i,j}$, and $cp^s_{i,j}$ are temperature-dependent, define $a(x^s_{i,j}) = -\Delta t \lambda^s_{i,j} / (\Delta z^2 \rho^s_{i,j} cp^s_{i,j})$, $\beta(x^s_{i,j}) = \Delta t\, a^s_{i,j} / \lambda^s_{i,j}$ and $\gamma = \Delta t\, v / \Delta l$. Then the nonlinear state space representation of subsystem s, deduced from the previous Equations (11.7)–(11.11), can be expressed as

$$\begin{cases} \mathbf{x}^s(k+1) = \mathbf{f}(\mathbf{x}^s(k)) \cdot \mathbf{x}^s(k) + \mathbf{g}(\mathbf{x}^s(k)) \cdot u^s(k) + \mathbf{D} \cdot \mathbf{x}^{s-1}_{n_{s-1}}(k) \\ y^s(k) = \mathbf{C} \cdot \mathbf{x}^s(k) \end{cases} \quad s = 1, 2, \ldots, N \tag{11.12}$$

where

$$\mathbf{x}^s = \left[\left(\mathbf{x}_1^s\right)^{\mathrm{T}}\ (\mathbf{x}_2^s)^{\mathrm{T}}\ \cdots\ (\mathbf{x}_{n_s}^s)^{\mathrm{T}}\right]^T;$$

$$\mathbf{x}_j^s = \left[x_{1,j}^s\ x_{2,j}^s\ \cdots\ x_{m,j}^s\right]^T, (j = 1, 2, \ldots, n_s) \tag{11.13}$$

is the state vector of subsystem s, y^s is the average temperature of the last column volumes of subsystem s, u^s is the input of subsystem s and is defined as follows:

$$\begin{cases} u^s = 2186.7 \times 10^{-6} \times \alpha \cdot \left(v/v_0\right)^b \times (F_s/F_0)^c,\ s \in \mathcal{C}_{\mathcal{W}} \\ u^s = 1, s \in \mathcal{C}_{\mathcal{A}} \end{cases} \tag{11.14}$$

where $\mathcal{C}_{\mathcal{W}}$ is the set of subsystems in which strips are cooled by water, and $\mathcal{C}_{\mathcal{A}}$ is the set of subsystems in which strips are cooled mainly through radiation. $\mathbf{f}(\mathbf{x}^s(k))$, $\mathbf{g}(\mathbf{x}^s(k))$, $\mathbf{D}$, and $\mathbf{C}$ are the coefficient matrices of subsystem s, and are defined as follows:

$$\mathbf{f}(\mathbf{x}^s(k)) = \begin{bmatrix} \mathbf{\Phi}_1\left(\mathbf{x}^s(k)\right) \cdot \mathbf{\Lambda} & \mathbf{0} & \cdots & \mathbf{0} \\ \mathbf{0} & \mathbf{\Phi}_2(\mathbf{x}^s(k)) \cdot \mathbf{\Lambda} & & \vdots \\ \vdots & & \ddots & \mathbf{0} \\ \mathbf{0} & \cdots & \mathbf{0} & \mathbf{\Phi}_{n_s}(\mathbf{x}^s(k)) \cdot \mathbf{\Lambda} \end{bmatrix} + \begin{bmatrix} (1-\gamma)\mathbf{I}_m & \mathbf{0} & \cdots & \mathbf{0} \\ \gamma \mathbf{I}_m & (1-\gamma)\mathbf{I}_m & \ddots & \vdots \\ \vdots & \ddots & \ddots & \mathbf{0} \\ \mathbf{0} & \cdots & \gamma \mathbf{I}_m & (1-\gamma)\mathbf{I}_m \end{bmatrix} \tag{11.15}$$

$$\mathbf{g}(\mathbf{x}^s(k)) = [\boldsymbol{\psi}_1(\mathbf{x}^s(k))^{\mathrm{T}}\ \cdots\ \boldsymbol{\psi}_{n_s}(\mathbf{x}^s(k))^{\mathrm{T}}]^{\mathrm{T}} \tag{11.16}$$

$$\mathbf{C} = m^{-1} \cdot \left[\mathbf{0}^{1\times m\cdot(n_s-1)}\ \mathbf{1}^{1\times m}\right] \tag{11.17}$$

$$\mathbf{D} = [\gamma \mathbf{I}_m\ \mathbf{0}^{m\times m\cdot(n_s-1)}]^{\mathrm{T}} \tag{11.18}$$

and

$$\mathbf{\Phi}_j(\mathbf{x}^s) = \begin{bmatrix} a\left(x_{1,j}^s\right) & \cdots & 0 \\ \vdots & \ddots & \vdots \\ 0 & \cdots & a(x_{m,j}^s) \end{bmatrix}$$

$$\boldsymbol{\psi}_j(\mathbf{x}^s) = \begin{bmatrix} \theta_{1,j}^s\left(x_{1,j}^s\right)(x_{1,j}^s - x_\infty) \cdot \beta(x_{1,j}^s) \\ \mathbf{0}^{(m-2)\times 1} \\ \theta_{m,j}^s(x_{m,j}^s)(x_{m,j}^s - x_\infty) \cdot \beta(x_{m,j}^s) \end{bmatrix}$$

$$\mathbf{\Lambda} = \begin{bmatrix} -1 & 1 & 0 & \cdots & 0 \\ 1 & -2 & 1 & \ddots & \vdots \\ 0 & \ddots & \ddots & \ddots & 0 \\ \vdots & \ddots & 1 & -2 & 1 \\ 0 & \cdots & 0 & 1 & -1 \end{bmatrix}$$

$$\mathbf{I}_m \in R^{m\times m}; (j = 1, 2, \ldots, n_s)$$

$$\begin{cases} \theta^s_{i,j}\left(x^s_{i,j}\right) = (x^s_{i,j}/x_0)^a, & s \in \mathcal{C}_\mathcal{W} \\ \theta^s_{i,j}(x^s_{i,j}) = h_{air}(x^s_{i,j}), & s \in \mathcal{C}_\mathcal{A} \end{cases}, (i = 1, 2, \ldots, m,\ j = 1, 2, \ldots, n_s) \tag{11.19}$$

It should be noticed that the notations $\mathbf{g}^s(\cdot)$ and u^s are different between the subsystems that belong to $\mathcal{C}_\mathcal{A}$ and the subsystems which belong to $\mathcal{C}_\mathcal{W}$, see (11.14) and (11.19). Since the current states of subsystems are not measurable inside the cooling section, an observer is designed to estimate the current states of each subsystem in the next subsection.

11.3.2 Design of Extended Kalman Filter

The nonlinear model of the overall system can be expressed as

$$\begin{cases} \mathbf{x}(k+1) = \mathbf{F}(\mathbf{x}(k))\mathbf{x}(k) + \mathbf{G}(\mathbf{x}(k))\mathbf{u}(k) + \overline{\mathbf{D}}\mathbf{x}^0(k) \\ \overline{\mathbf{y}}(k) = \overline{\mathbf{C}}\mathbf{x}(k) \end{cases} \tag{11.20}$$

where $\mathbf{x} = [(\mathbf{x}^1)^\mathrm{T}\ (\mathbf{x}^2)^\mathrm{T}\ \cdots\ (\mathbf{x}^N)^\mathrm{T}]^\mathrm{T}$ and $\mathbf{u} = [u^1\ u^2\ \cdots\ u^{n_s}]^\mathrm{T}$, $\mathbf{x}^0$ is the distribution of FT in the z-direction. $\overline{\mathbf{y}}$ is the output vector which includes the measurements of CT on the top and bottom surfaces. The expressions of $\mathbf{F}(\mathbf{x}(k))$, $\mathbf{G}(\mathbf{x}(k))$, and $\overline{\mathbf{D}}$ can be deduced easily from (11.12). Coefficient matrix $\overline{\mathbf{C}}$ is defined as

$$\overline{\mathbf{C}} = \begin{bmatrix} \mathbf{0}^{1\times(N-1)n_s m} & 1 & \mathbf{0}^{1\times(n_s m-1)} \\ \mathbf{0}^{1\times(N-1)n_s m} & \mathbf{0}^{1\times(n_s m-1)} & 1 \end{bmatrix} \tag{11.21}$$

This system is uniformly observable because each volume's temperature depends on its neighbors. Since the order of the system is inevitably high, the famous EKF is chosen to be more convenient to design in our case. The estimation of the strip temperature distribution can be expressed as

$$\hat{\mathbf{x}}(k+1) = \hat{\mathbf{x}}(k+1|k) + \mathbf{K}_{k+1}(\overline{\mathbf{y}}(k+1) - \overline{\mathbf{C}}\hat{\mathbf{x}}(k+1|k)) \tag{11.22}$$

The feedback coefficient $\mathbf{K}_{k+1}$ is deduced by the *difference Riccati equation*. This observer estimates the states of overall system at each control period, and transmits them to all subsystems.

11.3.3 Predictor

Since there are no manipulated variables in subsystem $s \in \mathcal{C}_\mathcal{A}$, a predictor is applied for estimating the future states $\mathbf{X}_s(k)$ where

$$\mathbf{X}_s(k) = \left[\mathbf{x}^s(k+1)\ \mathbf{x}^s(k+2)\ \cdots\ \mathbf{x}^s(k+P)\right]^\mathrm{T} \tag{11.23}$$

In the predictor, the prediction model is (11.12), and the measurable disturbance FT is assumed to be a constant during the estimation of $\mathbf{X}_1(k)$. When finishing estimating $\mathbf{X}_s(k)$, the predictor sends the estimation of $\mathbf{X}_s(k)$ to the downstream neighbors of subsystem s.

11.3.4 Local MPC Formulation

As for subsystem $s \in \mathcal{C}_{\mathcal{W}}$, the strip temperature is controlled by a local MPC. The local MPC is formulated based on neighborhood optimization and successive linearization of prediction model. The details of it are presented next.

Since the weighting matrices $\mathbf{Q}$ and $\mathbf{R}$ have block-diagonal forms in (11.6), the global performance index can be decomposed in terms of the local indices for each subsystem

$$J_s(k) = \sum_{i=1}^{P} \|\mathbf{r}_s(k+i) - \widehat{\mathbf{y}}^s(k+i|k)\|^2_{\mathbf{Q}_s} + \sum_{h=1}^{M} \|\Delta u^s(k+h-1|k)\|^2_{R_s},\ (s = 1, 2, \ldots, N) \tag{11.24}$$

The local control decision is computed by solving the local optimization problem $J_s(k)$ with local input/output variables and constraints. However, the optimal solution to the local optimization problem collectively is not equal to the global optimal control decision of the whole system. To enhance the global control performance, neighborhood optimization is adopted.

Define the set of the subsystems whose states are affected by the states of subsystem s as downstream neighborhood of subsystem s, and denote it by π_{-s}, $s \notin \pi_{-s}$. Similarly, define the set of subsystems whose states affect the states of subsystem s as upstream neighborhood of subsystem s, and denote it by π_{+s}, $s \notin \pi_{+s}$. Since the future states of the downstream neighbors are affected by the future inputs of subsystem s, the new performance index for each subsystem can be improved by

$$\min \overline{J}_s(k) = \sum_{j \in \{\pi_{-s}, s\}} J_j(k) \tag{11.25}$$

Note that the new performance index for the sth subsystem $\overline{J}_s(k)$ is composed not only of cost function of subsystem s but also of its downstream neighbors. Cooperation between subsystems is achieved by exchanging information between each subsystem and its neighbors in a distributed structure via network communication and by optimizing the local problem with the new performance index (11.25).

It should be noticed that model (11.12) is a nonlinear model. If the future evolution of each subsystem is predicted through it, the minimization of a quadratic index, subject to the nonlinear HSLC dynamic, would be a nonlinear optimization problem. This can be computationally demanding, depending on the states and constraints. To overcome this problem, the prediction model is linearized around the current operating point at each time step, and a linear MPC is designed for the resulting linear system. The idea of using time-varying models traces back to the early 1970s in the process-control field although it has been properly formalized only recently. Studies on linear parameter varying MPC schemes can be found in [128–131]. Among them, the works in [130, 131] are the closest to our approach.

In this case, the following prediction model is used to approximate the nonlinear model (11.12) at time instant k

$$\begin{cases} \mathbf{x}^s(i+1|k) = \mathbf{A}_s(k) \cdot \mathbf{x}^s(i|k) + \mathbf{B}_s(k) \cdot u_s(i|k) + \mathbf{D} \cdot \mathbf{x}^{s-1}_{n_{s-1}}(i|k) \\ y^s(i|k) = \mathbf{C} \cdot \mathbf{x}^s(i|k) \end{cases} \quad s = 1, 2, \ldots, N \tag{11.26}$$

where $\mathbf{A}_s(k) = \mathbf{f}(\mathbf{x}^s(k))$ and $\mathbf{B}_s(k) = \mathbf{g}(x^s(k))$. The $(s-1)$th subsystem and the $(s+1)$th subsystem are the upstream and the downstream neighbor of subsystem s, respectively. Assuming that $\mathbf{x}(k)$ is available, the local optimization problem for subsystem s at the sampling time instant k becomes

$$\min_{\Delta U_s(k)} \overline{J}_s(k) = \sum_{j\in\{s,s+1\}} \left(\sum_{i=1}^{P} \left\| \mathbf{r}_j(k+i) - \widehat{\mathbf{y}}^j(k+i|k) \right\|_{\mathbf{Q}_j}^2 + \sum_{h=1}^{M} \|\Delta u^j(k+h-1|k)\|_{R_j}^2 \right)$$

$$\begin{aligned}
\text{s.t. } & \mathbf{x}^j(i+1|k) = \mathbf{A}_j(k)\cdot \mathbf{x}^j(i|k) + \mathbf{B}_j(k)\cdot u^j(i|k) \\
& \qquad + \mathbf{D}\cdot \mathbf{x}_{n_{j-1}}^{j-1}(i|k), \quad j\in\{s,s+1\} \\
& \boldsymbol{u}_{\min}^s \le \boldsymbol{u}^s(k+h-1|k) \le \boldsymbol{u}_{\max}^s, \qquad h=1,\dots,M \\
& \Delta\boldsymbol{u}_{\min}^s \le \Delta\boldsymbol{u}^s(k+h-1|k) \le \Delta\boldsymbol{u}_{\min}^s, \quad h=1,\dots,M \\
& \mathbf{x}_{\min}^j \le \mathbf{x}^j(k+i|k) \le \mathbf{x}_{\min}^j, \ i=1,\dots,P, \ \ j\in\{s,s+1\}
\end{aligned} \tag{11.27}$$

where $\{\mathbf{u}_{\min}^s, \mathbf{u}_{\max}^s\}$,$\{\Delta\mathbf{u}_{\min}^s, \Delta\mathbf{u}_{\max}^s\}$, and $\{\mathbf{x}_{\min}^j, \mathbf{x}_{\max}^j\}(j\in\{s,s+1\})$ are boundaries of manipulated variables, increment of manipulated variables, and state vectors, respectively, and

$$\Delta\mathbf{U}_s(k) = \left[\Delta u^s(k) \ \ \Delta u^s(k+1) \ \cdots \ \Delta u^s(k+M)\right]^{\mathrm{T}} \tag{11.28}$$

Define that

$$\begin{aligned}
\mathbf{X}_{s,n_s}(k) &= \left[\mathbf{x}_{n_s}^{\mathrm{s}}(k+1) \ \mathbf{x}_{n_s}^{\mathrm{s}}(k+2) \ \cdots \ \mathbf{x}_{n_s}^{\mathrm{s}}(k+P)\right]^{\mathrm{T}} \\
\mathbf{U}_s(k) &= \left[u^s(k) \ \ u^s(k+1) \ \cdots \ u^s(k+M)\right]^{\mathrm{T}}
\end{aligned} \tag{11.29}$$

If sequences $\mathbf{X}_{s-1,n_{s-1}}(k)$ and $\mathbf{U}_{s+1}(k)$ are available to subsystem s, problem (11.27) can be recast as a quadratic program (QP). Then optimal control decision sequence $\Delta\mathbf{U}_s^*(k)$ of the subsystem s can be computed at time instant k by solving (11.27) for the current states. The first sample of $\mathbf{U}_s^*(k) = u^s(k-1)\mathbf{I}_{1\times M} + \Delta\mathbf{U}_s^*(k)$ is used to compute the optimal water flux set-point of subsystem s according to (11.14).

We remark that model (11.12) is linearized around an operating point, which, in general, is not an equilibrium point. When evaluating the online computational burden of the proposed scheme, one needs to account for the resources spent in computing the linear model (11.26) and translating (11.27) into a standard QP problem. Nevertheless, for the proposed application, complexity of problem (11.27) is reduced greatly comparing to the nonlinear model based MPC.

11.3.5 Iterative Algorithm

According to the neighborhood optimization, the local optimal control decision for each subsystem can be obtained by solving problem (11.27) if the local optimal control decision of its downstream neighbors and the future optimal states of its upstream neighbors are available, that is

$$\Delta U_s^*(k) = \arg\left\{ \min_{\Delta U_s(k)} \overline{J}_s(k)|_{U_j^*(k)(j\in\mathcal{N}_{-i},j\neq i),\mathbf{X}_h^*(k)(h\in\mathcal{N}_{+i},h\neq i)} \right\} \quad (s=1,\dots,N) \tag{11.30}$$

However, the local optimal control decision of its downstream neighbors and the future optimal states of its upstream neighbors are not available to subsystem s, and hence the estimations of them are used. To get an accurate solution of problem (11.27), an iterative algorithm is developed to seek the local optimal control decision for each subsystem at each sampling period.

Algorithm 11.1 Distributed MPC Algorithm

Step 1. *Initialization and Communication.* At the sampling instant k, the EKF sends the current states of the system to the corresponding subsystem. Each subsystem initializes the estimation of local optimal control decision and transmits it to its upstream neighbors. Set the iterative index $l=0$:

$$\boldsymbol{U}_s^{(l)}(k) = \hat{\boldsymbol{U}}_s(k), (s = 1, 2, \ldots, N)$$

Each subsystem calculates the estimate of local state sequence $\mathbf{X}_s^{(l)}(k)$ by (11.26) and transmits them to its downstream neighbors through network.

Step 2. *Subsystem Optimization.* Each subsystem that belongs to $\mathcal{C}_{\mathcal{W}}$ solves its local optimization problem described in (11.27) simultaneously to derive its control decision. That is

$$\Delta\boldsymbol{U}_s^{(l+1)}(k) = \arg\left\{\min_{\Delta\boldsymbol{U}_s(k)} \bar{J}_s(k)|_{\boldsymbol{U}_j^{(l)}(k)(j\in\mathcal{N}_{-i}),\ \mathbf{X}_h^{(l)}(k)(h\in\mathcal{N}_{+i})}\right\}$$

$$\boldsymbol{U}_s^{(l+1)}(k) = u^s(k-1)\mathbf{I}_{1\times M} + \Delta\boldsymbol{U}_s^{(l+1)}(k), s \in \mathcal{C}_{\mathcal{W}}$$

Set the optimal solution of each subsystem belonging to $\mathcal{C}_{\mathcal{A}}$

$$\boldsymbol{U}_s^{(l+1)}(k) = \begin{bmatrix}1 & 1 & \cdots & 1\end{bmatrix}^{\mathrm{T}},\ s \in \mathcal{C}_{\mathcal{A}}$$

Then, calculate the estimation of the local state sequence $\mathbf{X}_s^{(l)}(k)$ by (11.26).

Step 3. *Checking and Updating.* Each subsystem checks if its terminal iteration condition is satisfied, that is, for the given error tolerance $\varepsilon_s \in \mathbb{R}(s=1,\ldots,N)$, if there exists

$$\|\boldsymbol{U}_s^{(l+1)}(k) - \boldsymbol{U}_s^{(l)}(k)\| \le \varepsilon_s(s = 1, \ldots, N)$$

If all the terminal conditions are satisfied at iteration l^*, then stop the iteration, set the local optimal control decision for each subsystem $\boldsymbol{U}_s^*(k) = \boldsymbol{U}_s^{(l^*)}(k)$, and go to Step 4; otherwise, let $l=l+1$, each subsystem transmits the new information $\boldsymbol{U}_s^{(l)}(k)$ to its upstream neighbors and transmits $\mathbf{X}_s^{(l)}(k)$ to its downstream neighbors, and go to Step 2.

Step 4. *Assignment and Implementation.* Each subsystem computes the control law

$$u^{s*}(k) = \begin{bmatrix}1 & 0 & \cdots & 0\end{bmatrix} \boldsymbol{U}_s^*(k)\ (s = 1, \ldots, N)$$

and applies it to the corresponding subsystem.

Step 5. *Reassigning the Initial Estimation.* Set the initial estimate of the local optimal control decision for the next sampling time

$$\hat{\boldsymbol{U}}_s(k+1) = \boldsymbol{U}_s^*(k)\ (s = 1, \ldots, N)$$

Step 6. *Receding Horizon.* Move horizon to the next sampling time, that is, $k+1 \rightarrow k$, go to Step 1, and repeat the above steps.

The online optimization of HSLC, which is a large-scale nonlinear system, is converted into several small-scale systems via distributed computation; thus computational complexity is significantly reduced. In addition, information exchange among neighboring subsystems in a distributed structure via communication can improve control performance. Through this method, the whole temperature evolution of the strip is controlled online, which provides possibilities of producing many new types of steel with high quality (e.g., the multiphase steel). To prove the validation of the proposed strategy, both numerical simulations and experiments on a HSLC experimental apparatus are implemented in the next section.

11.4 Numerical Experiment

To test the validation of the proposed method, low-carbon C2 type steel is taken as an example. The parameters of C2 strip steel are shown in Table 11.1.

11.4.1 Validation of Designed Model

An experiment on full-scale industrial plant is performed with a strip of 3.51 mm in thickness to test the validation of the designed model. In the experiment, the spatial meshing chosen to validate the model is composed of 5 volumes of 0.7 mm in thickness and 30 volumes of 2.7 m in length, that is, $m=5$, $n=30$. The water fluxes in the main cooling section and in the fine cooling section equal to 150 $m^3/(s\ m^2)$ and 75 $m^3/(m^2\ s)$, respectively. The resulting prediction of CT and the measurement of CT are shown in Figure 11.5. The curve of predictive CT is very close to that of measurement. The phenomenon that the predictive curve is smoother than the measurement curve is caused by the second term in the right-hand side of the model (11.1).

Table 11.1 Thermal and physical properties of the strip

Item	Value	Units
Thermal conductivity $\lambda^s_{i,j}$	$\begin{cases} 56.43-\left(0.0363-c\left(v-v_0\right)\right)\times x^s_{0,j} \\ 56.43-(0.0363-c(v-v_0))\times x^s_{m,i} \end{cases}$	$W\ m^{-1}\ K^{-1}$
Thermal diffusivity $a(x^s_{i,j})$	$\begin{cases} 8.65+(5.0-8.65)(x^s_{i,j}-400)/250, & x^s_{i,j}\in[400,650) \\ 5.0+(2.75-5.0)(x^s_{i,j}-650)/50, & x^s_{i,j}\in[650,700) \\ 2.75+(5.25-2.75)(x^s_{i,j}-700)/100, & x^s_{i,j}\in[700,800) \\ 5.25+0.00225(x^s_{i,j}-800), & x^s_{i,j}\in[800,1000] \end{cases}$	$\times 10^{-6} m^2\ s^{-1}$
Temperature of ambient	$25+273.5$	K
Temperature of cooling water	$25+273.5$	K

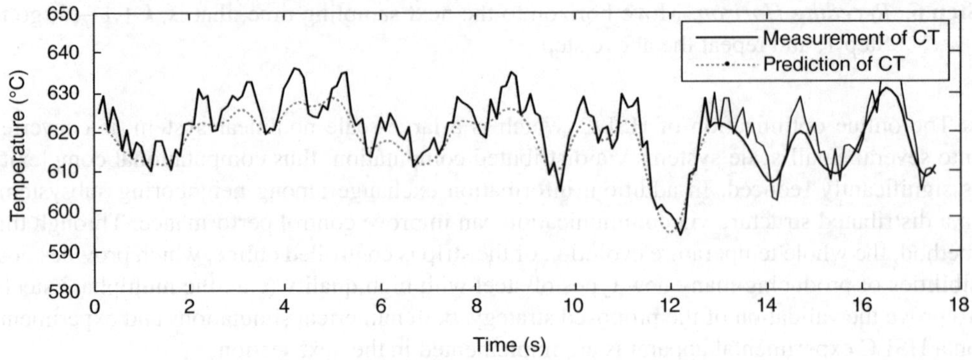

Figure 11.5 Comparison between the predictive CT and the measurement of CT

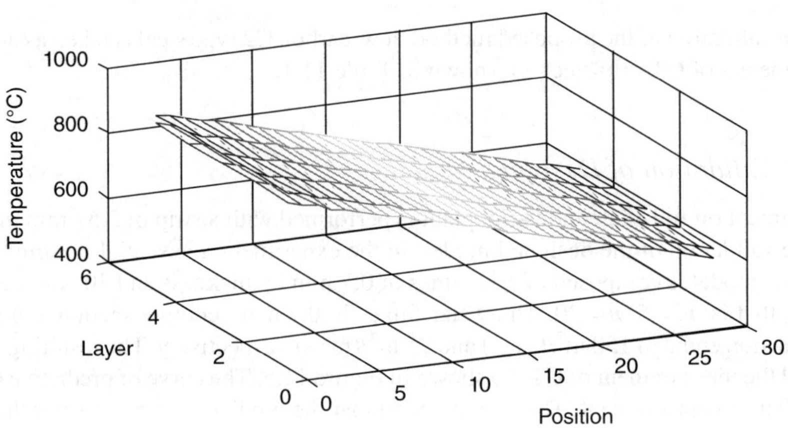

Figure 11.6 Initial states of process model and observer

11.4.2 Convergence of EKF

The convergence of EKF is verified through a simulation here. The initial states (the temperature of each volume) of process model and observer are shown in Figure 11.6, where the initial states of the observer are 30°C higher than those of the process model. The states of header groups are [1,1,1,1,1,1,0,0,0,1,1,1], where 1 stands for opening and 0 means closing. The coiling speed v equals to10.74 m s^{-1}. The water fluxes in the main cooling section and the fine cooling section are 200 m^3/(s m^2) and 150 m^3/(s m^2), respectively. FT is 870°C. For spatial reason, the temperatures of 2nd, 6th, 10th, and 14th subsystems at layers from top surface to central of strip are selected as examples to illustrate the convergence of EKF. Figure 11.7 shows that the temperatures observed by EKF are convergent to temperatures estimated by the process model.

11.4.3 Performance of DMPC Comparing with Centralized MPC

Since the main disturbance comes from FT, FT step signal is sent into a closed-loop system to compare the performance of the DMPC framework proposed and that of centralized MPC.

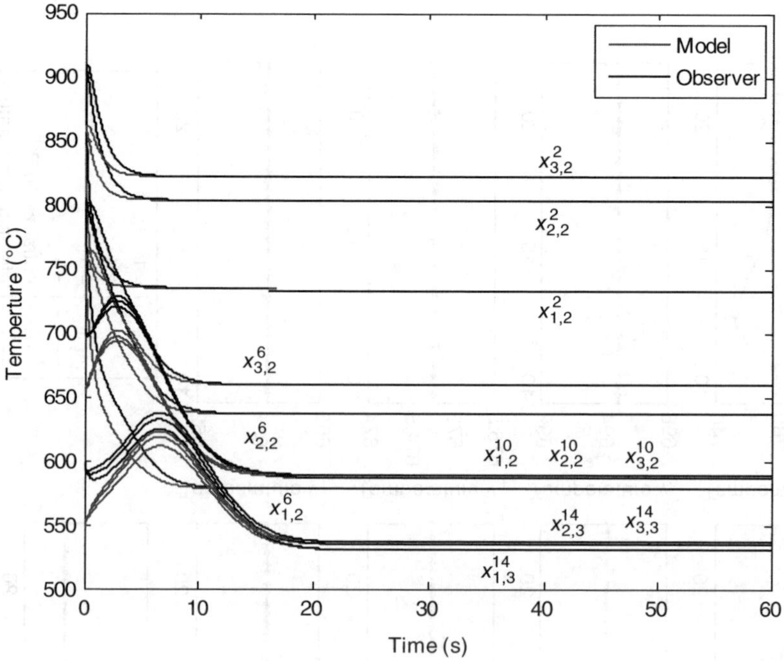

Figure 11.7 Comparison of temperatures estimated by process model and observer

The thickness of strip equals to 5 mm. Set the prediction horizon $P=15$, the control horizon $M=15$, and the control sampling period be 0.37 sec.

As shown in Figure 11.8, the disturbances coming from FT can be eliminated efficiently through DMPC. Figures 11.8 and 11.9 show that the performance and the manipulated variables of the closed-loop system with DMPC are close to those of the centralized MPC when iteration $l \geq 3$.

The time cost of centralized MPC and DMPC framework proposed, running in computers with a CPU of 1.8 G and a memory of 512 M, is illustrated in Table 11.2. It can be seen that the time consumed by DMPC proposed is quite less than that of centralized MPC. The maximum time cost of DMPC with $l=3$ is only 0.1192 sec, which is satisfied with the demand of online computation.

In Table 11.2, the time cost of constructing a system model is included in the time cost of DMPC and centralized MPC.

11.4.4 Advantages of the Proposed DMPC Framework Comparing with the Existing Method

Simulations are performed to illustrate the advantages of the proposed DMPC framework comparing with the existing method in industrial manufactory. Here, the existing method refers to the open-loop and closed-loop control introduced in Section 11.2. The cooling curves of each strip-point with the existing method and the proposed DMPC are shown in Figures 11.10 and 11.11, respectively. The existing method is able to control the CT well, while there is a rough approximation of cooling curve for each strip-point achieved by the

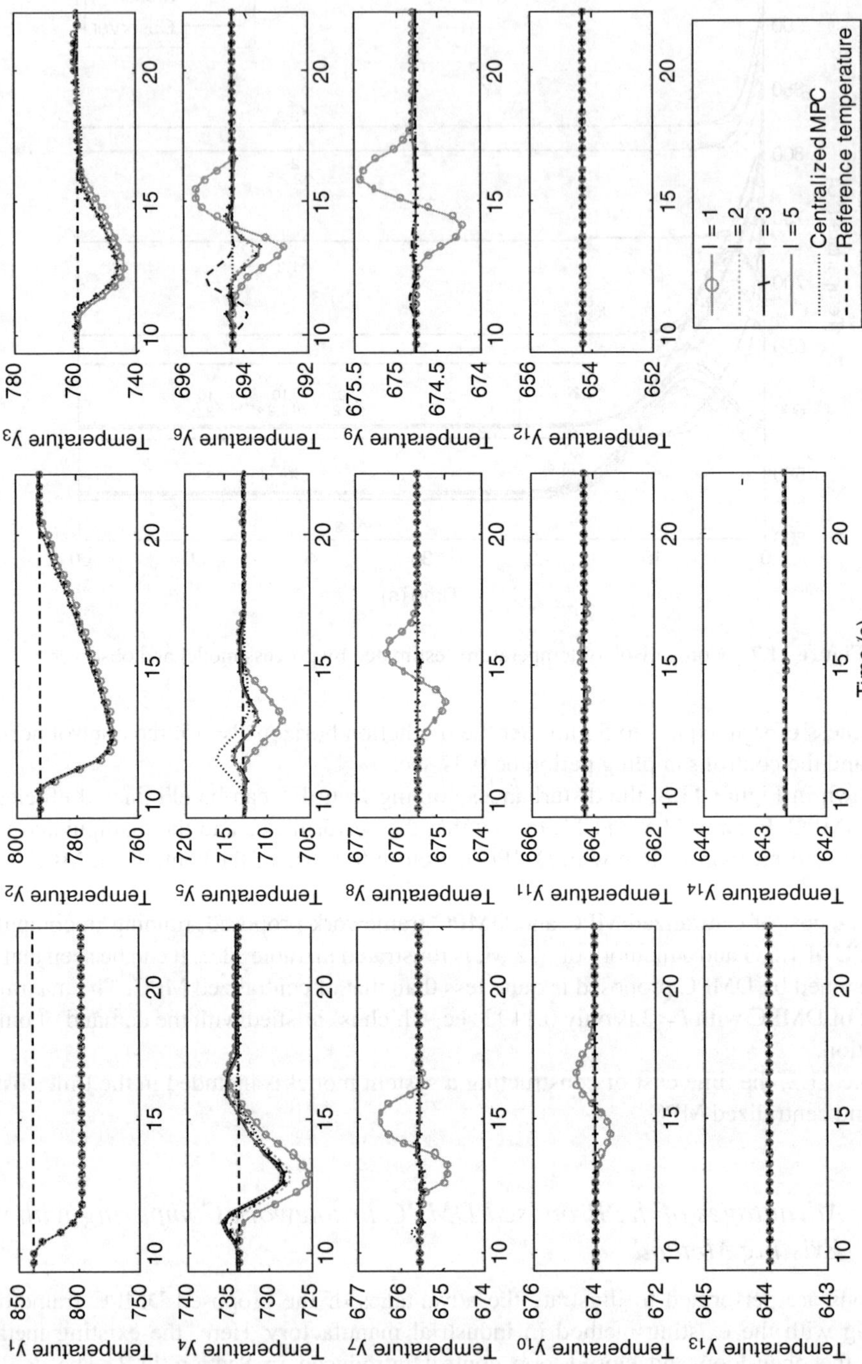

Figure 11.8 Performance comparison of different control strategies (centralized MPC and DMPC framework proposed)

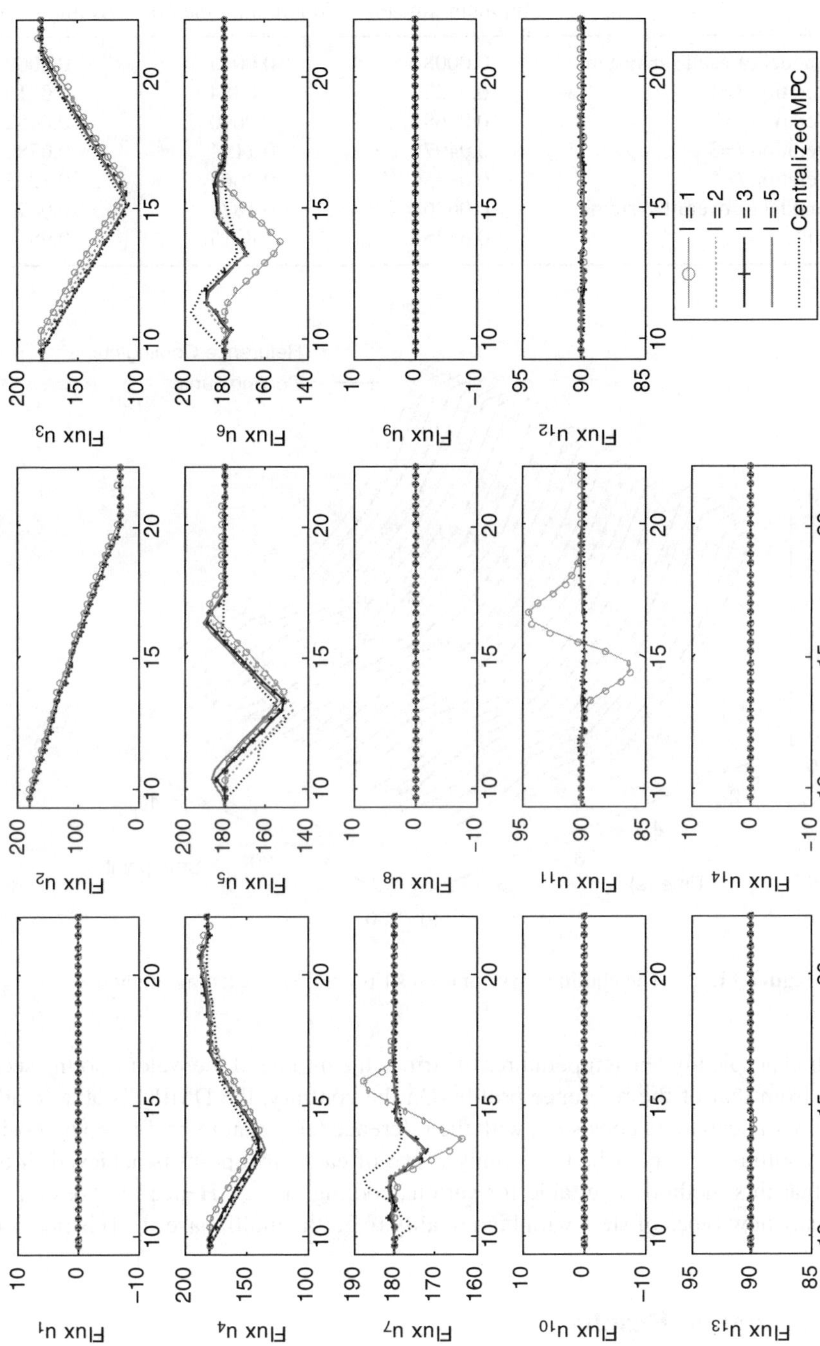

Figure 11.9 Flux of each header group with centralized MPC and DMPC framework proposed

Table 11.2 Computational burdens of DMPC and centralized MPC

Item	Minimum time (s)	Maximum time (s)	Average time (s)
Constructing model of each subsystem	0.0008	0.0012	0.0009
DMPC with iteration l=1	0.0153	0.0484	0.0216
DMPC with iteration l=2	0.0268	0.0690	0.0452
DMPC with iteration l=3	0.0497	0.1194	0.0780
DMPC with iteration l=5	0.0895	0.3665	0.1205
Constructing model of overall system	0.0626	0.1871	0.0890
Centralized MPC	0.6535	1.8915	0.9831

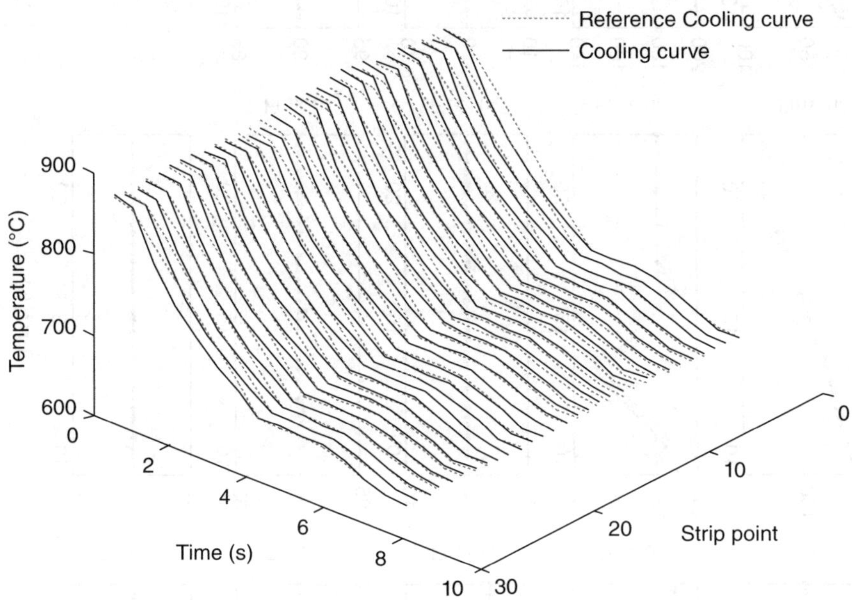

Figure 11.10 The cooling curve of each strip-point with existing method

existing method. Typically, the temperatures of strip at the middle of the water-cooling section are far away from that of the reference profile. On the contrary, the DMPC is able to adjust the temperature of strip to be consistent with the reference temperature profile at any position of the water-cooling section. A better cooling curve of each strip-point is achieved through it. It means that this method is suitable for various cooling curves. Hence the possibility of producing many new types of steel with high quality (e.g., the multiphase steel) is provided.

11.5 Experimental Results

To verify the validation of the method proposed, an experimental result is presented in this subsection. In the experiment, as shown in Figure 11.12, the DMPC framework is run in six

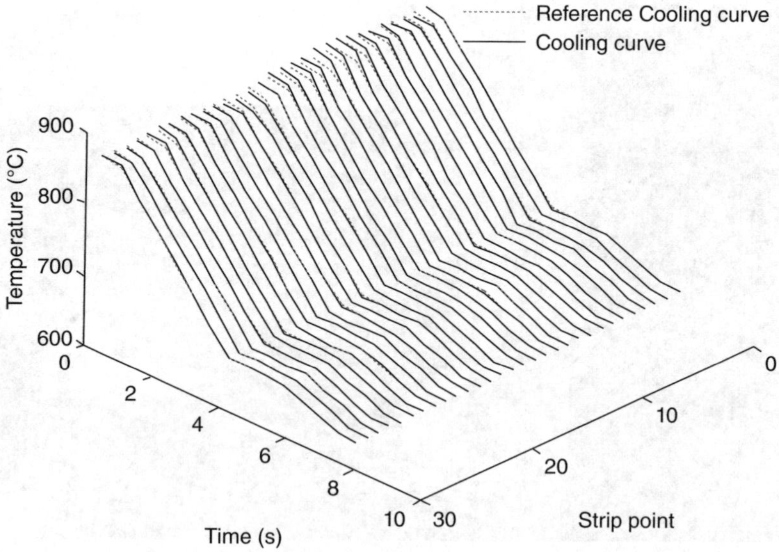

Figure 11.11 The cooling curve of each strip-point with the proposed DMPC framework

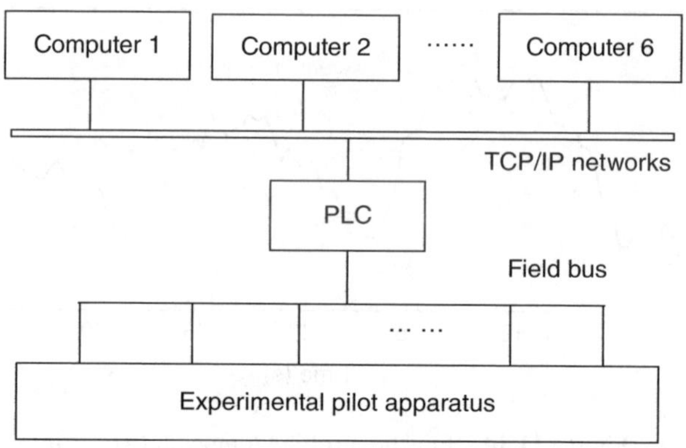

Figure 11.12 The structure of experimental system

computers (two for predictors and observer, the other four for local MPCs). The six computers cooperated with each other to derive the optimal inputs within the sampling period of 0.37 s, and then send the optimal inputs into a PLC that charges the field PI controllers. The runout table experimental apparatus, shown in Figure 11.13, which is a pilot apparatus, is used to test the performance of the DMPC framework.

The good performance of the proposed DMPC is further verified in Figures 11.14–11.16, which show the FT profile, the output of each closed-loop subsystem, and the water flux of each header group in the experiment, respectively. This method can not only control the CT

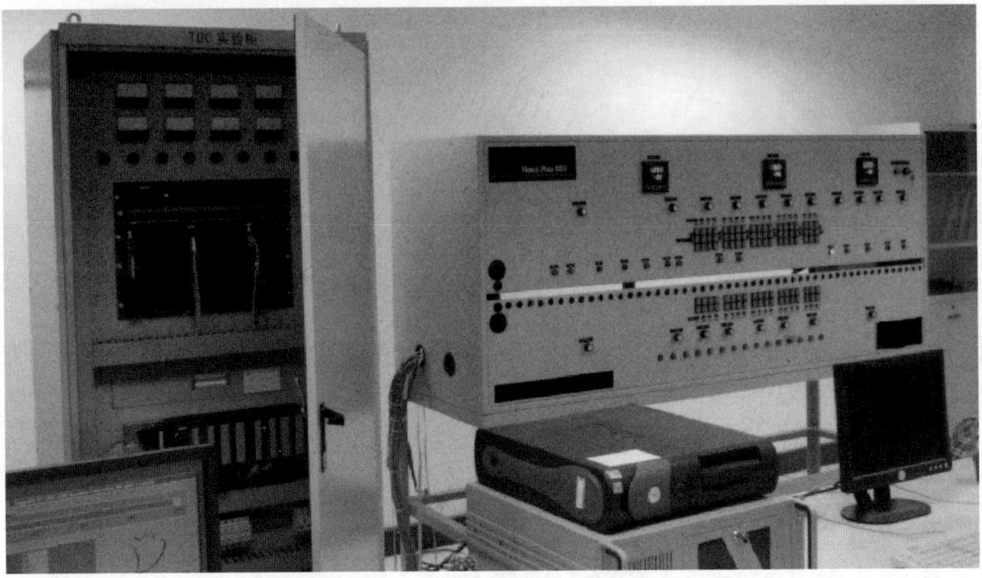

Figure 11.13 Runout table pilot apparatus

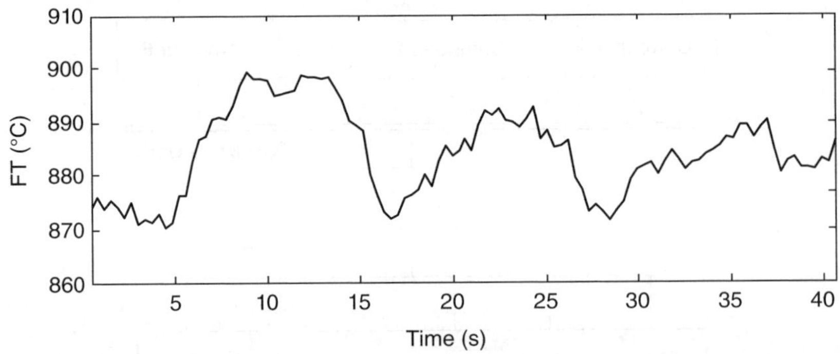

Figure 11.14 Finishing rolling temperature of strip

but also optimize the whole evolution of strip temperature online. The flexibility and precision of the control system are enhanced.

11.6 Conclusion

In the present study, a DMPC framework is designed for the HSLC process, in which the overall system is divided into several interconnected subsystems and each subsystem is controlled by a local MPC. First, the state space representation of each subsystem is developed using the finite volume method. Next, an observer based on EKF is designed to reconstruct the current

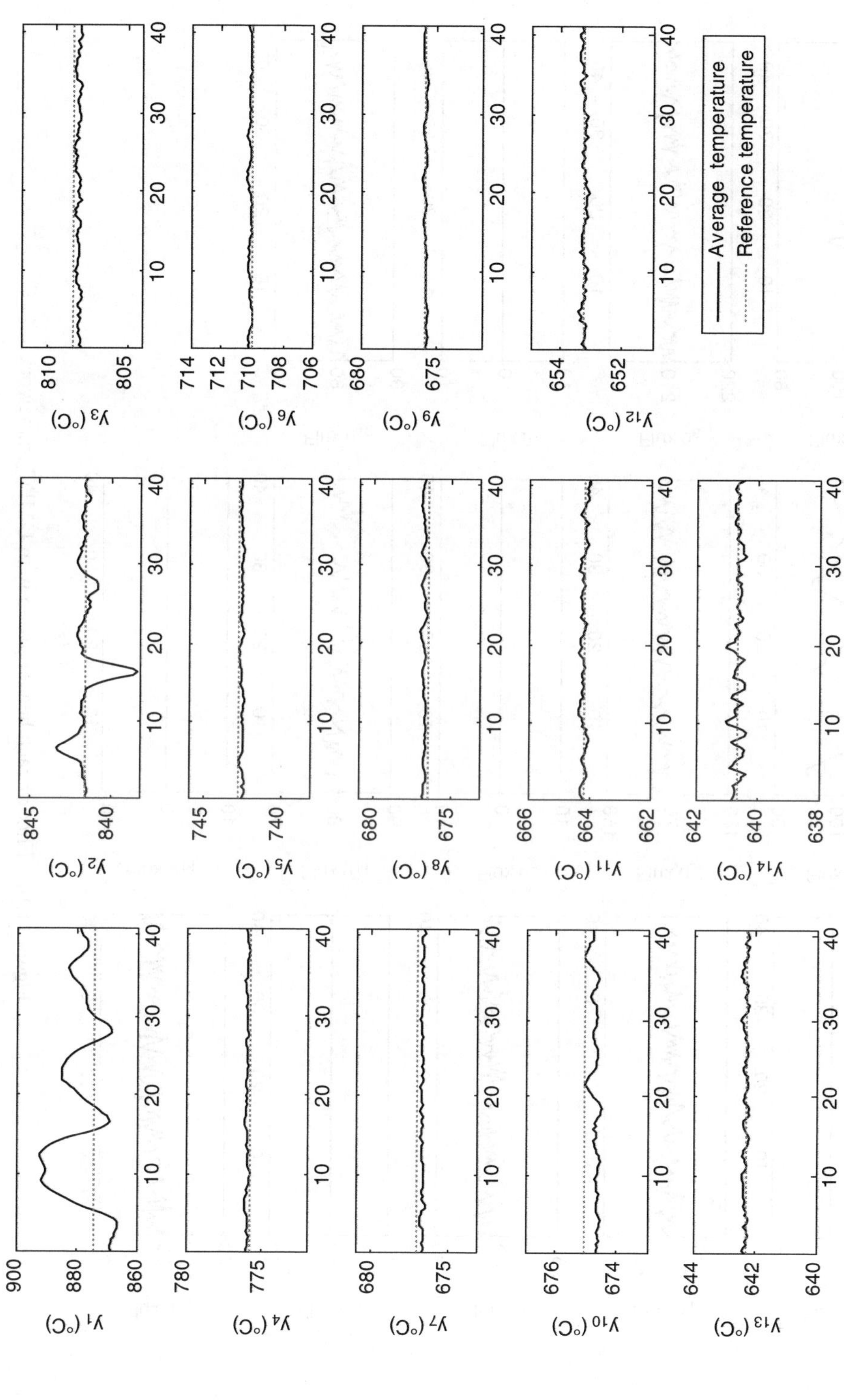

Figure 11.15 Output of each closed-loop subsystem with DMPC framework

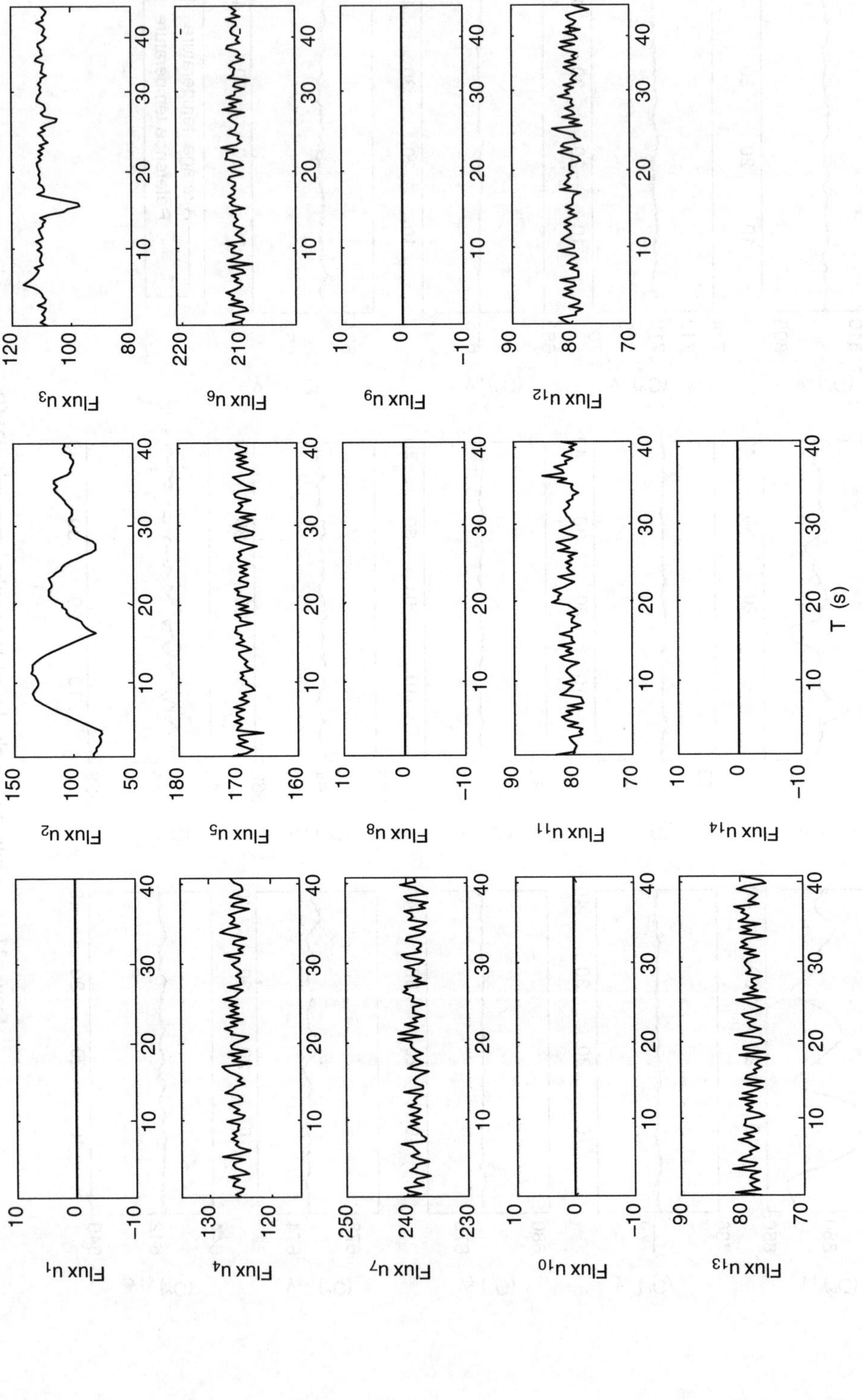

Figure 11.16 Flux of each header group with DMPC framework

temperature distribution of the strip. Then, the EKF sends current states to the corresponding local MPCs. In each local MPC, neighborhood optimization is adopted to enhance the global control performance. Furthermore, to overcome the computational obstacle of nonlinear model, the prediction model of each MPC is linearized around the current operating point at each step. Through this method, the online optimization of strip cooling curve is realized with a few computational burdens; both simulation and experiment results proved the efficiency of the proposed method. Following this method, the whole evolution procedure of strip temperature is controlled online with a relatively high precision, which provides possibilities of producing many new types of high-quality steel (e.g., the multiphase steel).

In HSLC, the aim is to obtain a uniform microstructure of the strip. Therefore, it is more reasonable that the temperature errors among different strip-points caused by FT are eliminated gradually along the rolling direction. However, in this work, these errors are eliminated mainly by the first several header groups, which is a problem to be solved in further works.

12

High-Speed Train Control with Distributed Predictive Control

12.1 Introduction

In recent years, the Chinese railway systems have gone through a massive phase of upgradation and expansion. More and more the China Railway High (CRH)-speed has been used for dedicated passenger lines. It is estimated that by the end of 2020, China will have 18,000 km dedicated passenger lines, with an operating speed of 350 km/h, which will cover almost the whole country.

The CRH consists of electric multiple units (EMUs), which include motor coaches and trailer coaches where the driving force of the CRH is distributed.

In the existing work, most researchers assumed that the couplers between adjacent coaches are stiff so that a complete train can be regarded as a rigid body [132]. In Ref. [133], the dynamics of high-speed train is modeled by a cascade of coaches connected by flexible couplers and subjected to rolling resistances, aerodynamic drag, and wind gust. However, this chapter assumes that the aerodynamic resistance acts on the leading coach only and the rolling resistance acts on every coach. It is more reasonable to assume that the aerodynamic resistance acts on every coach [134], but there still is a drawback that all the empirical constants are the same. In Ref. [135], robust adaptive control is proposed to track velocity and aerodynamic drag is taken as the uncertain variable and acts on each coach equally.

In Ref. [136], the aerodynamic drag in the real situation acts on every coach, and it mostly acts on the leading coach and the last coach. Since the aerodynamic drag is proportional to the square of the speed [136], its influence on high-speed train dynamic behavior becomes significant. Thus, it is very important to emend the dynamic model with real aerodynamics effects.

In this chapter, we establish a new spring–mass model with accurate parameters and a more real hypothesis. During the running of the EMUs, the in-train force is the most important thing for the safe driving. The in-train force is constrained strictly, besides the traction and the brake are limited by the real condition. Considering these constraints, we can use the model predictive control (MPC) method to handle constraints effectively [3]. In the existing research, the article in [137] proposes a cruise control of the longitudinal train longitudinal model based

Distributed Model Predictive Control for Plant-Wide Systems, First Edition. Shaoyuan Li and Yi Zheng.

on the global optimization, in which every subsystem controller has to connect with other subsystem controllers, which can result in high network load. The algorithm is complex and not convenient for engineering applications; the article in [138] improves the existing longitudinal multibody mass dynamic model, which optimizes the distribution of the train's aerodynamics, and then designs a MPC based on the global optimization. It comes across the same problem as the article in [137]; since the N-DMPC with information constraints can lower the high network load and high calculated load caused by the global optimization, the N-DMPC is adopted here in the control of the high-speed train.

12.2 System Description

The movement of EMUs is a very complicated process, which is affected by different forces. The driving of EMUs considers longitudinal movement. Influence of the longitudinal movement includes traction, rolling resistance, and brake force. There are many types of EMUs in the cavil's train systems, such as CRH1, CRH2, CRH3, CRH5, CRH380A, CRH380B, CRH380C, CRH380D, CRH6, and so on. The CRH2 is first used in the sixth time of improving speed of the Chinese railway systems in the year 2007 when the EMUs are first used in Chinese dedicated passenger lines. This type has been the largest amount among the imported other types of EMUs.

Different types of EMUs have different group organizations. In this chapter, we take the CRH2 as our research antetype.

12.3 N-DMPC for High-Speed Trains

12.3.1 Three Types of Force

12.3.1.1 The Traction of EMUs

EMUs are driven by tractions, which are generated by the electricity of the catenary [139]. The calculation of the traction uses the method of linear interpolation by the characteristics of tractions for the CRH2 given in Figure 12.1.

The traction is constrained by the adhesive force. When the force on the wheel rim is bigger than the adhesive force between wheels and rails, the wheel is idling. Thus, the adhesive force determines the upper bound of the traction.

12.3.1.2 The Resistance of EMUs

The resistance of EMUs includes two parts: basic resistance and additive resistance. The friction and concussion among components, surface and air, wheels, and rails are the main reasons of basic resistance. The additive resistance is caused by the path.

It is difficult to analyze the resistance theoretically because the resistance of the EMUs has a lot of parts. Generally, we find an approximate resistance formula for some EMUs from a lot of traction experiments which can be formulated by the following equation [139]:

$$w_0 = c_0 + c_1 v + c_2 v^2 \tag{12.1}$$

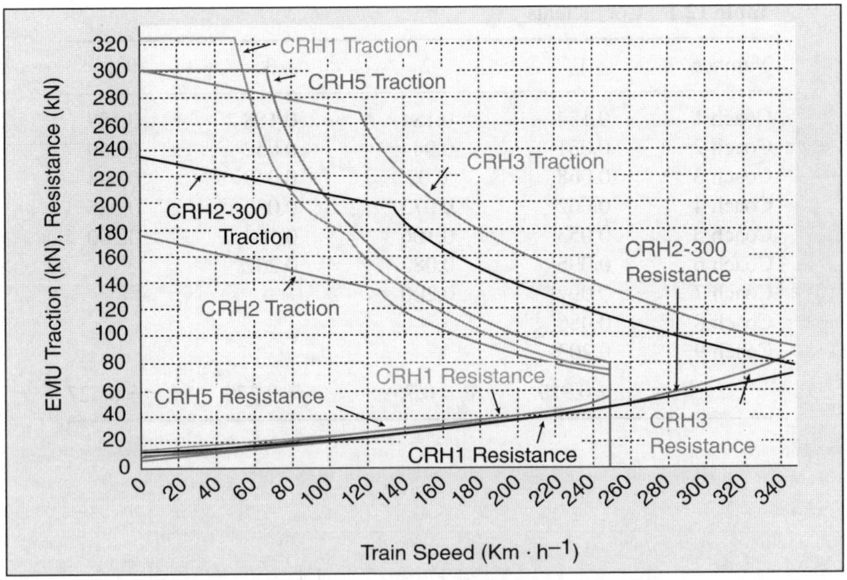

Figure 12.1 Traction characteristics of the CRH2 [140]

where w_0 is the basic resistance of EMUs (N/kN), v is the velocity of the EMUs (km/h), c_0, c_1, c_2 are the empirical constants, respectively, which are correlated with the type of the EMUs. The second order of (12.1) represents the aerodynamic drag. The rest two parts are considered to be rolling and mechanical resistance

$$c_2 = \frac{1}{2}\rho S_i C_i \tag{12.2}$$

where ρ is the air density, S_i is the biggest cross-sectional area of a train, and C_i is the aerodynamic drag coefficient.

Aerodynamic drag coefficient c_i is made up with head coach, middle coaches, and the last coach

$$C_i = \sum_{x=1}^{n} C_{ix} \tag{12.3}$$

where C_{ix} the xth coach aerodynamic drag coefficient and n is the number of cars. According to the book by Tian [136], we can get coefficients of different group organizations as shown in Table 12.1.

From Table 12.1, we can see that different coach has different aerodynamic resistance. The head coach and the last coach take the most part of the aerodynamic resistance of a whole train. In the multimass model, we must use different coefficients respectively.

12.3.1.3 The Brake of EMUs

The brake of EMUs is an external controlled force which is opposite to its current direction. Based on different generating mechanisms, the brake includes friction brake, power brake, and

Table 12.1 Coefficients

Number	1	2	3	4
Coach 1	0.184	0.184	0.188	0.189
Coach 2	0.081	0.044	0.052	0.069
Coach 3	0.068	0.07	0.049	0.094
Coach 4	0.062	0.073	0.083	0.085
Coach 5	0.053	0.106	0.104	0.200
Coach 6	0.095	0.082	0.202	
Coach 7	0.097	0.200		
Coach 8	0.086			
Coach 9	0.203			
	$c_i = 0.929$	$c_i = 0.759$	$c_i = 0.678$	$c_i = 0.627$

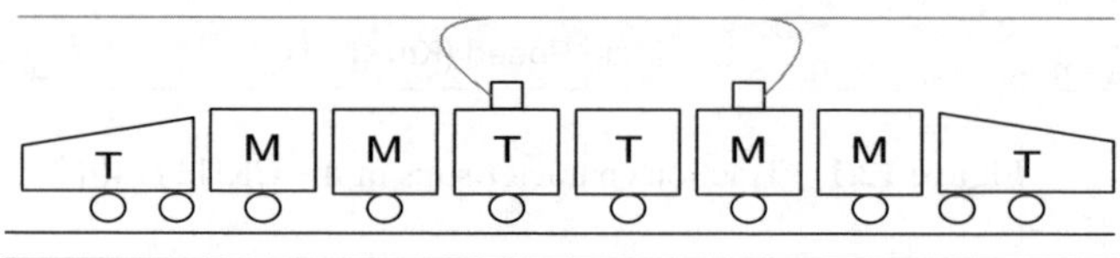

Figure 12.2 Groups of CRH2 ("M" means motor coach and "T" means trailer coach)

electromagnetic brake [139]. EMUs take the united brake with air brake and electric brake, which is called regenerative brake too. In the united brake system, electric brake has a higher priority than the air brake, which is only used in the emergency.

The calculation of the regenerative brake is similar to the calculation of the traction, which needs to refer to the characteristics of the traction with linear interpolation. Air brake is caused by the resistance torque, which is caused by the friction force between brake clips and brake discs.

12.3.2 The Force Analysis of EMUs

There are usually eight compartments in the CRH2 [141], four of which have self-traction locomotive, called motor coach, and the rest are carriages, called trailer coach. The diagram is shown in Figure 12.2.

To simplify the modeling, we only consider basic resistance and traction force shown in Figure 12.3.

Note that traction and braking actions are frequently taken by the control systems interchangeably. Traction and braking actions are combined in our method and represented by u_i where positive means the traction force and negative means the braking force. Based on this assumption, we can derive a spring–mass model as shown in Figure 12.4.

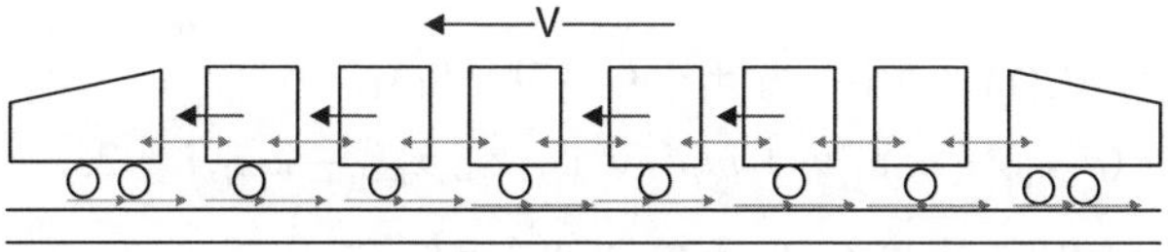

Figure 12.3 Analysis of the force of the CRH2

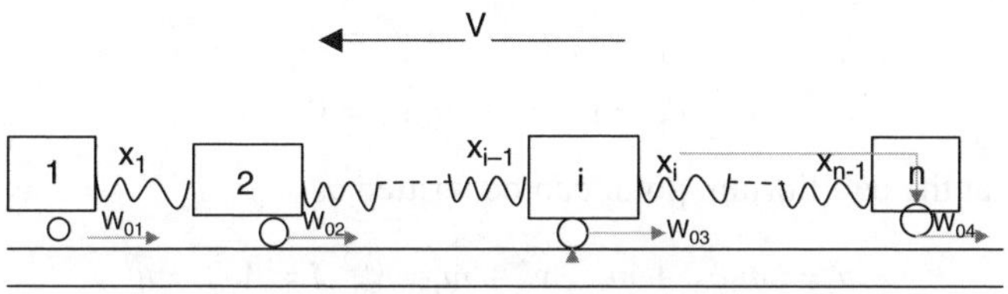

Figure 12.4 Spring–mass model

In-train force that is made by the couplers between two adjacent cars is described by an ideal spring model such that the restoring force of a couple is a function of the relative displacement x between two adjacent cars as

$$\begin{aligned} f_{in_i} &= k_i x_{in_i} + d_i \dot{x}_{in_i}, i = 1, \ldots, n-1, \\ f_{in_0} &= f_{in_n} = 0 \end{aligned} \tag{12.4}$$

where x_{in_i} is the relative displacement of the two adjacent cars, k_i is the stiffness coefficient, d_i is the damping coefficient, and n is the number of cars.

Based on the spring–mass model and Newton's second law of motion – force and acceleration, we can get the longitudinal dynamics of the EMUs [133] as

$$\begin{aligned} m_1 \dot{v}_1 &= u_1 - w_{01} - f_{in_1} \\ m_i \dot{v}_i &= u_i - w_{0i} + f_{in_{i-1}} - f_{in_i}, i = 2, \ldots, n-1 \\ m_n \dot{v}_n &= u_n - w_{0n} + f_{in_{n-1}} \\ \dot{x}_i &= v_i - v_{i+1}, i = 1, \ldots, n-1 \end{aligned} \tag{12.5}$$

where m_i is the mass of the ith coach, w_{0i} is the ith coach's basic resistance, f_{in_i} is the in-train force, v_i and $\dot{v}_i$ represent the velocity and acceleration of the ith coach in EMUs, respectively, u_i is the traction of the ith car that includes the braking force when u_i is negative, x_i is the relative displacement of the two adjacent cars, and n is the number of cars.

12.3.3 Model of CRH2

In the last section we get the force analysis of EMUs. From (12.5), we can get longitudinal dynamic model of the CRH2. Assuming damping effect of the couplers $d_i = 0$, substituting (12.1) into (12.5), we can get

$$\begin{aligned} m_1\dot{v}_1 &= u_1 - (m_1c_{01} + m_1c_{11}v_1 + m_ic_{21}v_1^2) - k_1x_1 \\ m_i\dot{v}_i &= u_i - (m_ic_{0i} + m_ic_{1i}v_i + m_ic_{2i}v_i^2) + k_{i-1}x_{i-1} - k_ix_i, i = 2, \dots, n-1 \\ m_n\dot{v}_n &= u_n - (m_nc_{0n} + m_nc_{1n}v_n + m_nc_{2n}v_n^2) + k_{n\text{-}1}x_{n\text{-}1} \\ \dot{x}_i &= v_i - v_{i+1}, i = 1, \dots, n-1 \end{aligned} \tag{12.6}$$

Assuming that the equilibrium state has a cruising speed

$$\begin{aligned} v_1^e &= v_2^e = \cdots = v_n^e = v_r \\ \dot{v}_1^e &= \dot{v}_2^e = \cdots = \dot{v}_n^e = 0 \end{aligned} \tag{12.7}$$

The inputs u_i^e at the equilibrium point can be written as

$$u_i^e = m_ic_{0i} + m_ic_{1i}v_i^e + m_ic_{2i}v_i^{e2}, i = 1, \dots, n \tag{12.8}$$

$$x_i = x_i^e + \delta x_i, v_i = v_i^e + \delta v_i, u_i = u_i^e + \delta u_i \tag{12.9}$$

Let

$$\begin{aligned} \mathbf{x} &= [\delta v_1, \delta v_2, \dots, \delta v_n, \delta x_1, \delta x_2, \dots, \delta x_{n-1}]^{\mathbf{T}} \\ \mathbf{u} &= [\delta u_1, \delta u_2, \dots, \delta u_n]^{\mathbf{T}} \end{aligned} \tag{12.10}$$

We can get the following linearized equations by substituting (12.9) into (12.6).

$$\begin{aligned} \dot{\mathbf{x}} &= \mathbf{A}\mathbf{x} + \mathbf{B}\mathbf{u} \\ \mathbf{y} &= \mathbf{C}\mathbf{x} \end{aligned} \tag{12.11}$$

where

$$\mathbf{A} = \begin{bmatrix} \mathbf{A}_{11} & \mathbf{A}_{12} \\ \mathbf{A}_{21} & \mathbf{A}_{22} \end{bmatrix}, \quad \mathbf{A}_{11} = -\text{diag}(c_{11} + c_{21}v_r, \dots, c_{1n} + c_{2n}v_r), \mathbf{A}_{22} = \mathbf{0}_{(n-1)\times(n-1)}$$

$$\mathbf{A}_{12} = \begin{bmatrix} -\frac{k_1}{m_1} & 0 & \cdots & 0 & 0 \\ \frac{k_1}{m_2} & -\frac{k_2}{m_2} & \cdots & 0 & 0 \\ \vdots & \vdots & \ddots & \vdots & \vdots \\ 0 & \cdots & 0 & \frac{k_{n-2}}{m_{n-1}} & -\frac{k_{n-1}}{m_{n-1}} \\ 0 & \cdots & 0 & 0 & \frac{k_{n-1}}{m_n} \end{bmatrix}$$

$$\mathbf{A}_{21} = \begin{bmatrix} 1 & -1 & 0 & \cdots & 0 & 0 \\ 0 & 1 & -1 & \cdots & 0 & 0 \\ \vdots & \vdots & \vdots & \ddots & \vdots & \vdots \\ 0 & 0 & 0 & \cdots & 1 & -1 \end{bmatrix}, \quad \mathbf{B} = \begin{bmatrix} \mathbf{B}_{11} \\ \mathbf{0}_{(n-1)\times n} \end{bmatrix},$$

$$\mathbf{B}_{11} = \text{diag}\left(\frac{1}{m_1}, \frac{1}{m_2}, \dots, \frac{1}{m_n}\right), \quad \mathbf{C} = \begin{bmatrix} \mathbf{I}_{n\times n} & \mathbf{0}_{(n-1)\times(n-1)} \end{bmatrix}$$

To represent the performance of a local subsystem of a coach, we define state variables of a local subsystem by the velocity of the coach and the relative displacement of the adjacent spring, as shown in (12.12):

$$\mathbf{Z}_i = \begin{bmatrix} v_i \\ x_i \end{bmatrix}, i = 1, \ldots, n-1 \tag{12.12}$$

With the definition of the state variables, we can rewrite the longitudinal dynamic model by n subsystems as follows:

$$\begin{bmatrix} \dot{v}_1 \\ \dot{x}_1 \end{bmatrix} = \begin{bmatrix} -(c_{11} + c_{21}v_r) & -k_1/m_1 \\ 1 & 0 \end{bmatrix} \begin{bmatrix} v_1 \\ x_1 \end{bmatrix} + \begin{bmatrix} 0 & 0 \\ -1 & 0 \end{bmatrix} \begin{bmatrix} v_2 \\ x_2 \end{bmatrix} + \begin{bmatrix} 1/m_1 \\ 0 \end{bmatrix} u_1 \tag{12.13}$$

$$\begin{bmatrix} \dot{v}_i \\ \dot{x}_i \end{bmatrix} = \begin{bmatrix} -(c_{1i} + c_{2i}v_r) & -k_i/m_i \\ 1 & 0 \end{bmatrix} \begin{bmatrix} v_i \\ x_i \end{bmatrix} + \begin{bmatrix} 0 & 0 \\ -1 & 0 \end{bmatrix} \begin{bmatrix} v_{i+1} \\ x_{i+1} \end{bmatrix} + \begin{bmatrix} 0 & k_i/m_i \\ 0 & 0 \end{bmatrix} \begin{bmatrix} v_{i-1} \\ x_{i-1} \end{bmatrix} + \begin{bmatrix} 1/m_i \\ 0 \end{bmatrix} u_i$$
$$i = 2, \ldots, n-1 \tag{12.14}$$

$$\dot{v}_n = -(c_{1n} + c_{2n}v_r) \times v_n + k_{n-1}/m_n \times x_{n-1} + 1/m_n \times u_n \tag{12.15}$$

In the real train system, there are n coaches with $n-1$ couplers. As the last coach, there is only the velocity v_n of the last coach without the relative displacement x_n. As a result, there is only one state variable in the local subsystem of the last coach shown in (12.15). To unite the representing forms, we assume an augmented variable x_n in the last coach. All the parameters related to the x_n are set up with zero as shown in (12.16).

$$\begin{bmatrix} \dot{v}_n \\ \dot{x}_n \end{bmatrix} = \begin{bmatrix} -(c_{1n} + c_{2n}v_r) & 0 \\ 0 & 0 \end{bmatrix} \begin{bmatrix} v_n \\ x_n \end{bmatrix} + \begin{bmatrix} 0 & k_{n-1}/m_{n-1} \\ 0 & 0 \end{bmatrix} \begin{bmatrix} v_{n-1} \\ x_{n-1} \end{bmatrix} + \begin{bmatrix} 1/m_n \\ 0 \end{bmatrix} u_n \tag{12.16}$$

The subsystem can be rewritten as follows:

$$\begin{aligned} \dot{\mathbf{z}}_1 &= \mathbf{A}_{11}\mathbf{z}_1 + \mathbf{B}_1 u_1 + \mathbf{A}_{12}\mathbf{z}_2 \\ \dot{\mathbf{z}}_i &= \mathbf{A}_{ii}\mathbf{z}_i + \mathbf{B}_i u_i + \mathbf{A}_{ii+1}\mathbf{z}_{i+1} + \mathbf{A}_{ii-1}\mathbf{z}_{i-1}\ i = 2, \ldots, n-1 \\ \dot{\mathbf{z}}_n &= \mathbf{A}_{nn}\mathbf{z}_n + \mathbf{B}_n u_n + \mathbf{A}_{nn-1}\mathbf{z}_{n-1} \end{aligned} \tag{12.17}$$

where $\mathbf{A}_{ii}$ is the matrix of the ith subsystem, which is not influenced by other subsystems, $\mathbf{A}_{ii+1}$ is the matrix of the ith subsystem, which is influenced by the $i+1$th coach, and $\mathbf{A}_{ii-1}$ is the matrix of the ith subsystem, which is influenced by the $i-1$th coach.

In Figure 12.5, we can see how subsystems communicate and work. To represent the system better, we have to propose some definitions.

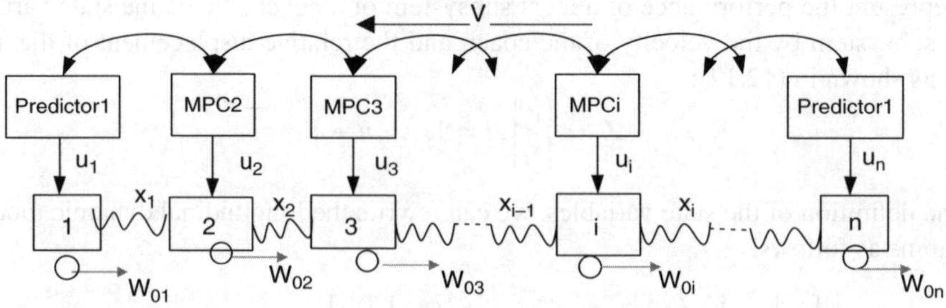

Figure 12.5 The structure diagram of the distributed model predictive control

Adjacent subsystems. If subsystems $\mathcal{S}_i$ and $\mathcal{S}_j$ interact with each other and the output of the subsystem $\mathcal{S}_i$ is the input of the subsystem $\mathcal{S}_j$, subsystem $\mathcal{S}_j$ is called the input-adjacent subsystem of the subsystem $\mathcal{S}_i$ and subsystem $\mathcal{S}_i$ is called the output-adjacent subsystem of the subsystem $\mathcal{S}_j$. By this way, subsystems $\mathcal{S}_i$ and $\mathcal{S}_j$ are called adjacent subsystems or neighbors.

Neighborhood of the subsystem. All the input (output) of neighbor's set of the subsystem $\mathcal{S}_i$ is defined as the input (output) of neighborhood $\mathcal{P}_{+i}$ of subsystem $\mathcal{S}_i$:

$$\mathcal{P}_{+i} = \{\mathcal{S}_i, \mathcal{S}_j | \mathcal{S}_j \text{ is the input neighbor of } \mathcal{S}_i\}$$

$$\mathcal{P}_{-i} = \{\mathcal{S}_i, \mathcal{S}_j | \mathcal{S}_j \text{ is the output neighbor of } \mathcal{S}_i\}$$

Neighborhood of the subsystem $\mathcal{S}_i$: all the set of neighbors of the subsystem $\mathcal{S}_i$ is shown as follows:

$$\mathcal{P}_i = \mathcal{P}_{+i} \cup \mathcal{P}_{-i}$$

Assumption 12.1

(a) subsystems' controllers act synchronously;
(b) during a sample time, the controllers contact others only once;
(c) there is one step delay during the communication course.

It is not rigorous for these three assumptions in the real systems. Normally the calculation time is much longer than the sample time, so the assumption (a) is not rigorous. The assumption (b) is to reduce the network communication between the controllers and improve the reliability of the algorithm. In the real systems, the instantaneous communication is not existence, so the assumption (c) is necessary.

Based on the definition of the neighborhood, the longitudinal dynamic model (12.17) can be rewritten as neighborhood subsystems

$$\dot{\mathbf{Z}}_{n1} = \begin{bmatrix} \dot{\mathbf{z}}_1 \\ \dot{\mathbf{z}}_2 \end{bmatrix} = \begin{bmatrix} \mathbf{A}_{11} & \mathbf{A}_{12} \\ \mathbf{A}_{21} & \mathbf{A}_{22} \end{bmatrix} \mathbf{Z}_{n1} + \begin{bmatrix} \mathbf{0} & \mathbf{0} \\ \mathbf{0} & \mathbf{A}_{23} \end{bmatrix} \mathbf{Z}_{n2} + \begin{bmatrix} \mathbf{B}_1 & \mathbf{0} \\ \mathbf{0} & \mathbf{B}_2 \end{bmatrix} \begin{bmatrix} u_1 \\ u_2 \end{bmatrix} \tag{12.18}$$

$$\dot{\mathbf{Z}}_{ni} = \begin{bmatrix} \dot{\mathbf{z}}_{i-1} \\ \dot{\mathbf{z}}_i \\ \dot{\mathbf{z}}_{i+1} \end{bmatrix} = \begin{bmatrix} \mathbf{A}_{i-1i-1} & \mathbf{A}_{i-1i} & \mathbf{0} \\ \mathbf{A}_{ii-1} & \mathbf{A}_{ii} & \mathbf{A}_{ii+1} \\ \mathbf{0} & \mathbf{A}_{i+1i} & \mathbf{A}_{i+1i+1} \end{bmatrix} \mathbf{Z}_{ni} + \begin{bmatrix} \mathbf{A}_{i-1i-2} & \mathbf{0} & \mathbf{0} \\ \mathbf{0} & \mathbf{0} & \mathbf{0} \\ \mathbf{0} & \mathbf{0} & \mathbf{0} \end{bmatrix} \mathbf{Z}_{ni-1}$$
$$+ \begin{bmatrix} \mathbf{0} & \mathbf{0} & \mathbf{0} \\ \mathbf{0} & \mathbf{0} & \mathbf{0} \\ \mathbf{0} & \mathbf{0} & \mathbf{A}_{i+1i+2} \end{bmatrix} \mathbf{Z}_{ni+1} + \begin{bmatrix} \mathbf{B}_{i-1i-1} & \mathbf{0} & \mathbf{0} \\ \mathbf{0} & \mathbf{B}_{ii} & \mathbf{0} \\ \mathbf{0} & \mathbf{0} & \mathbf{B}_{i+1i+1} \end{bmatrix} \begin{bmatrix} u_{i-1} \\ u_i \\ u_{i+1} \end{bmatrix}$$
$$i = 2, \ldots, n-1 \tag{12.19}$$

$$\dot{\mathbf{Z}}_{nn} = \begin{bmatrix} \dot{\mathbf{z}}_{n-1n-1} \\ \dot{\mathbf{z}}_{nn} \end{bmatrix} = \begin{bmatrix} \mathbf{A}_{n-1n-1} & \mathbf{A}_{n-1n} \\ \mathbf{A}_{nn-1} & \mathbf{A}_{nn} \end{bmatrix} \mathbf{Z}_{nn}$$
$$+ \begin{bmatrix} \mathbf{A}_{n-1n-2} & 0 \\ 0 & 0 \end{bmatrix} \mathbf{Z}_{nn-1} + \begin{bmatrix} \mathbf{B}_{n-1} & \mathbf{0} \\ \mathbf{0} & \mathbf{B}_n \end{bmatrix} \begin{bmatrix} u_{n-1} \\ u_n \end{bmatrix} \tag{12.20}$$

where Z_{ni} is state variable of the ith neighborhood subsystem in this system. Equations (12.18)–(12.20) can be simplified as follows:

$$\begin{aligned} \dot{\mathbf{Z}}_{n1} &= \hat{\mathbf{A}}_{11}\mathbf{Z}_{n1} + \hat{\mathbf{A}}_{12}\mathbf{Z}_{n2} + \hat{\mathbf{B}}_1\hat{\mathbf{U}}_1 \\ \dot{\mathbf{Z}}_{ni} &= \hat{\mathbf{A}}_{ii}\mathbf{Z}_{ni} + \hat{\mathbf{A}}_{ii-1}\mathbf{Z}_{ni-1} + \hat{\mathbf{A}}_{ii+1}\mathbf{Z}_{ni+1} + \hat{\mathbf{B}}_i\hat{\mathbf{U}}_i \\ & i = 2, \ldots, n-1 \\ \dot{\mathbf{Z}}_{nn} &= \hat{\mathbf{A}}_{nn}\mathbf{Z}_{nn} + \hat{\mathbf{A}}_{nn-1}\mathbf{Z}_{nn-1} + \hat{\mathbf{B}}_n\hat{\mathbf{U}}_n \end{aligned} \tag{12.21}$$

12.3.4 Performance Index

To guarantee that the train would work at the given velocity and at the same time the traction force would stay the smallest, we propose a global optimization performance index as

$$J(k) = \sum_{i=1}^{P} \|\mathbf{y}(k+i|k) - \mathbf{y}_r(i)\|^2_{\mathbf{Q}_i} + \sum_{i=1}^{M} \|\mathbf{u}(k+i-1|k)\|^2_{\mathbf{R}_i} \tag{12.22}$$

where P is the optimization horizon, M is the control horizon, and Q_i and R_i are the weight matrices.

Similarly, we can get a networked optimization performance index with information constraints as follows:

$$\begin{aligned} \overline{J}_i(k) &= \sum_{j \in \mathcal{N}_i^{\text{out}}} J_i(k) \\ &= \sum_{j \in \mathcal{N}_i^{\text{out}}} \left[\sum_{l=1}^{P} \left\| \hat{\mathbf{y}}_j(k+l|k) - \mathbf{y}_j^r(k+l|k) \right\|^2_{\mathbf{Q}_j} + \sum_{l=1}^{M} \|\Delta\mathbf{u}_j(k+l-1|k)\|^2_{\mathbf{R}_j} \right] \end{aligned} \tag{12.23}$$

12.3.5 *Optimization Problem*

According to the N-DMPC introduced in Chapter 7, the optimization problem of each subsystem-based MPC can be written as per the following quadratic programming (QP):

$$\min J_i(k) = \mathbf{U}_i^{\mathbf{T}}\mathbf{H}\mathbf{U}_i + 2\mathbf{f}^{\mathbf{T}}\mathbf{U}_i$$

$$\begin{cases} u_{\min} - u^e \le U_i \le u_{\max} - u^e \\ x_{\min} \le \hat{\mathbf{A}}x_i(k) + \hat{\mathbf{B}}\mathbf{U}_i \le x_{\max} \end{cases}$$

where

$$\mathbf{H} = \tilde{\mathbf{B}}_i^{\mathbf{T}}\tilde{\mathbf{Q}}\tilde{\mathbf{B}}_i + \tilde{\mathbf{R}}$$

Head coach:

$$f = -\tilde{\mathbf{B}}_i^{\mathbf{T}} \times \tilde{\mathbf{Q}} \times (\mathbf{Y}_r - \tilde{\mathbf{A}}_{ii}\mathbf{Z}_{ni}(k-1) - \tilde{\mathbf{A}}_{ii-1}\mathbf{Z}_{ni-1}(k-1)) + \tilde{u}^e$$

Middle coach:

$$f = -\tilde{\mathbf{B}}_i^{\mathbf{T}} \times \tilde{\mathbf{Q}} \times (\mathbf{Y}_r - \tilde{\mathbf{A}}_{ii}\mathbf{Z}_{ni}(k-1) - \tilde{\mathbf{A}}_{ii+1}\mathbf{Z}_{ni+1}(k-1) - \tilde{\mathbf{A}}_{ii-1}\mathbf{Z}_{ni-1}(k-1)) + \tilde{u}^e$$

Tail coach:

$$f = -\tilde{\mathbf{B}}_i^{\mathbf{T}} \times \tilde{\mathbf{Q}} \times (\mathbf{Y}_r - \tilde{\mathbf{A}}_{ii}\mathbf{Z}_{ni}(k-1) - \tilde{\mathbf{A}}_{ii+1}\mathbf{Z}_{ni+1}(k-1)) + \tilde{u}^e$$

$$\tilde{\mathbf{A}}_{ii} = \begin{bmatrix} \hat{\mathbf{C}}_i\hat{\mathbf{A}}_{ii} \\ \hat{\mathbf{C}}_i\hat{\mathbf{A}}_{ii}^2 \\ \vdots \\ \hat{\mathbf{C}}_i\hat{\mathbf{A}}_{ii}^P \end{bmatrix}, \tilde{\mathbf{A}}_{ii+1} = \begin{bmatrix} \hat{\mathbf{C}}_i\hat{\mathbf{A}}_{ii+1} & \mathbf{0} & \mathbf{0} & \mathbf{0} \\ \hat{\mathbf{C}}_i\hat{\mathbf{A}}_{ii}\hat{\mathbf{A}}_{ii+1} & \hat{\mathbf{C}}_i\hat{\mathbf{A}}_{ii+1} & \mathbf{0} & \mathbf{0} \\ \vdots & \vdots & \vdots & \vdots \\ \hat{\mathbf{C}}_i\hat{\mathbf{A}}_{ii}^{P-1}\hat{\mathbf{A}}_{ii+1} & \hat{\mathbf{C}}_i\hat{\mathbf{A}}_{ii}^{P-2}\hat{\mathbf{A}}_{ii+1} & \cdots & \hat{\mathbf{C}}_i\hat{\mathbf{A}}_{ii+1} \end{bmatrix}$$

$$\tilde{\mathbf{A}}_{ii+1} = \begin{bmatrix} \hat{\mathbf{C}}_i\hat{\mathbf{A}}_{ii+1} & \mathbf{0} & \mathbf{0} & \mathbf{0} \\ \hat{\mathbf{C}}_i\hat{\mathbf{A}}_{ii}\hat{\mathbf{A}}_{ii+1} & \hat{\mathbf{C}}_i\hat{\mathbf{A}}_{ii+1} & \mathbf{0} & \mathbf{0} \\ \vdots & \vdots & \vdots & \vdots \\ \hat{\mathbf{C}}_i\hat{\mathbf{A}}_{ii}^{P-1}\hat{\mathbf{A}}_{ii+1} & \hat{\mathbf{C}}_i\hat{\mathbf{A}}_{ii}^{P-2}\hat{\mathbf{A}}_{ii+1} & \cdots & \hat{\mathbf{C}}_i\hat{\mathbf{A}}_{ii+1} \end{bmatrix},$$

$$\tilde{\mathbf{B}}_i = \begin{bmatrix} \hat{\mathbf{C}}_i\mathbf{B}_i & \mathbf{0} & \mathbf{0} & \mathbf{0} \\ \hat{\mathbf{C}}_i\hat{\mathbf{A}}_{ii}\mathbf{B}_i & \hat{\mathbf{C}}_i\hat{\mathbf{B}}_i & \mathbf{0} & \mathbf{0} \\ \vdots & \vdots & \vdots & \vdots \\ \hat{\mathbf{C}}_i\hat{\mathbf{A}}_{ii}^{P-1}\hat{\mathbf{B}}_i & \cdots & \cdots & \sum_{i=1}^{P-M+1} \hat{\mathbf{C}}_i\hat{\mathbf{A}}_{ii}^{i-1}\hat{\mathbf{B}}_i \end{bmatrix}$$

12.4 Simulation Results

We take the half of CRH2 EMUs as the simulation model as shown in Figure 12.6.

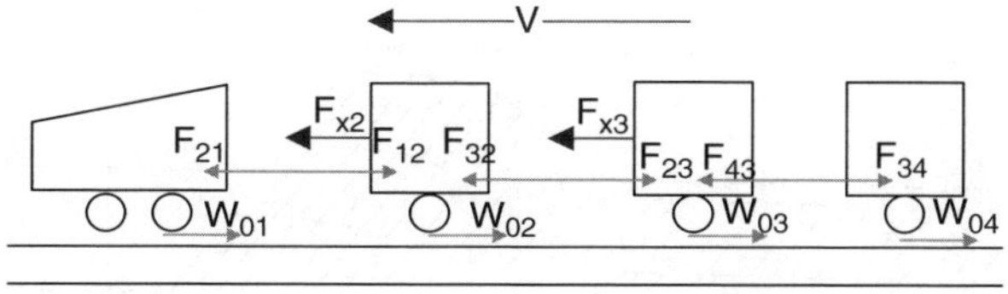

Figure 12.6 Half of the CRH2 EMUs schematic diagram

12.4.1 Parameters of CRH2

The EMU's configuration in the simulation is according to [141]; the mass of the first coach m_1 is 42.8 ton, the second coach m_2 is 48 ton, the third coach m_3 is 46.5 ton, and the fourth coach m_4 is 42 ton. The resistance coefficient c_0 is 0.8806 N/kN/kg, c_1 is 0.007444 N/kN m/s kg, and c_2 is 0.0001143 N/kN m^2/s^2 kg. The optimization horizon P is 4, the control horizon M is 2, the sample time T_s is 20 s, and the spring coefficient k_i is 10488 kN/m.

12.4.2 Simulation Matrix

Substituting parameters we can get four neighborhood subsystems:

$$\mathbf{A}_{11} = \begin{bmatrix} -(c_1 + c_2 v_r) & -k_1/m_1 \\ 1 & 0 \end{bmatrix}, \quad \mathbf{A}_{12} = \begin{bmatrix} 0 & 0 \\ -1 & 0 \end{bmatrix}$$

$$\mathbf{B}_1 = \begin{bmatrix} 1/m_1 \\ 0 \end{bmatrix}, \quad \mathbf{C}_1 = \begin{bmatrix} 1 & 0 \end{bmatrix}$$

$$\mathbf{A}_{22} = \begin{bmatrix} -(c_1 + c_2 v_r) & -k_1/m_2 \\ 1 & 0 \end{bmatrix}, \quad \mathbf{A}_{23} = \begin{bmatrix} 0 & 0 \\ -1 & 0 \end{bmatrix}$$

$$\mathbf{A}_{21} = \begin{bmatrix} 0 & k_1/m_2 \\ 0 & 0 \end{bmatrix}, \quad \mathbf{B}_2 = \begin{bmatrix} 1/m_2 \\ 0 \end{bmatrix}, \quad \mathbf{C}_2 = \begin{bmatrix} 1 & 0 \end{bmatrix}$$

$$\mathbf{A}_{33} = \begin{bmatrix} -(c_1 + c_2 v_r) & -k_1/m_3 \\ 1 & 0 \end{bmatrix}, \quad \mathbf{A}_{34} = \begin{bmatrix} 0 & 0 \\ -1 & 0 \end{bmatrix}$$

$$\mathbf{A}_{32} = \begin{bmatrix} 0 & k_1/m_3 \\ 0 & 0 \end{bmatrix}, \quad \mathbf{B}_3 = \begin{bmatrix} 1/m_3 \\ 0 \end{bmatrix}, \quad \mathbf{C}_3 = \begin{bmatrix} 1 & 0 \end{bmatrix}$$

$$\mathbf{A}_{44} = \begin{bmatrix} -(c_1 + c_2 v_r) & 0 \\ 0 & 0 \end{bmatrix}, \quad \mathbf{A}_{43} = \begin{bmatrix} 0 & k_1/m_4 \\ 0 & 0 \end{bmatrix}$$

$$\mathbf{B}_4 = \begin{bmatrix} 1/m_4 \\ 0 \end{bmatrix}, \quad \mathbf{C}_4 = \begin{bmatrix} 1 & 0 \end{bmatrix}$$

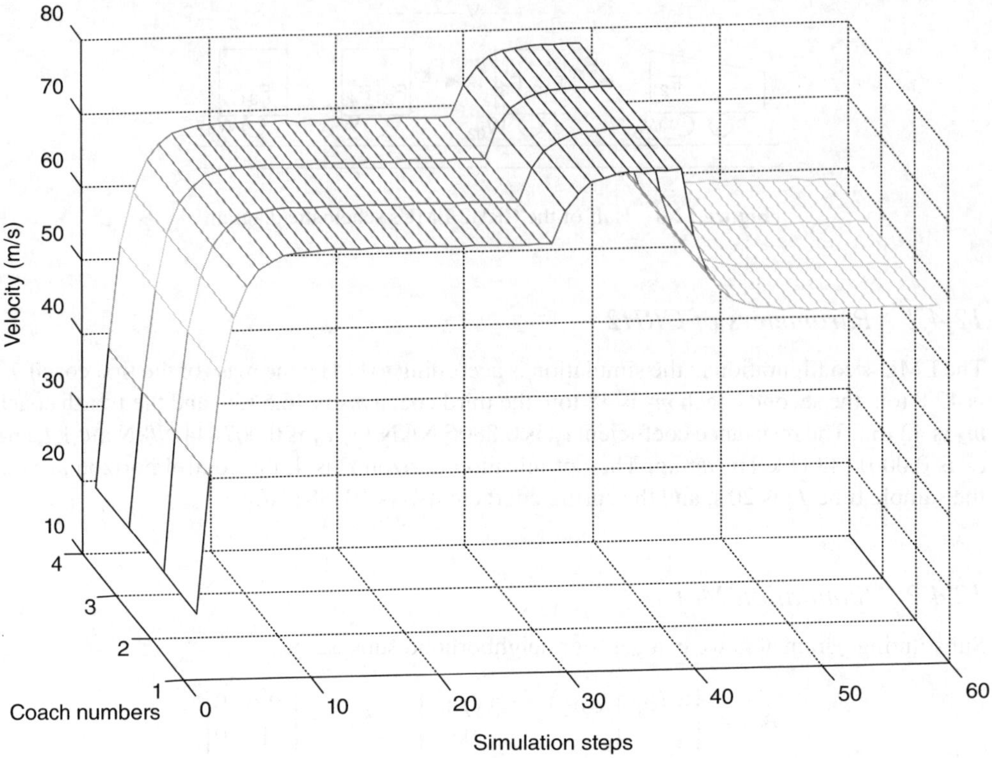

Figure 12.7 Velocity track

Based on the Part Two, we can get parameters of neighborhood subsystems:

$$\hat{\mathbf{A}}_{11} = \begin{bmatrix} \mathbf{A}_{11} & \mathbf{A}_{12} \\ \mathbf{A}_{21} & \mathbf{A}_{22} \end{bmatrix}, \ \hat{\mathbf{A}}_{12} = \begin{bmatrix} \mathbf{0} & \mathbf{0} \\ \mathbf{0} & \mathbf{A}_{23} \end{bmatrix}; \ \hat{\mathbf{A}}_{22} = \begin{bmatrix} \mathbf{A}_{11} & \mathbf{A}_{12} & \mathbf{0} \\ \mathbf{A}_{21} & \mathbf{A}_{22} & \mathbf{A}_{23} \\ \mathbf{0} & \mathbf{A}_{32} & \mathbf{A}_{33} \end{bmatrix}, \ \hat{\mathbf{A}}_{23} = \begin{bmatrix} \mathbf{0} & \mathbf{0} & \mathbf{0} \\ \mathbf{0} & \mathbf{0} & \mathbf{0} \\ \mathbf{0} & \mathbf{0} & \mathbf{A}_{34} \end{bmatrix};$$

$$\hat{\mathbf{A}}_{33} = \begin{bmatrix} \mathbf{A}_{22} & \mathbf{A}_{23} & \mathbf{0} \\ \mathbf{A}_{32} & \mathbf{A}_{33} & \mathbf{A}_{34} \\ \mathbf{0} & \mathbf{A}_{43} & \mathbf{A}_{44} \end{bmatrix}, \ \hat{\mathbf{A}}_{32} = \begin{bmatrix} \mathbf{A}_{21} & \mathbf{0} & \mathbf{0} \\ \mathbf{0} & \mathbf{0} & \mathbf{0} \\ \mathbf{0} & \mathbf{0} & \mathbf{0} \end{bmatrix}, \ \hat{\mathbf{A}}_{44} = \begin{bmatrix} \mathbf{A}_{33} & \mathbf{A}_{34} \\ \mathbf{A}_{43} & \mathbf{A}_{44} \end{bmatrix}, \ \hat{\mathbf{A}}_{43} = \begin{bmatrix} \mathbf{A}_{32} & \mathbf{0} \\ \mathbf{0} & \mathbf{0} \end{bmatrix}$$

12.4.3 Results and Some Comments

Figure 12.7 shows that the velocity of every coach can track the reference velocity accurately. Figure 12.8 shows that all the traction force is in the appropriate range. Figure 12.9 shows that the forces between coaches are in the constraint range. To explain the performance of global optimization, decentralized optimization, and neighborhood optimization, we take a four-coach (T-M-M-T) CRH2 as an example by three strategies respectively. Simulation results are shown in Figures 12.10–12.13.

The traction of the second coach is shown in Figure 12.12.

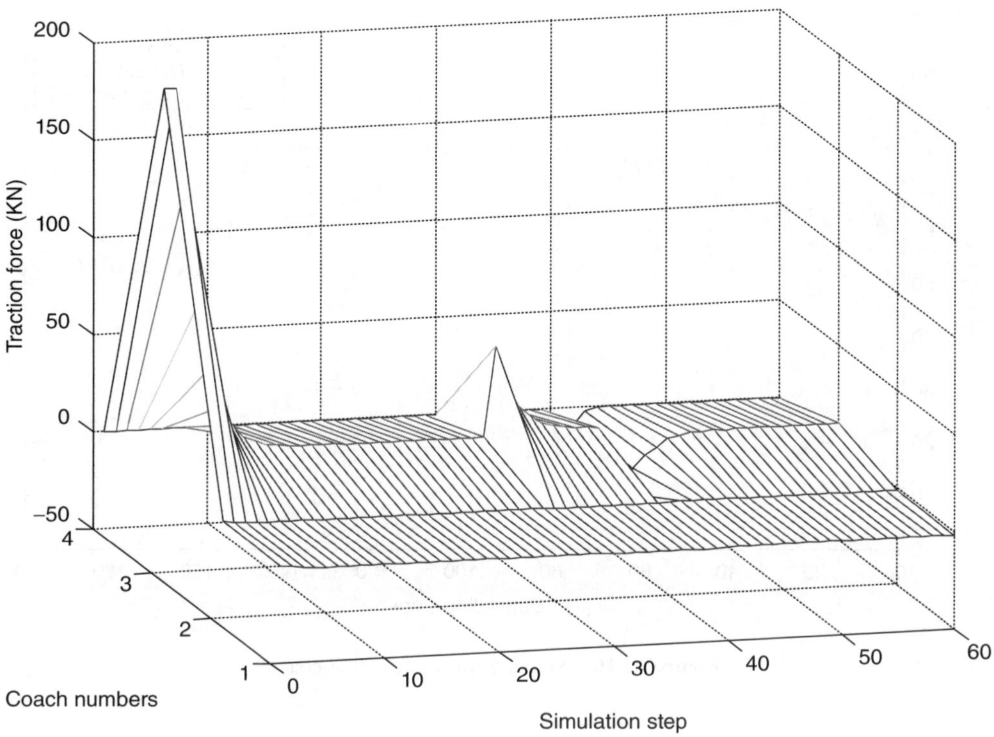

Figure 12.8 Driving force optimal scheduling

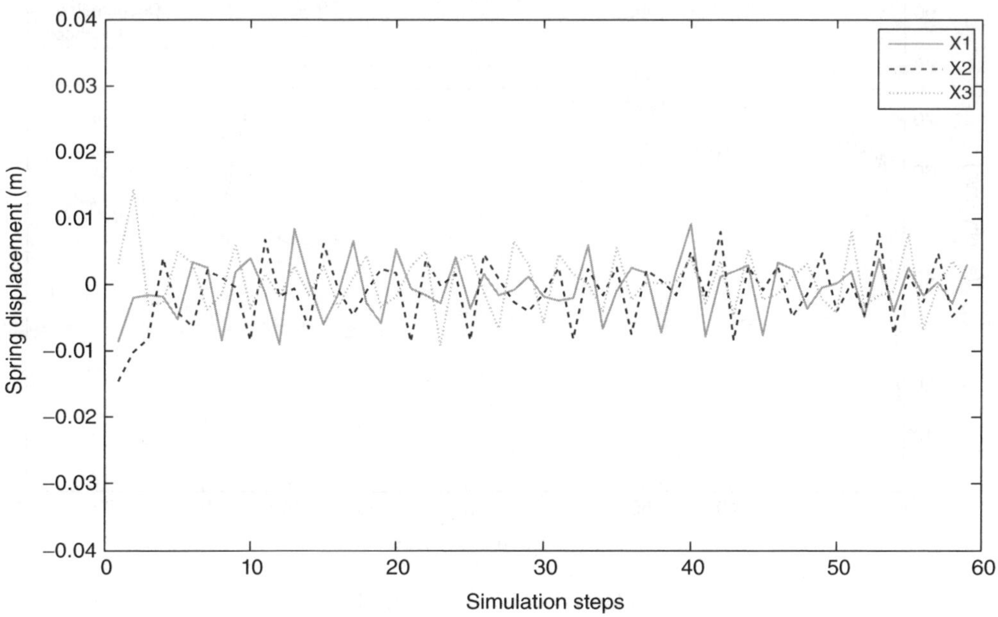

Figure 12.9 The relative displacement of the coaches

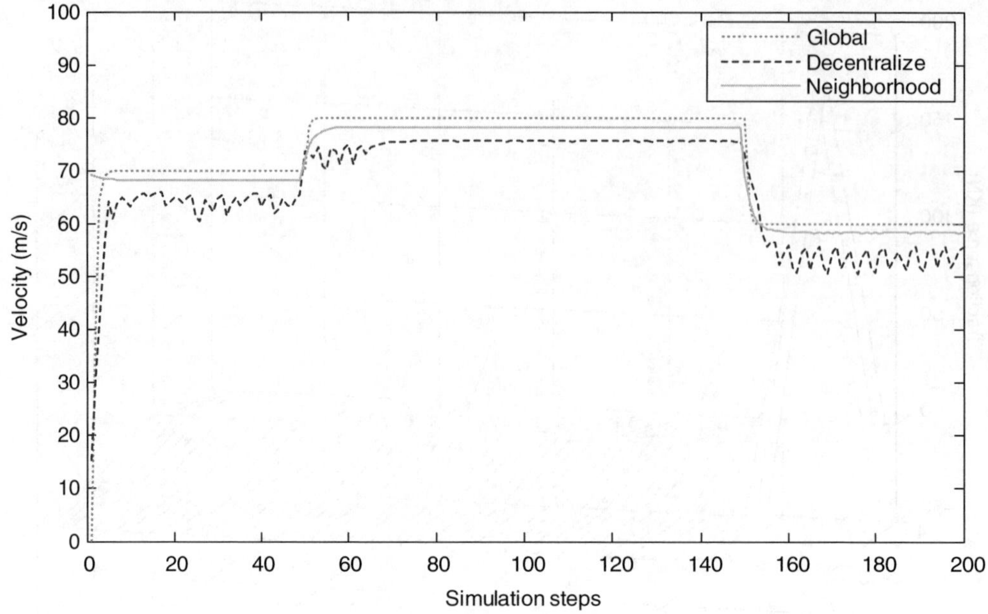

Figure 12.10 Velocity track of the first coach

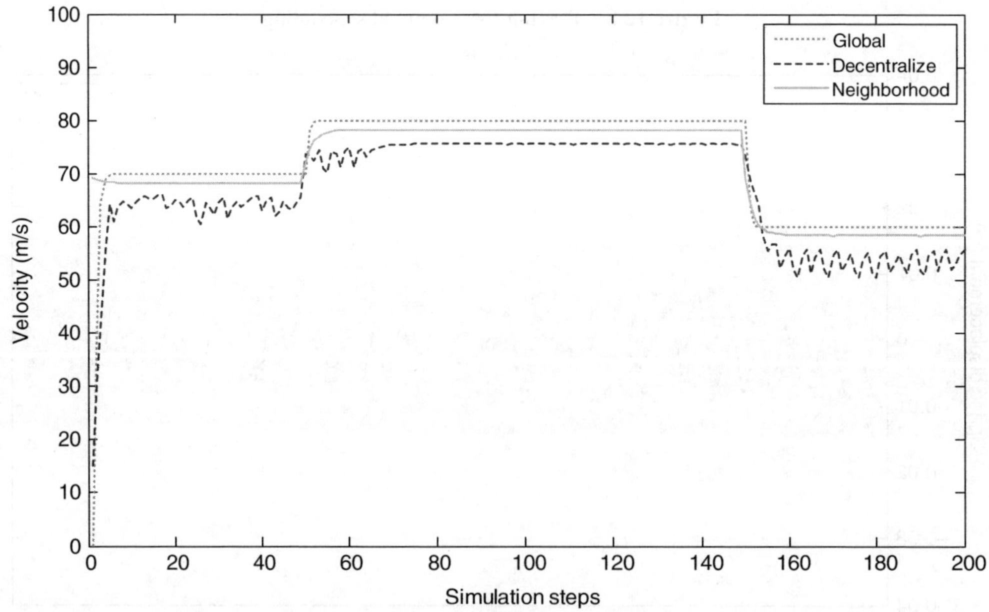

Figure 12.11 Velocity track

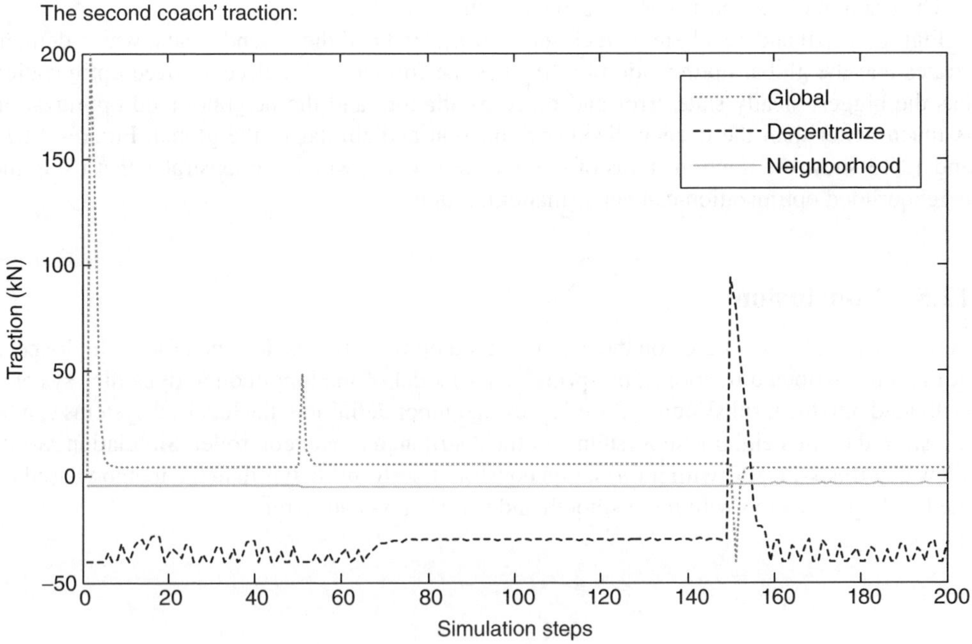

Figure 12.12 Driving force optimal scheduling of the second coach

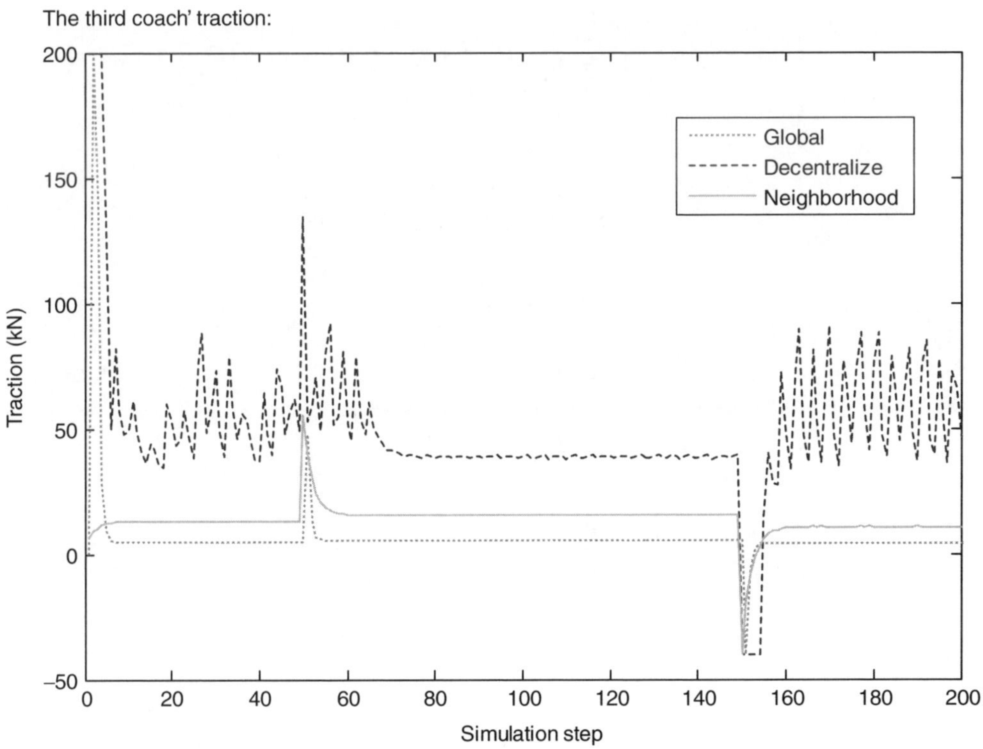

Figure 12.13 Driving force optimal scheduling of the third coach

The traction of the third coach is shown in Figure 12.13.

Figures 12.10 and 12.11 show track status of the first and the second coach, which demonstrates that the global optimization has the best performance, the decentralized optimization has the biggest steady-state error and fierce oscillation, and the neighborhood optimization is much better than the decentralized optimization and similar to the global. Figures 12.12 and 12.13 show the traction status of the powered coach, which demonstrates that using the neighborhood optimization can get a smaller traction.

12.5 Conclusion

In this chapter, DMPC based on the neighborhood optimization is designed for the high-speed train with distributed traction. The spring–mass model of the longitudinal dynamics systems is divided into local subsystems. Based on the neighbor definition, the local subsystems can be transferred to the neighbor subsystems for the distributed model controller. Simulation results show that the N-DMPC with information constraint performs better than the method based on the local optimization with more smooth and less steady-state error.

13

Operation Optimization of Multitype Cooling Source System Based on DMPC

13.1 Introduction

With the rapid growth of energy consumption, various kinds of cooling and storage devices are widely used in air-conditioning system in giant buildings so as to realize energy saving and emission reduction. The main characteristic of the subsystem is a diversity of cooling devices, which generally include conventional electric refrigerators, ice storage system, ground-source heat pump system, and combined cooling heating and power.The load features of these cooling devices are not identical and the types of energy they consume are complex.

To improve energy utilization efficiency and reduce cost, there is an urgent need to provide a new kind of economic scheduling strategy that makes full use of the time-of-use pricing, which is currently widely used in China. The conventional electric refrigerators and the ice storage tank can satisfy at least 90% of cooling load. The ice storage systems are widely applied because of the autocontrol of ice storage and ice-making. Cooperating between the conventional electric refrigerator and the ice storage tank is the key to reduce the cost and improve the economic effect.

This chapter proposes a hierarchically distributed MPC strategy, which builds an economic model of electric refrigerator and ice storage tank, and gets steady power states and optimal set-point of each electric refrigerator and ice storage tanks under optimal conditions by using mixed integer programming. Then it uses DMPC to make sure that each electric refrigerators and ice storage tanks can track the upper optimal set-point as soon as possible with guaranteeing that the total power can track predictive load.

13.2 Structure of Joint Cooling System

The typical cooling system includes dual mode refrigerator, ice storage tank, exchanger, and some other components. Normal electric refrigerator just has the air-condition mode. Dual

Distributed Model Predictive Control for Plant-Wide Systems, First Edition. Shaoyuan Li and Yi Zheng.

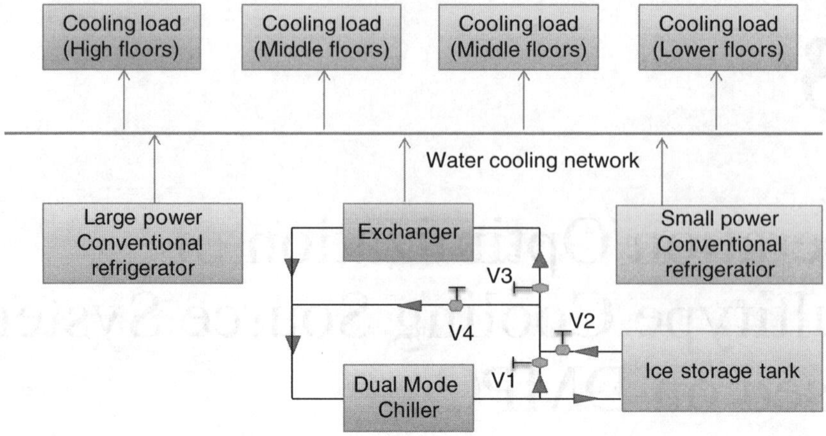

Figure 13.1 Structure of joint cooling system

operating mode refrigerator has two modes and it is able to either make ice or directly supply cold.Ice storage system has four working modes including electric refrigerator ice-making mode, electric refrigerator cooling mode, ice storage cooling mode, and electric refrigerator ice storage cooling mode. We can use the valves to change the mode of the multitype cooling source system.

A typical multitype cooling system is shown in Figure 13.1. The valves between the exchangers, ice storage, and the dual mode electric refrigerator are used to control the switch of the working mode. The low-temperature water from the conventional electric refrigerators and the exchanger after the ice storage tank supplies cooling loads to customers via the cold water network. The purpose of the whole air-conditioning system is to ensure the minimum cost while the predictive load is satisfied by scheduling the power of conventional electric refrigerators and ice storage tanks.

13.3 Control Strategy of Joint Cooling System

This chapter proposes a conventional electric refrigerator and dual mode electric refrigerator control strategy, which can be divided into two levels – upper and bottom levels. The upper level optimal strategy uses predictive load data and the steady power data of the air-mode and ice-mode electric refrigerators, and the ice storage tank to get the lowest cost of electricity. The bottom dynamic optimization uses MPC to decide the real-time power reference, according to the dynamic process of starting the electric refrigerator. The expected power is the steady reference of each electric refrigerator from the upper strategy. This chapter uses the DMPC to cooperate the control variables to avoid the large calculation of the central control mode. Different MPCs from subsystems can share the data from the communicators to track the total expected power. Figure 13.2 depicts the structure of a whole control strategy of a joint cooling system, where DR refers to dual mode refrigerator and CR refers to conventional refrigerator. The single-pole double-throw switch in the dynamic optimization level is used to choose the air-mode electric refrigerator controlled by the DMPC, then the disabled electric refrigerator

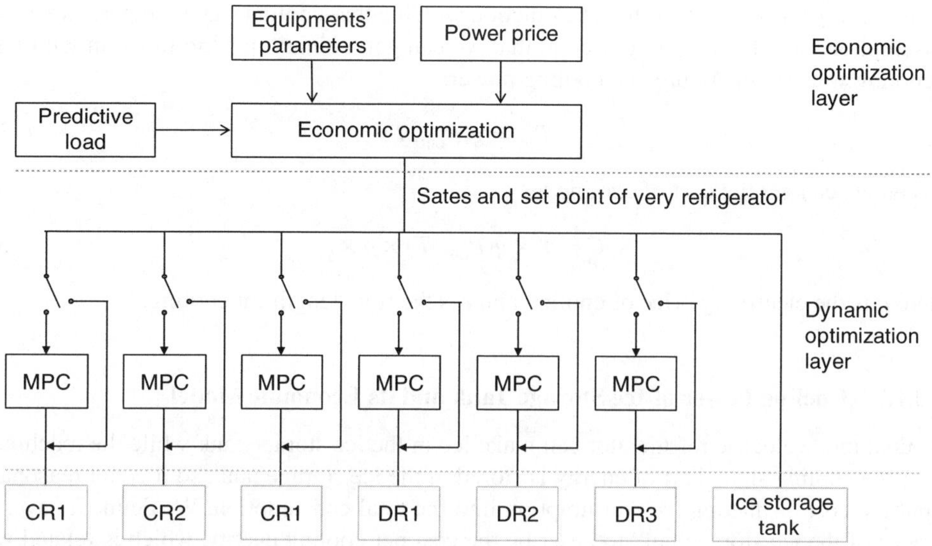

Figure 13.2 Control strategy of joint cooling system

or ice-mode electric refrigerator is neglected. The virtual network is the data transmit channel for MPC subsystems.

13.3.1 Economic Optimization Strategy

13.3.1.1 Economic Models of Conventional and Dual Operating Mode Refrigerators

Each refrigerator has two working states – on and off – so we choose the on–off state $Y\{0,1\}$ to describe the electric refrigerator working state for the dual operating mode electric refrigerator air condition and ice-making condition.

The cooling power is described by the continuous variables. Gordon *et al.* [142] introduced the physical model of the centrifugal electric refrigerator and mentioned that the coefficient of performance (COP) of the centrifugal refrigerator can be presented by a binary function with the temperature of the input cooling water in the electric refrigerator T and cooling power of the electric refrigerator P_{out}, which is shown as follows:

$$\text{COP} = \frac{P_{\text{out}}}{P_{\text{in}}} = f(P_{\text{out}}, T) \tag{13.1}$$

where P_{in} is the power consumption of the electric refrigerator. Then, we can get the relationship between the power consumption and the cooling power of the electric refrigerator as

$$P_{\text{in}} = \frac{P_{\text{out}}}{f(P_{\text{out}}, T)} \tag{13.2}$$

In the real application, relation (13.2) is difficult to use as the specific algorithm to optimize the fractional function. Therefore, we need a new function for the electric refrigerator.

We use the regression analysis to get a function with input as cooling water temperature, power consumption, and the cooling power so that we can get a binary quadric function with input as cooling water temperature and cooling power:

$$P_{\text{in}} = g(P_{\text{out}}, T) \tag{13.3}$$

Then we can get the cost of cooling as

$$C = Y \times g(P_{\text{out}}, T) \times p \times t \tag{13.4}$$

where p is the electricity price of cooling and t is the time length for cooling.

13.3.1.2 Cooling Power of Ice Storage Tank and its Economic Model

The dual mode electric refrigerator can make ice in the ice storage tank while the machine is under the cooling state. Part of energy is stored in the ice storage tank, so it is not reasonable to put the cost of making ice consumption into the total consumption. We define an average price p^s of the ice storage tank to describe the cost per cooling energy, which is related with cooling energy, the power price, and ice-making condition of dual mode electric refrigerator:

$$p^s = \frac{p^{\text{init}} W^{\text{init}} + pT^{\text{in}} P - p^{\text{init}} T^{\text{out}} P^s Y^s}{W^{\text{init}} + T^{\text{in}} P - T^{\text{out}} P^s Y^s} \tag{13.5}$$

where p^{init} is the initial value of average price, W^{init} is the initial value of energy reserves in ice storage tank, T^{in} is time of making ice, T^{out} is soaking time, P is total power of the dual mode electric refrigerator, P^s is the cooling power for the ice storage, and Y^s is the start or stop sign of ice storage. From the definition of average price of ice in the ice storage tank, we can see that the average price has nothing to do with the ice storage tank soaking, without considering the decreasing ice product with the time consumption. It is only related to the ice storage tank while making ice.

13.3.1.3 Objective Function

The objective function is to minimize the electricity power from the sum of power cost of supplying cold in the air mode and ice storage tank, shown as follows:

$$\min_{Y_i, Y^s, P_{\text{out},i}, P^s} J = pt \sum Y_i g_i(P_{\text{out},i}, T_i) + p^s t P^s Y^s \tag{13.6}$$

where $P_{\text{out},i}$ is the supplying cold power of the ith electric refrigerator. The cost of dual mode electric refrigerator in the ice mode is not included in the objective function in the current optimization period, so the ice mode of the dual mode electric refrigerator does not impact the optimal value of the optimization problem. In the application, we can use the starting and stopping sign solved by the optimization problem to start or stop the ice-making for the dual mode electric refrigerator, which means any two dual mode electric refrigerators cannot have a state such that one is ice mode and the other is air mode to make sure that the working state of the electric refrigerators has the same mode.

13.3.1.4 Constraint Conditions

The main constraints include the output power of refrigerators and the power balance between refrigerator and predictive load:

$$\begin{cases} Y_i P_{\text{out},i} + P^s Y^s = P^p \\ \alpha_i P_{i,\text{rated}} \le P_{\text{out},i} \le P_{i,\text{rated}} \end{cases} \tag{13.7}$$

where, P^p is predictive load, $\alpha_i \in (0, 1)$ is the minimum power ratio of the ith electric refrigerator, and $P_{i,\text{rated}}$ is the rated output power of the ith electric refrigerator.

According to the objective function and constraint conditions form, we can see that the economic optimization problem is made up of continuous variables and integer variables. It is a mixed programming problem, which can be solved by the branch and bound tool in the MATLAB. The solution is $Y_i, Y^s, P_{\text{out},i}, P^s$.

13.3.2 Design of Distributed Model Predictive Control in Multitype Cold Source System

Bottom control system can be divided into N subsystems according to the number of air-mode electric refrigerators, which is made up with a MPC and the plant electric refrigerator. Different MPCs from subsystems can share the data from the communicators to track the total expected power.

13.3.2.1 Model of Subsystems

Browne and Bansal [143] introduce the dynamic process of the starting state of the electric refrigerator. From the curves of the starting process of an electric refrigerator, we can see that the real-time output power can be seen as a step response of a first-order system, for which the response time is four times of the time constant in the first-order system. Then, we can estimate the time constant from the starting time of the electric refrigerator. Considering the delay in the real control system, the process transfer function of tracking the electric refrigerator cooling power reference is

$$H(s) = \frac{1}{1+\tau s} e^{-\tau_d s} \tag{13.8}$$

where τ is time constant of the first-order inertial element and τ_d is the delay time. Then, the discrete space state for the sth subsystem is

$$\begin{cases} \mathbf{x}_s(k+1) = \mathbf{A}_s \mathbf{x}_s(k) + \mathbf{B}_s u_s(k) \\ p_s(k) = \mathbf{C}_s \mathbf{x}_s(k) \end{cases} \tag{13.9}$$

$$s = 1, 2, \ldots, N$$

where $u(k)$ is the output power reference of electric refrigerator at time k, which is the optimal variable in the following optimal problem, $p_s(k)$ is the real output power of the electric refrigerator at the time $k+1$, n_d is the times between the delay time of electric refrigerator and discrete system sampling time, $\mathbf{x}_s(k) = [p_s(k)\ p_s(k+1)\ \cdots\ p_s(k+n_d)]^{\mathrm{T}}$ is the state variable, $\mathbf{A} \in \mathbb{R}^{(n_d+1)\times(n_d+1)}$ and $\mathbf{B} \in \mathbb{R}^{(n_d+1)\times 1}$ are decided by the sampling time and the time constant of electric refrigerator. Let the sampling time be Δt, we can obtain

$$\mathbf{A}_s = \begin{bmatrix} 0 & 1 & \cdots & 0 \\ 0 & 0 & \ddots & \vdots \\ \vdots & \ddots & \ddots & 1 \\ 0 & \cdots & 0 & e^{-\Delta t/\tau} \end{bmatrix}$$

$$\mathbf{B}_s = \begin{bmatrix} 0 & \cdots & 0 & 1 - e^{-\Delta t/\tau} \end{bmatrix}^{\mathrm{T}}$$

$$\mathbf{C}_s = \begin{bmatrix} 1 & 0 & \cdots & 0 \end{bmatrix} \tag{13.10}$$

13.3.2.2 MPC Model of Subsystems

In each subsystem, the real-time reference of supplying the cold power by an electric refrigerator is decided by the current MPC and other MPCs of other subsystems.

For the subsystem s ($s = 1, 2, \ldots, N$), MPC coordinate strategy includes the tracking of the upper level optimal reference $\mathbf{r}_s$ in the future P sampling time and it is not the normal MPC strategy. In this chapter, we introduce a sum of square of the difference between the total predictive power and all expected powers to track total expected power $\mathbf{r}$, which makes the objective functions coupling from different subsystems. Above all, the MPC objective function of the subsystem s is

$$\min_{\mathbf{U}_s} J_s = Q\sum_{i=1}^{P} (r_s(k+i) - \hat{p}_s(k+i|k))^2 + R\sum_{i=1}^{P}\left(r(k+i) - \hat{p}_s(k+i|k) - \sum_{\substack{j=1 \\ j\neq s}}^{N} \hat{p}_j(k+i|k) \right)^2 \tag{13.11}$$

where $\mathbf{U}_s = [u(k)\ u(k+1)\ \cdots\ u(k+P-1)]^{\mathrm{T}}$ is the reference sequences of the electric refrigerator output power in the predictive horizon. $Q, R \in [0, 1]$ are the weights used to measure the predictive load tracking level for a single electric refrigerator and all electric refrigerators, respectively. Each system solves the self-optimal question with not only the output in the predictive horizon based on the model itself but also the tracking level of the total expected power based on the other subsystems. Based on the last section, we can get the predictive model as

$$\begin{cases} \hat{\mathbf{x}}_s(k+i|k) = \mathbf{A}_s^i \mathbf{x}_s(k) + \sum_{j=1}^{i} (\mathbf{A}_s^{i-j}\mathbf{B}_s u_s(k+j-1)) \\ \hat{p}_s(k+i|k) = \mathbf{C}_s \hat{\mathbf{x}}_s(k+i|k) \end{cases}$$

$$i = 1, 2, \ldots, P \tag{13.12}$$

Above all, for the MPC controller of the sth subsystem the optimal problem form is as follows:

Problem 13.1

$$\min_{\mathbf{U}_s} \quad J_s = Q\sum_{i=1}^{P}(r_s(k+i) - \hat{p}_s(k+i|k))^2$$
$$+R\sum_{i=1}^{P}\left(r(k+i) - \hat{p}_s(k+i|k) - \sum_{\substack{j=1\\ j\neq s}}^{N}\hat{p}_j(k+i|k)\right)^2$$
$$\text{s.t.} \quad \hat{\mathbf{x}}_s(k+i|k) = \mathbf{A}_s{}^i\mathbf{x}_s(k) + \sum_{j=1}^{i}(\mathbf{A}_s{}^{i-j}\mathbf{B}_s u_s(k+j-1))$$
$$\hat{p}_s(k+i|k) = \mathbf{C}_s\hat{\mathbf{x}}_s(k+i|k)$$
$$u_{s,\min} \leq u_s(k+i) \leq u_{s,\max}$$
$$p_{s,\min} \leq \hat{p}_s(k+i|k) \leq p_{s,\max}$$
$$i = 1, 2, \ldots, P \tag{13.13}$$

where $\{u_{s,\min}, u_{s,\max}\}$, $\{p_{s,\min}, p_{s,\max}\}$ are border constraints of the control variables. Based on solving the optimal problem, we can get the optimal reference of the electric refrigerator cooling power. Put the first data from the optimal results into the electric refrigerator and calculate the next receding horizon MPC problem.

13.3.2.3 Iterative Algorithm

Considering that each subsystem needs other subsystems' predictive models to solve the local optimization, this chapter uses a general iterative algorithm.

Distributed MPC algorithm is shown as follows:

Algorithm 13.1

Step 1. *Initialization.* At the time k, according to the upper level optimization, we can get the ideal electric refrigerator power $\mathbf{r}_s$ and initialize the $\mathbf{U}_s$ of each subsystem in the predictive horizon, let the iterative number be $l = 0$.

$$\mathbf{U}_s^l = [u_s^l(k+1) \; u_s^l(k+2) \; \cdots \; u_s^l(k+P)]^T$$
$$= [r_s(k+1) \; r_s(k+2) \; \cdots \; r_s(k+P)]^T$$
$$s = 1, 2, \ldots, N$$

Step 2. *Model prediction and data transmission.* According to the predictive model (13.12), we can calculate each subsystem's predictive output in the control horizon:

$$\hat{\mathbf{p}}_s = [\hat{p}_s(k+1|k) \; \hat{p}_s(k+2|k) \; \cdots \; \hat{p}_s(k+P|k)]^{\mathrm{T}}$$

and transmit the data to other $N-1$ subsystem MPC controllers.

Step 3. *Subsystem optimization*. Each subsystem solves the optimal Problem 13.1, and then gets the solution $\mathbf{U}_s^{l+1}$.

Step 4. *Check and update*. Check all the subsystems' conditions of convergence, which is for the given

$$\varepsilon_s \in \mathbb{R}(s = 1, 2, \ldots, N,)$$

$$\|\mathbf{U}_s^{l+1} - \mathbf{U}_s^l\| < \varepsilon_s \quad (s = 1, 2, \ldots, N)$$

If all the subsystems can satisfy the conditions of convergence, then let the optimal variable of each subsystem be $\mathbf{U}_s^* = \mathbf{U}_s^{(l*)}$, go to Step 5; otherwise, let $\mathbf{U}_s^{l+1} = \alpha \mathbf{U}_s^l + (1 - \alpha)\mathbf{U}_s^l (s = 1, 2, \ldots, N)$, α is the constant between 0 and 1, which is decided by the need of subsystems, $l = l + 1$. Go to Step 2.

Step 5. *Control variables decision*. At the time k, apply $u_s^* = \begin{bmatrix} 1 & 0 & \cdots & 0 \end{bmatrix} \mathbf{U}_s^* (s = 1, 2, \ldots, N)$ to the subsystem.

Step 6. *Receding optimization*. Move to the next sampling time, $k + 1 \to k$, return Step 1, repeat.

By the distributed predictive control algorithm, the optimal problem for the globally supplying cold dynamic system can be divided into each subsystem made by electric refrigerator and corresponding MPC, which can reduce the complexity and calculation of the control problem and improve the response speed. To testify the effectiveness of the two-level control strategy, this chapter applies the method to a cooling system of a super-high building.

13.4 Results and Analysis of Simulation

To validate the scheduling algorithm, this chapter applies the above method to the low-district cooling system of a super-high building in Shanghai, China. The main cold sources of the system are three conventional centrifugal refrigerators, three dual operating mode centrifugal refrigerators, an ice storage tank, a ground-source heat pump, and lithium-bromide absorption refrigerators. To simplify the system, considering high-efficiency and low-output power of ground-source heat pump and lithium-bromide absorption refrigerators, we preferentially let them loaded fully and jointly dispatch conventional electric refrigerators and ice storage system to satisfy the rest load.

There are three conventional electric refrigerators. For two of them, the cooling power is 3900 kW, while for the third one the cooling power is 2150 kW. There are three dual operating mode electric cooling refrigerators. The rated cooling power of dual operating mode centrifugal refrigerators in the air mode is 6392 kW and in the ice-making mode is 3868 kW. We use the regression analysis to get a function with input as cooling water temperature, power consumption, and the cooling power so that we can get a binary quadric function with input as cooling water temperature and cooling power. Power unit is kilowatt and temperature unit is degree centigrade, as shown in Table 13.1.

For the ice mode of electric refrigerator, to make ice as quickly as possible, refrigerators work under the condition of rated power in the ice mode. Then, the cooling power *cold* and

Table 13.1 Power consumption functions of refrigerators under air conditioning operation

Electric refrigerators in the air-mode rated power (kW)	Binary quadratic function
3900	$P_{in} = 146.7 + 0.033P_{out} - 7.3T + 0.17T^2 + 3.5 \times 10^{-3}P_{out}T + 3.2 \times 10^{-6}P_{out}^2$
2150	$P_{in} = 88.9 + 0.042P_{out} - 4.6T + 0.10T^2 + 3.3 \times 10^{-3}P_{out}T + 1.8 \times 10^{-6}P_{out}^2$
6329	$P_{in} = 270.9 + 0.030P_{out} - 9.1T + 0.19T^2 + 3.9 \times 10^{-3}P_{out}T + 4.7 \times 10^{-6}P_{out}^2$

the consumption power *elec* are the unit and hypo-two function of the temperature T of input cooling water:

$$\begin{cases} cold = -1.87T^2 + 17.63T + 5239.50 \\ elec = -0.50T^2 + 23.07T + 766.22 \end{cases} \tag{13.14}$$

The valley price of the building is shown in Table 13.2.

During the simulation, total predictive cooling load sampling time is 30 min and the dynamic optimization period is 2.5 min. That means every 30 min the upper control systems calculate the optimal cooling power for each electric refrigerator and ice storage, and bottom systems update the power reference of each electric refrigerator every 2.5 min to track the steady reference made by the upper level and the predictive load. The dynamic model time constant and delay time of each electric refrigerator in air mode are shown in Table 13.3.

The simulation time is 05:00–11:00, which contains the peak, average, and the valley price of the power. Besides, the load is from valley to the peak according to the time from morning to forenoon. It is meaningful to make a simulation during this time period. During the whole

Table 13.2 Time-of use power price

Time (h)	Price of power (Yuan/kWh)
22:00–06:00	0.234
06:00–08:00	0.706
08:00–11:00	1.037
11:00–13:00	0.706
13:00–15:00	1.037
15:00–18:00	0.706
18:00–21:00	1.037
21:00–22:00	0.706

Table 13.3 Dynamic parameters of refrigerators under air conditioning operation

Electric refrigerators in the air-mode power (kW)	Time constant (min)	Delay time (min)
2150	3.0	2.5
3900	4.0	5.0
6329	5.0	7.5

simulation period, the input cooling water temperature of each electric refrigerator is stable and constant. The input cooling water temperature of air-mode electric refrigerator is 32°C, and in the ice mode it is 30°C. The maximum volume of ice storage tank is 50,000 kWh and the lower limit is set to 5000 kWh.The time constant of first-order model for the ice storage tank is 2 min. The average price of the ice storage is 0.0576 RMB/kWh according to the full power of an electric refrigerator based on the valley electricity price.

First, to testify the economic effect of the upper level optimization, this chapter chooses a normal schedule strategy as a comparison, which makes ice in the ice storage tank during the price valley and use electric refrigerator to supply cooling during the normal price and the peak price. The start order is the same as the size of the electric refrigerator power.

The predictive load curve and the price of two strategies are shown in Figures 13.3 and 13.4, respectively. The total cost of the upper economic optimization strategy is 3970 Yuan in 6 h, and the comparison is 4053 RMB. The proposed strategy saves about 2% of the total money

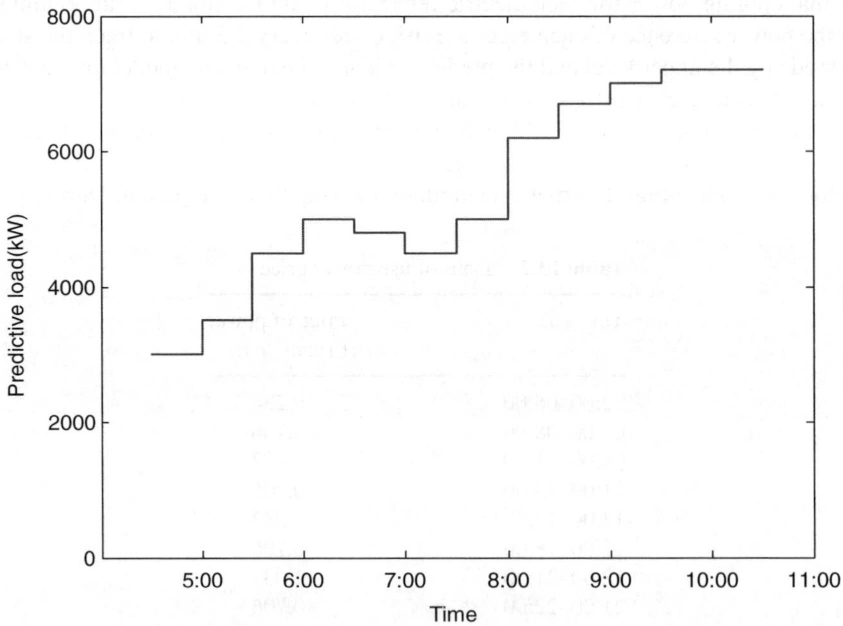

Figure 13.3 Load forecasting curve

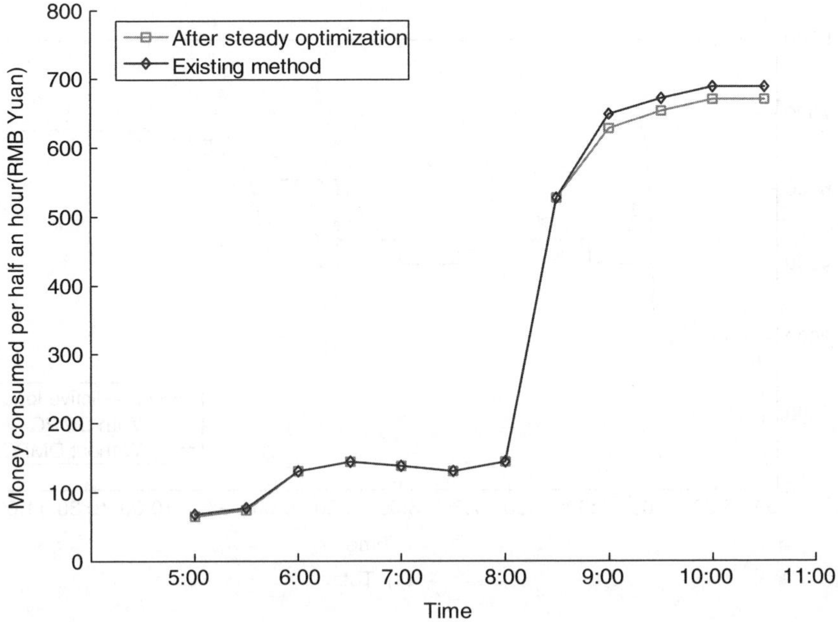

Figure 13.4 Comparison of power cost in every half an hour between the two strategies

in 6 h. In Figure 13.4, two curves have overlapped during the time 6:30–8:300. The reason is that two strategies are same; they all use ice storage tank. The comparison of two strategies are obvious while the ice storage is not the main source during the time 5:00–6:00 and 9:00–11:00, and the results show that the upper economic optimization is effective.

Second, to test the bottom dynamic optimization, we compare the strategy that only gets the upper optimization and the strategy that has upper optimization and the bottom dynamic optimization on the predictive load data between the time 05:00 and 11:00. We get curves of total expected load of electric refrigerators in air mode and real-time supplying cold source power by two different strategies, as shown in Figure 13.5. Because of the 6329 kW electric refrigerator off, curves only include 3900 and 2150 kW electric refrigerators.

From Figure 13.5, we can see that each electric refrigerator has an ahead-of-time process with off and on, which makes the real-time power track the references quickly by using the bottom dynamic optimization. This shows the advantage of the predictive algorithm in the system including the delay part. The curves of the total cost of power do not include the peak at the time 6:00 because of the small time constant of ice storage tank, which makes that the reference can be reached quickly. At time 6:00, the ice storage tank supplies all the predictive loads; the power of the electric refrigerators does not disappear because of the system inertance. The power of the electric refrigerators decreases ahead of time and reduces the peak when we use DMPC strategy.

To have a quantitative description of the difference of different strategies for the tracking level of the power of the electric refrigerators in air mode, this chapter calculates the sum of squares of bias during 6 h; the data in Table 13.4 show a good performance of the dynamic optimization.

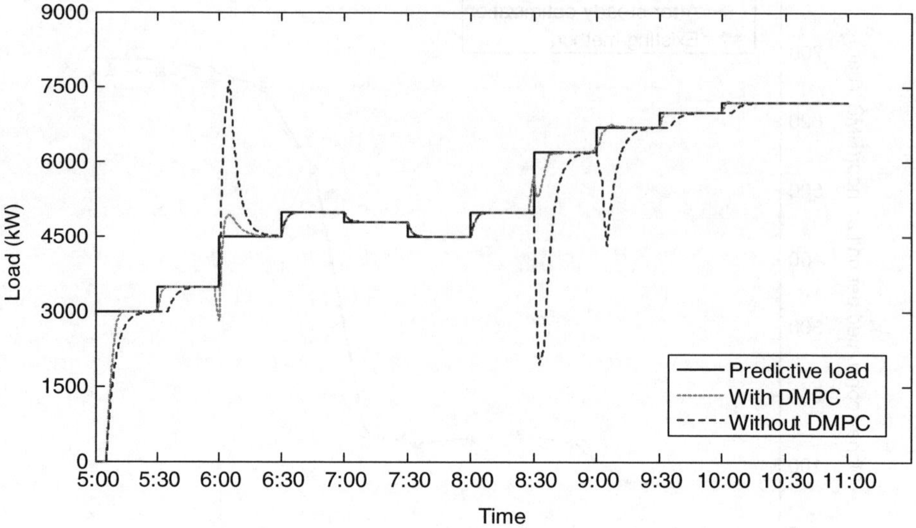

(a) Total

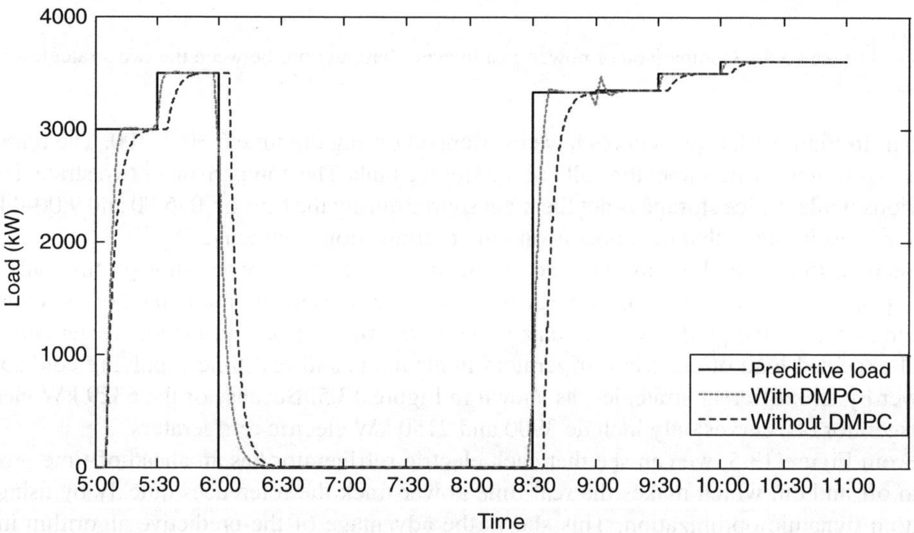

(b) First electric refrigerator with 3900 kW

Figure 13.5 Simulation results of dynamic performance optimization

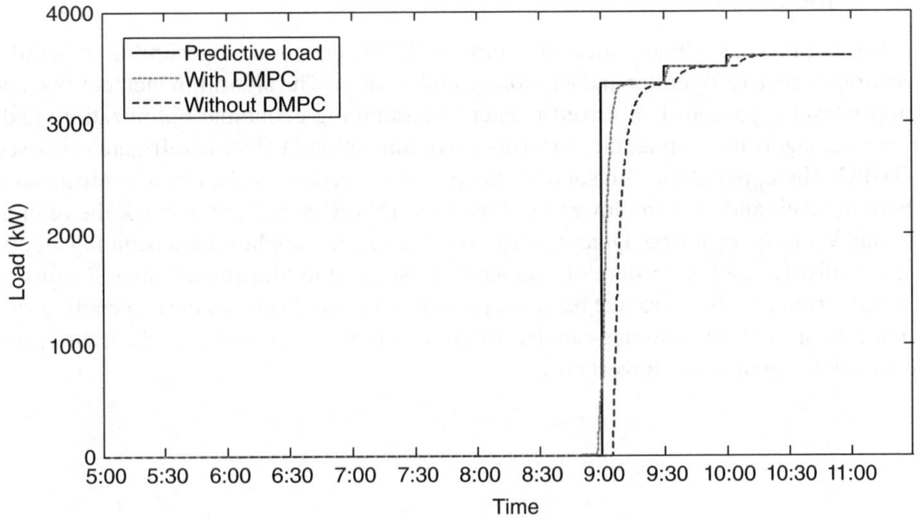

(c) Second electric refrigeratorwith 3900 kW

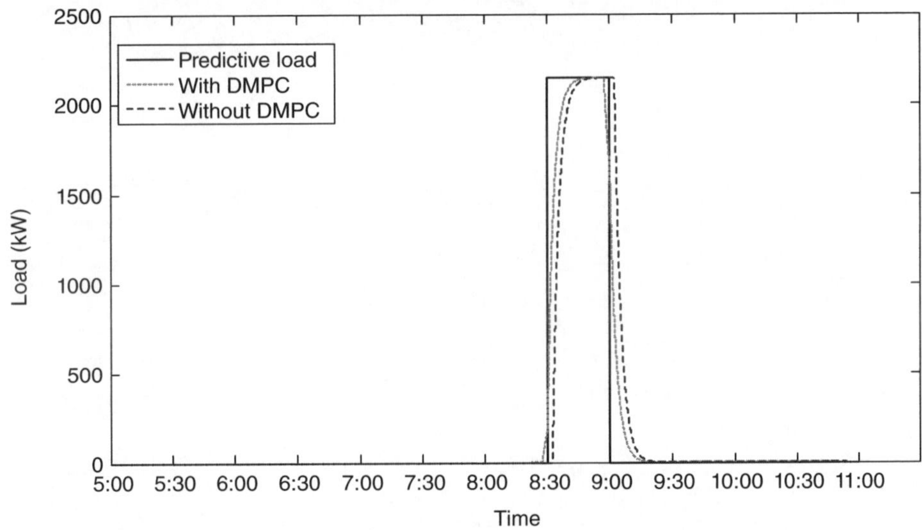

(d) First electric refrigerator with 2150 kW

Figure 13.5 (*Continued*)

Table 13.4 Effect of dynamic optimization

Load	DMPC	Non-DMPC
Total	3.40×10^{7}	1.03×10^{8}
First electric refrigerator	4.91×10^{7}	1.11×10^{8}
Second electric refrigerator	8.71×10^{6}	3.82×10^{7}
Third electric refrigerator	8.41×10^{6}	2.06×10^{7}

13.5 Conclusion

This chapter proposes a hierarchical structure of DMPC economic algorithm to schedule the convention electric refrigerator and ice storage tank system. The algorithm includes two levels: the upper level algorithm is a mixed integer programming economic optimization; and the bottom level algorithm is dynamic performance optimization of electric refrigerators based on the DMPC. The upper algorithm supplies the power references of the electric refrigerators in the bottom level, and then the lower level uses the DMPC algorithm to track the references above quickly to speed up the dynamic response. The chapter applies the scheduling algorithm in the low-district cooling system of a super-high building in Shanghai, China. Results show that the algorithm is able to reflect basic characteristics of cold sources in such cooling system and put forward optimum strategy under different conditions of load and electricity price so as to realize the goal of the lowest cost.

References

[1] C. E. Garcia and M. Morari, "Internal model control. A unifying review and some new results," Industrial and Engineering Chemistry Process Design and Development, vol. 21, pp. 308–323, 1982.
[2] Y. G. Xi, Predictive Control, Beijing: National Defense Industry Press, pp. 92–97, 1993.
[3] S. J. Qin and T. A. Badgwell, "A survey of industrial model predictive control technology," Control Engineering Practice, vol. 11, pp. 733–764, 2003.
[4] J. Richalet, A. Rault, J. Testud, and J. Papon, "Model predictive heuristic control: applications to industrial processes," Automatica, vol. 14, pp. 413–428, 1978.
[5] C. R. Cutler and B. L. Ramaker, "Dynamic matrix control – a computer control algorithm," in Proceedings of the Joint Automatic Control Conference, 1980, pp. Wp5–B.
[6] J. Richalet, "Industrial applications of model based predictive control," Automatica, vol. 29, pp. 1251–1274, 1993.
[7] S. Li, Y. Zheng, and B. Wang, "Steady-state target calculation for constrained predictive control systems based on goal programming," Asia-Pacific Journal of Chemical Engineering, vol. 3, pp. 648–655, 2008.
[8] H. Peng, K. Nakano, and H. Shioya, "Nonlinear predictive control using neural nets-based local linearization ARX model – Stability and industrial application," IEEE Transactions on Control Systems Technology, vol. 15, pp. 130–143, 2007.
[9] S. J. Qin and T. A. Badgwell, "An overview of nonlinear model predictive control applications," in Nonlinear Model Predictive Control, F. Allgower and A. Zheng, Eds., Berlin: Birkhauser, 2000, pp. 369–392.
[10] M. Xu, S. Li, and W. Cai, "Cascade generalized predictive control strategy for boiler drum level," ISA Transactions, vol. 44, pp. 399–411, 2005.
[11] Y. Zheng, S. Li, and H. Qiu, "Networked coordination-based distributed model predictive control for large-scale system," IEEE Transactions on Control Systems Technology, vol. 21, pp. 991–998, 2012.
[12] Y. Zheng, S. Li, and X. Wang, "Horizon-varying model predictive control for accelerated and controlled cooling process," IEEE Transactions on Industrial Electronics, vol. 58, pp. 329–336, 2011.
[13] S. Lee, S. Kumara, and N. Gautam, "Market-based model predictive control for large-scale information networks: completion time and value of solution," IEEE Transactions on Automation Science and Engineering, vol. 5, pp. 630–640, 2008.
[14] J. H. Lee, M. S. Gelormino, and M. Morarih, "Model predictive control of multi-rate sampled-data systems: a state-space approach," International Journal of Control, vol. 55, pp. 153–191, 1992.
[15] D. M. Prett and R. Gillette, "Optimization and constrained multivariable control of a catalytic cracking unit," in Proceedings of the Joint Automatic Control Conference, New York: AIChE, 1980.
[16] C. Cutler and R. Hawkins, "Constrained multivariable control of a hydrocracker reactor," in Proceedings of the American Control Conference, June 10–12, Minneapolis, MN, USA, 1987, pp. 1014–1020.
[17] R. K. Mehra, R. Rouhani, J. Eterno, J. Richalet, and A. Rault, "Model algorithmic control: review and recent developments," in Proceedings of the Engineering Foundation Conference on Chemical Process Control II, January 18–23, 1981, the Cloister, Sea Island, GA, USA, 1982, pp. 287–310.

Distributed Model Predictive Control for Plant-Wide Systems, First Edition. Shaoyuan Li and Yi Zheng.

[18] M. Soliman, O. Malik, and D. T. Westwick, "Multiple model predictive control for wind turbines with doubly fed induction generators," IEEE Transactions on Sustainable Energy, vol. 2, pp. 215–225, 2011.
[19] Y. Zheng, S. Li, and X. Wang, "Distributed model predictive control for plant-wide hot-rolled strip laminar cooling process," Journal of Process Control, vol. 19, pp. 1427–1437, 2009.
[20] Y. Zheng, S. Li, and X. Wang, "Optimization target resetting distributed model predictive control for accelerated cooling process," in 10th World Congress on Intelligent Control and Automation, July 6–8, Beijing, 2012, pp. 2764–2769.
[21] Y. Zheng, N. Li, and S. Li, "Hot-rolled strip laminar cooling process plant-wide temperature monitoring and control," Control Engineering Practice, vol. 21, pp. 23–30, 2013.
[22] H. Liu and S. Li, "Speed control for PMSM servo system using predictive functional control and extended state observer," IEEE Transactions on Industrial Electronics, vol. 59, pp. 1171–1183, 2012.
[23] S. Kouro, P. Cortés, R. Vargas, U. Ammann, and J. Rodríguez, "Model predictive control – a simple and powerful method to control power converters," IEEE Transactions on Industrial Electronics, vol. 56, pp. 1826–1838, 2009.
[24] M. M'Saad, L. Dugard, and S. Hammad, "A suitable generalized predictive adaptive controller case study: control of a flexible arm," Automatica, vol. 29, pp. 589–608, 1993.
[25] R. Negenborn, Multi-agent model predictive control with applications to power networks, TRAIL Thesis Series T2007/14, TRAIL Research School, The Netherlands, 2008.
[26] J. M. Maciejowski, Predictive Control with Constraints, Englewood Cliffs, NJ: Prentice-Hall, 2000.
[27] R. Scattolini, "Architectures for distributed and hierarchical Model Predictive Control – A review," Journal of Process Control, vol. 19, pp. 723–731, 2009.
[28] P. D. Christofides, R. Scattolini, D. Muñoz de la Peña, and J. Liu, "Distributed model predictive control: a tutorial review and future research directions," Computers and Chemical Engineering, vol. 51, pp. 21–41, 2013.
[29] S. Li, Y. Zhang, and Q. Zhu, "Nash-optimization enhanced distributed model predictive control applied to the Shell benchmark problem," Information Sciences, vol. 170, pp. 329–349, 2005.
[30] I. Alvarado, D. Limon, D. Muñoz de la Peña, J. M. Maestre, M. A. Ridao, H. Scheu, W. Marquardt, R. R. Negenborn, B. De Schutter, F. Valencia, and J. Espinosa, "A comparative analysis of distributed MPC techniques applied to the HD-MPC four-tank benchmark," Journal of Process Control, vol. 21, pp. 800–815, 2011.
[31] B. T. Stewart, S. J. Wright, and J. B. Rawlings, "Cooperative distributed model predictive control for nonlinear systems," Journal of Process Control, vol. 21, pp. 698–704, 2011.
[32] W. Al-Gherwi, H. Budman, and A. Elkamel, "Selection of control structure for distributed model predictive control in the presence of model errors," Journal of Process Control, vol. 20, pp. 270–284, 2010.
[33] A. J. del Real, A. Arce, and C. Bordons, "An integrated framework for distributed model predictive control of large-scale power networks," IEEE Transactions on Industrial Informatics, vol. 10, pp. 197–209, 2014.
[34] D. Jia and B. H. Krogh, "Distributed model predictive control," in Proceedings of the American Control Conference, vol. 4, June 25–27, Arlington, VA, USA, 2001, pp. 2767–2772.
[35] S. Leirens, C. Zamora, R. Negenborn, and B. De Schutter, "Coordination in urban water supply networks using distributed model predictive control," in Proceedings of the American Control Conference, June 30–July 2, Baltimore, MD, 2010, pp. 3957–3962.
[36] P.-D. Moroşan, R. Bourdais, D. Dumur, and J. Buisson, "Building temperature regulation using a distributed model predictive control," Energy and Buildings, vol. 42, pp. 1445–1452, 2010.
[37] B. T. Stewart, A. N. Venkat, J. B. Rawlings, S. J. Wright, and G. Pannocchia, "Cooperative distributed model predictive control," Systems and Control Letters, vol. 59, pp. 460–469, 2010.
[38] W. Al-Gherwi, H. Budman, and A. Elkamel, "A robust distributed model predictive control algorithm," Journal of Process Control, vol. 21, pp. 1127–1137, 2011.
[39] L. Giovanini, "Game approach to distributed model predictive control," IET Control Theory and Applications, vol. 5, pp. 1729–1739, 2011.
[40] P.-D. Moroşan, R. Bourdais, D. Dumur, and J. Buisson, "A distributed MPC strategy based on Benders' decomposition applied to multi-source multi-zone temperature regulation," Journal of Process Control, vol. 21, pp. 729–737, 2011.
[41] H. Scheu and W. Marquardt, "Sensitivity-based coordination in distributed model predictive control," Journal of Process Control, vol. 21, pp. 715–728, 2011.
[42] A. Ferramosca, D. Limon, I. Alvarado, and E. F. Camacho, "Cooperative distributed MPC for tracking," Automatica, vol. 49, pp. 906–914, 2013.
[43] H. Junbiao, S. K. Solanki, and J. Solanki, "Coordinated predictive control of a wind/battery microgrid system," IEEE Journal of Emerging and Selected Topics in Power Electronics, vol. 1, pp. 296–305, 2013.

[44] Z. Yi, L. Shaoyuan, and Q. Hai, "Networked coordination-based distributed model predictive control for large-scale system," IEEE Transactions on Control Systems Technology, vol. 21, pp. 991–998, 2013.
[45] E. Camponogara, D. Jia, B. H. Krogh, and S. Talukdar, "Distributed model predictive control," IEEE Control Systems, vol. 22, pp. 44–52, 2002.
[46] M. Vaccarini, S. Longhi, and M. R. Katebi, "Unconstrained networked decentralized model predictive control," Journal of Process Control, vol. 19, pp. 328–339, 2009.
[47] Y. Zheng, S. Li, and N. Li, "Distributed model predictive control over network information exchange for large-scale systems," Control Engineering Practice, vol. 19, pp. 757–769, 2011.
[48] A. N. Venkat, I. A. Hiskens, J. B. Rawlings, and S. J. Wright, "Distributed MPC strategies with application to power system automatic generation control," IEEE Transactions on Control Systems Technology, vol. 16, pp. 1192–1206, 2008.
[49] P. Giselsson, M. D. Doan, T. Keviczky, B. D. Schutter, and A. Rantzer, "Accelerated gradient methods and dual decomposition in distributed model predictive control," Automatica, vol. 49, pp. 829–833, 2013.
[50] M. Farina and R. Scattolini, "Distributed predictive control: a non-cooperative algorithm with neighbor-to-neighbor communication for linear systems," Automatica, vol. 48, pp. 1088–1096, 2012.
[51] W. B. Dunbar, "Distributed receding horizon control of dynamically coupled nonlinear systems," IEEE Transactions on Automatic Control, vol. 52, pp. 1249–1263, 2007.
[52] W. B. Dunbar and R. M. Murray, "Distributed receding horizon control for multi-vehicle formation stabilization," Automatica, vol. 42, pp. 549–558, 2006.
[53] Z. Yi and S. Li, "The stabilization of coordinated distributed model predictive control," Large Scale Complex Systems Theory and Applications, vol. 13, pp. 182–187, 2013.
[54] Y. Zhang and S. Li, "Networked model predictive control based on neighbourhood optimization for serially connected large-scale processes," Journal of Process Control, vol. 17, pp. 37–50, 2007.
[55] Y. Zheng, S. Li, J. Wu, and X. Zhang, "Stabilized neighborhood optimization based distributed model predictive control for distributed system," in 31st Chinese Control Conference (CCC), July 25–27, Hefei, China, 2012, pp. 4212–4217.
[56] S. Li, Y. Zheng, and Z. Lin, "Impacted-region optimization for distributed model predictive control systems with constraints," IEEE Transactions on Automation Science and Engineering, vol. PP, pp. 1–14, 2014.
[57] Y. Zheng and S. Li, Distributed Predictive Control for Building Temperature Regulation with Impact-Region Optimization, Cape Town, South Africa, 2014, pp. 12074–1279.
[58] X. Du, Y. Xi, and S. Li, "Distributed model predictive control for large-scale systems," in Proceedings of the American Control Conference, vol. 4, June 25–27, Arlington, VA, 2001, pp. 3142–3143.
[59] A. Richards and J. P. How, "Robust distributed model predictive control," International Journal of Control, vol. 80, pp. 1517–1531, 2007.
[60] A. N. Venkat, J. B. Rawlings, and S. J. Wright, "Distributed model predictive control of large-scale systems," in Assessment and Future Directions of Nonlinear Model Predictive Control, Berlin, Heidelberg: Springer, 2007, pp. 591–605.
[61] E. Camponogara, D. Jia, B. H. Krogh, and S. Talukdar, "Distributed model predictive control," IEEE Control Systems Magazine, vol. 22, pp. 44–52, 2002.
[62] T. Keviczky, F. Borrelli, and G. J. Balas, "Decentralized receding horizon control for large scale dynamically decoupled systems," Automatica, vol. 42, pp. 2105–2115, 2006.
[63] L. Magni and R. Scattolini, "Stabilizing decentralized model predictive control of nonlinear systems," Automatica, vol. 42, pp. 1231–1236, 2006.
[64] M. Mercangöz and F. J. Doyle Iii, "Distributed model predictive control of an experimental four-tank system," Journal of Process Control, vol. 17, pp. 297–308, 2007.
[65] J. Richalet, A. Rault, J. Testud, and J. Papon, "Algorithmic control of industrial processes," in Proceedings of the 4th IFAC Symposium on Identification and System Parameter Estimation, 1976, pp. 1119–1167.
[66] S. Joe Qin, "Control performance monitoring – a review and assessment," Computers and Chemical Engineering, vol. 23, pp. 173–186, 1998.
[67] C. Cutler, A. Morshedi, and J. Haydel, "An industrial perspective on advanced control," in AICHE Annual Meeting, New York: AIChE, 1983.
[68] C. E. Garcia and A. Morshedi, "Quadratic programming solution of dynamic matrix control (QDMC)," Chemical Engineering Communications, vol. 46, pp. 73–87, 1986.
[69] D. Q. Mayne, J. B. Rawlings, C. V. Rao, and P. O. Scokaert, "Constrained model predictive control: stability and optimality," Automatica, vol. 36, pp. 789–814, 2000.

[70] V. Nevistić and J. A. Primbs, "Finite receding horizon linear quadratic control: a unifying theory for stability and performance analysis," Technical Report, CaltechCDSTR:1997.001, Pasadena, CA: California Institute of Technology, 1997.
[71] H. Michalska and D. Q. Mayne, "Robust receding horizon control of constrained nonlinear systems," IEEE Transactions on Automatic Control, vol. 38, pp. 1623–1633, 1993.
[72] H. Michalska, "A new formulation of receding horizon stabilising control without terminal constraint on the state," European Journal of Control, vol. 3, pp. 2–14, 1997.
[73] H. Chen and F. Allgöwer, "Nonlinear model predictive control schemes with guaranteed stability," in Nonlinear Model Based Process Control, Berlin: Springer, 1998, pp. 465–494.
[74] D. Bao-Cang, Modern Predictive Control: Boca Raton, FL: CRC Press, 2010.
[75] L. Ljung, System Identification, Berlin: Springer, 1998.
[76] Y. Zhu, Multivariable System Identification for Process Control, Amsterdam: Elsevier, 2001.
[77] B. Huang and S. L. Shah, Performance Assessment of Control Loops: Theory and Applications, Berlin: Springer, 1999.
[78] J.-W. Lee, W. Hyun Kwon, and J. Choi, "On stability of constrained receding horizon control with finite terminal weighting matrix," Automatica, vol. 34, pp. 1607–1612, 1998.
[79] D. D. Siljak, Decentralized Control of Complex Systems, New York: Courier Dover Publications, 2011.
[80] G. Weiss, Multiagent Systems: A Modern Approach to Distributed Artificial Intelligence, Cambridge, MA: MIT Press, 1999.
[81] H. Sildir, Y. Arkun, B. Cakal, D. Gokce, and E. Kuzu, "Plant-wide hierarchical optimization and control of an industrial hydrocracking process," Journal of Process Control, vol. 23, pp. 1229–1240, 2013.
[82] Y. Xi, Dynamic Large-Scale Systems Introduction (in Chinese), Beijing: National Defense Industry Press, 1988.
[83] S. Skogestad and I. Postlethwaite, Multivariable Feedback Control: Analysis and Design, vol. 2, New York: Wiley, 2007.
[84] M. Morari, Robust Process Control, Zurich: Morari, 1989.
[85] W. Hu, W.-J. Cai , and G. Xiao, "Relative gain array for MIMO processes containing integrators and/or differentiators," in 11th International Conference on Control Automation Robotics and Vision, December 7–10, Singapore, 2010, pp. 231–235.
[86] E. Bristol, "On a new measure of interaction for multivariable process control," IEEE Transactions on Automatic Control, vol. 11, pp. 133–134, 1966.
[87] R. Wood and M. Berry, "Terminal composition control of a binary distillation column," Chemical Engineering Science, vol. 28, pp. 1707–1717, 1973.
[88] C. E. Garcia, D. M. Prett, and M. Morari, "Model predictive control: theory and practice – a survey," Automatica, vol. 25, pp. 335–348, 1989.
[89] X. Du, Y. Xi, and S. Li, "Distributed model predictive control for large-scale systems," in Proceedings of the American Control Conference, vol. 4, June 25–27, Arlington, VA, 2001, pp. 3142–3143.
[90] N. Motee and B. Sayyar-Rodsari, "Optimal partitioning in distributed model predictive control," in Proceedings of the American Control Conference, vol. 6, June 4–6, 2003, pp. 5300–5305.
[91] N. H. El-Farra and P. D. Christofides, "Coordinating feedback and switching for control of hybrid nonlinear processes," AIChE Journal, vol. 49, pp. 2079–2098, 2003.
[92] N. H. El-Farra, A. Gani, and P. D. Christofides, "Fault-tolerant control of process systems using communication networks," AIChE Journal, vol. 51, pp. 1665–1682, 2005.
[93] P. Mhaskar, A. Gani, and P. D. Christofides, "Fault-tolerant control of nonlinear processes: performance-based reconfiguration and robustness," International Journal of Robust and Nonlinear Control, vol. 16, pp. 91–111, 2006.
[94] R. J. Patton, R. N. Clark, and P. M. Frank, Issues of Fault Diagnosis for Dynamic Systems, Berlin: Springer, 2000.
[95] J. Nash, "Non-cooperative games," Annals of Mathematics, vol. 54, pp. 286–295, 1951.
[96] Y. Xi, "New design method for discrete-time multi-variable predictive controllers," International Journal of Control, vol. 49, pp. 45–56, 1989.
[97] D. M. Prett and M. Morari, The Shell Process Control Workshop, Amsterdam: Butterworth-Heinemann, 1987.
[98] A. N. Venkat, J. B. Rawlings, and S. J. Wright, "Stability and optimality of distributed model predictive control," in CDC-ECC'05: 44th IEEE Conference on Decision and Control, 2005 and 2005 European Control Conference, December 12–15, 2005, pp. 6680–6685.

[99] M. Moradzadeh, R. Boel, and L. Vandevelde, "Voltage coordination in multi-area power systems via distributed model predictive control," IEEE Transactions on Power Systems, vol. 28, pp. 1–9, 2012.
[100] R. Scattolini, "Architectures for distributed and hierarchical model predictive control – A review," Journal of Process Control, vol. 19, pp. 723–731, 2009.
[101] M. R. Katebi and M. A. Johnson, "Predictive control design for large-scale systems," Automatica, vol. 33, pp. 421–425, 1997.
[102] A. Mukhopadhyay and S. Sikdar, "Implementation of an on-line run-out table model in a hot strip mill," Journal of Materials Processing Technology, vol. 169, pp. 164–172, 2005.
[103] S. Guan, X. Wang, and T. Chai, "An expert controller for the laminar cooling process of hot rolled slab," presented at the Proceedings of the American Control Conference, vol. 4, June 24–26, Philadelphia, PA, 1998, pp. 2165–2169.
[104] Y. Zheng, S. Li, and X. Wang, "An approach to model building for accelerated cooling process using instance-based learning," Expert Systems with Applications, vol. 37, pp. 5364–5371, 2010.
[105] Z. J. Wang, Q. D. Wu, and T. Y. Chai, "Optimal-setting control for complicated industrial processes and its application study," Control Engineering Practice, vol. 12, pp. 65–74, 2004.
[106] P. Zhou, T. Chai, and H. Wang, "Intelligent optimal-setting control for grinding circuits of mineral processing process," IEEE Transactions on Automation Science and Engineering, vol. 6, pp. 730–743, 2009.
[107] L. Tang, X. Wang, and J. Liu, "Color-coating production scheduling for coils in inventory in steel industry," IEEE Transactions on Automation Science and Engineering, vol. 5, pp. 544–549, 2008.
[108] N. Jin, S. Zhou, T.-S. Chang, and H. Hsun-Hau Huang, "Identification of influential functional process variables for surface quality control in hot rolling processes," IEEE Transactions on Automation Science and Engineering, vol. 5, pp. 557–562, 2008.
[109] Z. Wang, Q. Wu, and T. Chai, "Optimal-setting control for complicated industrial processes and its application study," Control Engineering Practice, vol. 12, pp. 65–74, 2004.
[110] H. S. Ko, J.-S. Kim, T.-W. Yoon, M. Lim, D. R. Yang, and I. S. Jun, "Modeling and predictive control of a reheating furnace," in Proceedings of the American Control Conference, vol. 4, June 28–30, Chicago, IL, 2000, pp. 2725–2729.
[111] H. Fang and Z. Lin, "Stability analysis for linear systems under state constraints," IEEE Transactions on Automatic Control, vol. 49, pp. 950–955, 2004.
[112] T. Hu and Z. Lin, Control Systems with Actuator Saturation: Analysis and Design, Berlin: Springer, 2001.
[113] J. B. Rawlings and K. R. Muske, "The stability of constrained receding horizon control," IEEE Transactions on Automatic Control, vol. 38, pp. 1512–1516, 1993.
[114] Y. Zheng, S. Li, and H. Qiu, "Networked coordination-based distributed model predictive control for large-scale system," IEEE Transactions on Control Systems Technology, vol. 21, pp. 991–998, 2013.
[115] P. Mhaskar, N. H. El-Farra, and P. D. Christofides, "Predictive control of switched nonlinear systems with scheduled mode transitions," IEEE Transactions on Automatic Control, vol. 50, pp. 1670–1680, 2005.
[116] D. Wu and Z. Li, "Effect of thermomechanical controlled processing on the microstructure and mechanical properties of Fe-C-Mn-Si multiphase steels," ISIJ International, vol. 46, pp. 1059–1066, 2006.
[117] Q.-Y. Sha, G.-Y. Li, L.-F. Qiao, and P.-Y. Yan, "Effect of cooling rate and coiling temperature on precipitate in ferrite of a Nb–V–Ti microalloyed strip steel," International Journal of Iron and Steel Research, vol. 14, pp. 316–319, 2007.
[118] J. Wang, G.-D. Wang, and X.-H. Liu, "Hot strip laminar cooling control model," International Journal of Iron and Steel Research, vol. 11, pp. 13–17, 2004.
[119] B. Han, Z.-P. Zhang, X.-H. Liu, and G.-D. Wang, "Element tracking strategies for hot strip laminar cooling control," International Journal of Iron and Steel Research, vol. 12, pp. 18–21, 2005.
[120] G. Dian-Yao, X. Jian-Zhong, P. Liang-Gui, W. Guo-Dong, and L. Xiang-Hua, "Self-learning and its application to laminar cooling model of hot rolled strip," International Journal of Iron and Steel Research, vol. 14, pp. 11–14, 2007.
[121] H. Xie, Z. Jiang, X. Liu, G. Wang, A. Tieu, M. Yang, and K. Manabe, "Application of fuzzy control of laminar cooling for hot rolled strip," Journal of Materials Processing Technology, vol. 187, pp. 715–719, 2007.
[122] E. Camponogara and M. L. de Lima, "Distributed optimization for MPC of linear networks with uncertain dynamics," IEEE Transactions on Automatic Control, vol. 57, pp. 804–809, 2012.
[123] H. F. Scherer, M. Pasamontes, J. L. Guzmán, J. D. Álvarez, E. Camponogara, and J. E. Normey-Rico, "Efficient building energy management using distributed model predictive control," Journal of Process Control, vol. 24, pp. 740–749, 2014.
[124] A. N. Venkat, "Distributed model predictive control: theory and applications," Citeseer, 2006.

[125] J. ZárateFlórez, J. Martinez, G. Besançon, and D. Faille, "Decentralized-coordinated model predictive control for a hydro-power valley," Mathematics and Computers in Simulation, vol. 91, pp. 108–118, 2013.
[126] S. Latzel, "Advanced automation concept of runout table strip cooling for hot strip and plate mills," in Conference Record of the 2000 IEEE Industry Applications Conference., 2000, pp. 2604–2612.
[127] R. D. Pehlke, A. Jeyarajan, and H. Wada, "Summary of thermal properties for casting alloys and mold materials," NASA STI/Recon Technical Report N, vol. 83, p. 36293, 1982.
[128] L. Chisci, P. Falugi, and G. Zappa, "Gain-scheduling MPC of nonlinear systems," International Journal of Robust and Nonlinear Control, vol. 13, pp. 295–308, 2003.
[129] Z. Wan and M. V. Kothare, "Efficient scheduled stabilizing model predictive control for constrained nonlinear systems," International Journal of Robust and Nonlinear Control, vol. 13, pp. 331–346, 2003.
[130] P. Falcone, F. Borrelli, J. Asgari, H. E. Tseng, and D. Hrovat, "Predictive active steering control for autonomous vehicle systems," IEEE Transactions on Control Systems Technology, vol. 15, pp. 566–580, 2007.
[131] T. Keviczky and G. J. Balas, "Flight test of a receding horizon controller for autonomous UAV guidance," in Proceedings of the American Control Conference, June 8–10, 2005, pp. 3518–3523.
[132] S. Gordon and D. Lehrer, "Coordinated train control and energy management control strategies," in Proceedings of the 1998 ASME/IEEE Joint Railroad Conference, 1998, 1998, pp. 165–176.
[133] X. Zhang and Z. Qiu, "CRH train traction calculation model and algorithm based on automatic constant speed," Bridges, vol. 10, pp. 506–512, 2014.
[134] C.-D. Yang and Y.-P. Sun, "Mixed H_2/H cruise controller design for high speed train," International Journal of Control, vol. 74, pp. 905–920, 2001.
[135] S. Qi and S. Yongduan, "Adaptive control and optimal power/brake distribution of high speed trains with uncertain nonlinear couplers," in 29th Chinese Control Conference, July 29–31, Beijing, 2010, pp. 1966–1971.
[136] H. Tian, Train Aerodynamics, Beijing: China Railway Publishing House, pp. 268–303, 2007.
[137] X. Zhuan and X. Xia, "Cruise control scheduling of heavy haul trains," IEEE Transactions on Control Systems Technology, vol. 14, pp. 757–766, 2006.
[138] X. Zhuan and X. Xia, "Optimal scheduling and control of heavy haul trains equipped with electronically controlled pneumatic braking systems," IEEE Transactions on Control Systems Technology, vol. 15, pp. 1159–1166, 2007.
[139] J. Peng, Traction and Brake of EMUs, Beijing: China Railway Publishing House, 2009.
[140] H. Dong, B. Ning, B. Cai, and Z. Hou, "Automatic train control system development and simulation for high-speed railways," IEEE Circuits and Systems Magazine, vol. 10, pp. 6–18, 2010.
[141] S. Zhang, The CRH2 EMU, Beijing: China Railway Press, 2008.
[142] J. Gordon, K. C. Ng, and H. T. Chua, "Centrifugal chillers: thermodynamic modelling and a diagnostic case study," International Journal of Refrigeration, vol. 18, pp. 253–257, 1995.
[143] M. Browne and P. Bansal, "Transient simulation of vapour-compression packaged liquid chillers," International Journal of Refrigeration, vol. 25, pp. 597–610, 2002.

Index

Distributed Model Predictive Control for Plant-Wide Systems, First Edition. Shaoyuan Li and Yi Zheng.